Milk Processing and Quality Management

Other books in the Society of Dairy Technology series:

Probiotic Dairy Products (ISBN 978-1-4051-2124-8)
Fermented Milks (ISBN 978-0632-06458-8)
Brined Cheeses (ISBN 978-1-4051-2460-7)
Structure of Dairy Products (ISBN 978-1-4051-2975-6)
Dairy Powders and Concentrated Milk Products (ISBN 978-1-4051-5764-3)
Cleaning-in-Place: Dairy, Food and Beverage Operations (ISBN 978-1-4051-5503-8)

Milk Processing and Quality Management

Edited by

Dr Adnan Y. Tamime
Dairy Science and Technology Consultant
Ayr, UK

WILEY-BLACKWELL
A John Wiley & Sons, Ltd., Publication

SDT

This edition first published 2009

Blackwell Publishing was acquired by John Wiley & Sons in February 2007. Blackwell's publishing programme has been merged with Wiley's global Scientific, Technical, and Medical business to form Wiley-Blackwell.

Registered office
John Wiley & Sons Ltd, The Atrium, Southern Gate, Chichester, West Sussex, PO19 8SQ, United Kingdom

Editorial offices
9600 Garsington Road, Oxford, OX4 2DQ, United Kingdom
350 Main Street, Malden, MA 02148-5020, USA

For details of our global editorial offices, for customer services and for information about how to apply for permission to reuse the copyright material in this book please see our website at www.wiley.com/wiley-blackwell.

Library of Congress Cataloging-in-Publication Data

Milk processing and quality management / edited by Adnan Y. Tamime.
p. cm. – (Society of Dairy Technology series)
Includes bibliographical references and index.
ISBN 978-1-4051-4530-5 (hardback : alk. paper) 1. Dairy processing–Quality control. I. Tamime, A. Y.
SF250.5.M56 2008
637′.1–dc22

2008026000

A catalogue record for this book is available from the British Library.

Set in 10/12.5pt Times New Roman by Aptara® Inc., New Delhi, India
Printed in Singapore by Utopia Press Pte Ltd

1 2009

Contents

Preface to Technical Series xiii
Preface xiv
Contributors xv

1 On-Farm Hygienic Milk Production **1**
M.M.M. VISSERS AND F. DRIEHUIS

1.1 Introduction 1
1.2 Sources of microbial contamination of bulk tank milk 2
1.2.1 Background 2
1.2.2 Mastitis 3
1.2.3 Environment 5
1.2.4 Milking equipment 7
1.2.5 Microbial growth during milk storage 8
1.3 Control of microbial contamination of bulk tank milk 8
1.3.1 Good farming practice 8
1.3.2 Animal health management 9
1.3.3 Control of feed 9
1.3.4 Facility hygiene 11
1.3.5 Milking operations 12
1.3.6 Milking machine design and operations 13
1.3.7 Bulk tank design and operations 14
1.3.8 Identification of effective control measures 14
1.4 Future developments in handling of the milk on the farm 15
1.4.1 Concentration of milk 15
1.4.2 Heat treatment of the milk 16
1.4.3 In/online monitoring of bulk tank milk quality 16
1.5 Conclusions 16
References 17

2 Properties and Constituents of Cow's Milk **23**
T. HUPPERTZ AND A.L. KELLY

2.1 Introduction 23
2.2 Milk composition 24
2.3 Milk constituents 25

2.3.1 Lactose 25
2.3.2 Milk salts 28
2.3.3 Lipids 29
2.3.4 Proteins 31
2.3.5 Indigenous milk enzymes 36
2.4 Heat-induced changes in physicochemical properties of milk 37
2.4.1 pH 37
2.4.2 Buffering capacity 38
2.4.3 Creaming 39
2.5 Heat-induced changes in processing characteristics of milk 39
2.5.1 Rennet-induced coagulation 39
2.5.2 Acid-induced coagulation 40
2.5.3 Heat-induced coagulation 40
2.6 Relationship between the quality of raw milk and that of products 41
2.7 Conclusions 43
References 43

3 Microbiology of Raw and Market Milks 48
V. TOUCH AND H.C. DEETH

3.1 Introduction 48
3.2 Microflora of raw milk 48
3.2.1 Spoilage organisms 48
3.2.2 Pathogenic organisms 50
3.3 Microflora of pasteurised milk 51
3.3.1 Spoilage organisms 51
3.3.2 Pathogenic organisms 54
3.4 Microflora of UHT milk 55
3.4.1 Heat-resistant spore-formers 55
3.4.2 Post-sterilisation contaminants 56
3.5 Microflora of ESL milk 57
3.6 Sources of contamination 58
3.6.1 Raw milk 58
3.6.2 Pasteurised milk 60
3.6.3 UHT milk 61
3.7 Measures to reduce bacterial contamination of raw and market milks 61
3.7.1 Cleaning and sanitation 61
3.7.2 Cooling the milk during storage 62
3.7.3 Addition of carbon dioxide 62
3.7.4 Applications of bio-preservatives 62
3.7.5 Thermal treatments 64
3.7.6 Non-thermal treatments 65
3.7.7 Multitarget attack/integrative approaches 65
3.8 Conclusion 65
References 66

4 Quality Control **72**
J. BELLOQUE, R. CHICÓN AND I. RECIO

4.1 Introduction 72
4.2 Quality control of raw milk 72
4.3 Quality control of processed milk 73
4.4 Methods of analysis 74
4.4.1 Analysis of abnormal milk 75
4.4.2 Microbial analysis 75
4.4.3 Mastitic milk – somatic cell count (SCC) 77
4.4.4 Testing for residues compounds 78
4.5 Major components analysis 79
4.5.1 Protein composition analysis 82
4.5.2 Fat composition analysis 83
4.5.3 Analysis of other milk components 83
4.5.4 Analysis of degradation products formed during milk storage 84
4.5.5 Evaluation of heat load 84
4.6 Analysis of fraudulent addition of ingredients and authentication 86
4.6.1 Addition of water 86
4.6.2 Addition of whey proteins 86
4.6.3 Addition of non-dairy proteins 89
4.6.4 Addition of reconstituted milk 89
4.6.5 Substitution of milk from a different species of mammals 90
4.6.6 Substitution of milk with different feeding or geographical origin 90
4.6.7 Mislabelling of fortified and functional milks 90
4.7 Conclusion 91
Acknowledgement 92
References 92
Further Reading 99
Web Pages of Interest 100

5 Current Legislation of Market Milks **101**
M. HICKEY

5.1 Introduction 101
5.2 EU legislation 102
5.2.1 Access to EU legislation 102
5.2.2 Compositional requirements 102
5.2.3 Amendment to existing compositional requirements 105
5.2.4 Hygiene and food safety requirements 105
5.3 UK legislation 115
5.3.1 Background 115
5.3.2 Present legislation on composition 115
5.3.3 Present legislation on hygiene 116
5.3.4 The Dairy UK Code of Practice for HTST pasteurisation 118
5.3.5 Permitted range of packaging sizes for milk in the UK 118

5.4 Irish legislation 119
5.4.1 Introduction 119
5.4.2 Present legislation on hygiene 119
5.4.3 Present legislation on composition 120
5.4.4 Range of packaging sizes for fresh drinking milk in Ireland 120
5.5 USA legislation 120
5.5.1 Background 120
5.5.2 Historical details 121
5.5.3 The Pasteurised Milk Ordinance (PMO) and the Code of Federal Regulations 122
5.5.4 Definitions and standards for milk in the USA 122
5.5.5 Sale of raw milk for direct consumption in the USA 124
5.6 The international perspective – Codex Alimentarius 125
5.6.1 What is Codex Alimentarius? 125
5.6.2 Codex Alimentarius Commission (CAC) membership and structures 125
5.6.3 Codex Alimentarius standards 126
5.6.4 Codex Alimentarius standards and milk for human consumption 128
5.7 Conclusions and possible future developments 132
References 133

6 The Safety of Raw Liquid Milk **139**
M. O'MAHONY, S. FANNING AND P. WHYTE

6.1 Introduction 139
6.2 Implication of milk in human disease 139
6.3 Microbial hazards in milk 140
6.3.1 *Salmonella* species 140
6.3.2 *Campylobacter* species 141
6.3.3 Verocytotoxigenic *Escherichia coli* 141
6.3.4 *Enterotoxigenic* Staphylococcus aureus 142
6.3.5 *Listeria monocytogenes* 143
6.3.6 *Mycobacterium bovis* 144
6.3.7 *Brucella abortus* and *Brucella melitensis* 145
6.4 Chemical contaminants and residues 146
6.4.1 Industrially derived contaminants 146
6.4.2 Biologically derived contaminants 147
6.4.3 Pesticides and residues of plant health agrichemicals 147
6.4.4 Residues of animal remedies 148
6.5 Poorly understood and emerging hazards 149
6.5.1 Bovine spongiform encephalopathy (BSE) 149
6.5.2 *Mycobacterium avium* subsp. *paratuberculosis* 150
6.5.3 Enterobacter sakazakii 151
6.5.4 Antimicrobial resistance 152
6.6 Risk management strategies to ensure safety of liquid milk 156
6.6.1 Pre-harvest risk management 156

6.6.2 Harvest phase risk management 156
6.6.3 Post-harvest risk management 157
6.7 Summation 158
References 158

7 Heat Treatment of Milk **168**
M.J. LEWIS AND H.C. DEETH

7.1 Introduction 168
7.2 Milk composition 169
7.3 Reaction kinetics 170
7.4 Principles of heat transfer 170
7.5 Thermisation and tyndallisation 171
7.6 Pasteurisation 171
7.6.1 Introduction and principles 171
7.6.2 Methods of pasteurisation 174
7.6.3 Factors affecting the quality of pasteurised milk 177
7.7 Sterilisation – safety and spoilage considerations 180
7.8 In-container sterilisation 181
7.9 UHT processing 183
7.9.1 Introduction and principles 183
7.9.2 Methods of UHT processing 185
7.9.3 Factors affecting the quality of UHT milk 190
7.9.4 Heat stability, sediment formation and fouling 193
7.10 High-temperature processing (extended shelf life) 195
7.11 Reconstituted and recombined milk products 197
7.12 Conclusions 198
7.13 Appendix 199
7.13.1 Some heat transfer properties 199
7.13.2 Definitions and equations 199
References 200

8 Novel Methods of Milk Processing **205**
M. VILLAMIEL, M.A.I. SCHUTYSER AND P. DE JONG

8.1 Introduction 205
8.2 Microwaves 205
8.2.1 Operating principle 207
8.2.2 Effects on product properties 207
8.2.3 State of the art and opportunities 209
8.3 High Pressure 209
8.3.1 Operating principle 210
8.3.2 Effect on product properties 210
8.3.3 State of the art and opportunities 212
8.4 Pulsed electric fields 213
8.4.1 Operating principle 213

8.4.2 Effect on product properties 214
8.4.3 State of the art and opportunities 215
8.5 Ultrasound 216
8.5.1 Operating principle 216
8.5.2 Effects on product properties 216
8.5.3 State of the art and opportunities 217
8.6 Microfiltration 218
8.6.1 Operating principle 219
8.6.2 Effect on product properties 219
8.6.3 State of the art and opportunities 220
8.7 Innovative steam injection – a novel heating method 220
8.7.1 Operating principle 221
8.7.2 Effect on product properties 221
8.7.3 State of the art and opportunities 222
8.8 Combined technologies 222
8.8.1 Effect on product properties 224
8.8.2 Preservation in the presence of antimicrobials 224
8.8.3 State of the art and opportunities 225
8.9 Computer models for design and implementation of novel milk-processing methods 226
8.9.1 Optimisation of milk-processing methods with models 226
8.9.2 Availability of models 227
References 228

9 Hygiene Practices in Liquid Milk Dairies 237
A.J. VAN ASSELT AND M.C. TE GIFFEL

9.1 Introduction 237
9.2 Principal hazards 237
9.2.1 Raw milk 237
9.2.2 Pasteurised milk 238
9.2.3 Sterilised milk 240
9.3 Hygienic processing 240
9.3.1 General 240
9.3.2 Building construction 241
9.3.3 Zoning 241
9.3.4 Processing equipment 242
9.3.5 Cleaning 243
9.3.6 Packaging materials 244
9.3.7 Processing aids 244
9.4 Monitoring and control 244
9.4.1 In-line detection of microorganisms 245
9.4.2 Optimisation of CIP procedures 245
9.5 Concluding remarks 246
References 249

10 Hazard Analysis (Appraisal) Critical Control Point (HACCP) in Milk Processing – A Practical Overview **253**
A. SAYLER

10.1 Introduction 253
10.2 General aspects of the benefits of the HACCP system 254
10.3 HACCP definitions 255
10.4 Implementation and maintenance of the HACCP programme 257
10.4.1 Commitment by management 259
10.4.2 Letter of endorsement of an HACCP programme 259
10.5 Steps to HACCP implementation 260
10.5.1 Assemble the HACCP team 260
10.5.2 Describe the food and its distribution 262
10.5.3 Intended use and consumers of the food (see form above) 263
10.5.4 Develop a flow diagram that describes the process 263
10.5.5 Verify the flow diagram 263
10.6 Hazard components 263
10.6.1 Biological hazards 264
10.6.2 Chemical hazards 266
10.6.3 Physical hazards 267
10.6.4 Potential hazards identified by source 267
10.7 Prerequisite programme 269
10.7.1 Background 269
10.7.2 Prerequisite examples and forms 277
10.8 The principles of an HACCP plan 277
10.8.1 Principle 1 – conduct hazard analysis 277
10.8.2 Principle 2 – determine CCPs 283
10.8.3 Principle 3 – establishments of critical limits 283
10.8.4 Principle 4 – establish monitoring procedures 284
10.8.5 Principle 5 – establish corrective actions 285
10.8.6 Principle 6 – establish verification procedure 287
10.8.7 Principle 7 – records 289
10.9 Internal and external auditing in the HACCP process 291
10.10 Overview and summary 293

11 Sensory Profiling of Market Milks **294**
E. MOLINA, L. AMIGO AND A. QUIRÓS

11.1 Introduction 294
11.2 Sensory properties of market milks 295
11.2.1 Sensory evaluation of milk 295
11.2.2 Sensory attributes of flavour defects 297
11.2.3 Other factors influencing the sensory properties 303
11.3 Different types of market milks 305
11.3.1 Raw and heat processed milk 306
11.3.2 Whole and fat-reduced milk 306

11.3.3 Enriched and modified milks 309
11.3.4 Other types of milks 311
11.4 Conclusion 312
References 313

Index 317

Preface to Technical Series

For more than 60 years, the Society of Dairy Technology (SDT) has sought to provide education and training in the dairy field, disseminating knowledge and fostering personal development through symposia, conferences, residential courses, publications, and its journal, the *International Journal of Dairy Technology* (previously known as the *Journal of the Society of Dairy Technology*).

In recent years, there have been significant advances in our understanding of milk systems, probably the most complex natural food available to man. Improvements in process technology have been accompanied by massive changes in the scale of many milk/dairy-processing operations, and the manufacture of a wide range of dairy and other related products.

The Society has now embarked on a project with Blackwell Publishing to produce a Technical Series of dairy-related books to provide an invaluable source of information for practicing dairy scientists and technologists, covering the range from small enterprises to modern large-scale operation. This fifth volume in the series, on *Milk Processing and Quality Management* under the editorship of Dr Adnan Y. Tamime, provides timely and comprehensive guidance on the processing of liquid milks. The economic production of liquid milk is of vital importance to the dairy industry, for example in the UK half of the ex-farm milk is processed for the liquid milk market, almost 90% being pasteurised while the remainder is more severely heat treated to provide long-life products. Attention to detail is essential if the consumer is to be provided with a safe product that meets taste and shelf life expectations.

Andrew Wilbey
Chairman of the Publications Committee, SDT

Preface

Given the recent developments in dairy technology, it has become apparent that revision of some of the SDT publications (e.g. *Pasteurisation Manual*) is overdue. Although there have been some technical developments (i.e. with the exception of automation) in pasteurisation of liquid milk over the past couple decades, consumption of liquid milk is rather high worldwide. It is important to note that in certain parts of Europe and majority of developing countries, sterilised, extended shelf life (ESL) and ultra-heat treatment (UHT) of liquid milk products are widely produced, and this sector of the dairy industry is highly profitable and represents a large proportion (e.g. >50%) of the processed milk in any dairying country.

Milks Processing and Quality Management is another book proposed within the Technical Series of The Society of Dairy Technology (SDT). Numerous scientific data are available in journals and books that have been published since the early 1990s, and the primary aim of this text is to detail the manufacturing methods (i.e. pasteurisation, sterilisation, ESL and UHT), scientific aspects, quality control (i.e. hygiene and analytical methods), safety of raw milk consumption and properties of all these liquid milk products in one publication.

The authors, who are all specialists in these products, have been chosen from around the world. There is no doubt that the book will have an international recognition by dairy scientists, students, researchers and dairy operatives, and will become an important component of the Technical Series promoted by the Society of Dairy Technology.

Adnan Y. Tamime

Contributors

Editor

Dr Adnan Y. Tamime
Dairy Science & Technology Consultant
24 Queens Terrace
Ayr KA7 1DX
Scotland
UK
Tel: +44 (0)1292 265498
Fax: +44 (0)1292 265498
Mobile: +44 (0)7980 278950
E-mail: adnan@tamime.fsnet.co.uk

Contributors

Dr Lourdes Amigo
Instituto de Fermentaciones Industriales (CSIC)
Juan de la Cierva 3
28006 Madrid
Spain
Tel: +34-91-5622900 Ext. 220
Fax: +34-91-5644853
E-mail: amigo@ifi.csic.es

Mr Arjan J. van Asselt
Project Manager Process
Hygiena & Efficiency
NIZO Food Research B.V.
P.O. Box 20
6710 BA Ede
The Netherlands
Tel: +31 318 659 646
Fax: +31 318 650 400
E-mail: Arjan.van.Asselt@nizo.nl

Dr Josefina Belloque
Instituto de Fermentaciones Industriales (CSIC)
Consejo Superior de Investigaciones Científicas
Juan de la Cierva 3
28006 Madrid
Spain
Tel: +34 91 5622900
Fax: +34 91 5644853
E-mail: belloque@ifi.csic.es

Dr Rosa Chicón
Instituto de Fermentaciones Industriales (CSIC)
Consejo Superior de Investigaciones Científicas
Juan de la Cierva 3
28006 Madrid
Spain

Tel: +34 91 5622900
Fax: +34 91 5644853
E-mail: rosa.chicon@ifi.csic.es

Dr H.C. Deeth
School Land, Crop and Food Sciences
The University of Queensland
Brisbane 4072
Australia
Tel: +61 (0)7 3347 9191
Fax: +61 (0)7 3365 1177
E-mail: h.deeth@uq.edu.au

Dr Frank Driehuis
NIZO Food Research
Department of Health and Safety
Kerhemseweg 2
P.O. Box 20
6710 BA Ede
The Netherlands
Tel: +31 318 659511
Fax: +31 318 650400
E-mail: Frank.Driehuis@nizo.nl

Prof Séamus Fanning
Centre for Food Safety
School of Agriculture, Food Science and Veterinary Medicine
College of Life Sciences
University College Dublin
Belfield
Dublin 4
Republic of Ireland
Tel: +353 1 716 6082
Fax: +353 1 716 6091
E-mail: sfanning@ucd.ie

Mr Michael Hickey
Michael Hickey Associates, Dairy and Food Consultancy
Derryreigh
Creggane
Charleville
Co. Cork
Ireland
Tel: +353 (0)63 89392
Mobile: +353 (0)87 2385653
E-mail: mfhickey@oceanfree.net

Dr Thom Huppertz
NIZO food research
P.O. Box 20
6710 BA Ede
The Netherlands
Tel: +31 318 659600
Fax: +31 318 650400
E-mail: Thom.Huppertz@nizo.nl

Dr Peter de Jong
Principal Scientist
Division Manager Processing
NIZO Food Research B.V.
PO Box 20
6710 BA Ede
The Netherlands
Kernhemseweg 2
6718 ZB Ede
The Netherlands
Tel: +31(0)318 659 575
Fax: +31 318 650 400
E-mail: peter.de.jong@nizo.nl

Prof Alan L. Kelly
University College Cork
Department of Food and Nutritional Sciences
Cork
Ireland
Tel: +353 21 4903405
Fax: +353 21 4270213
E-mail: a.kelly@ucc.ie

Dr Mike J. Lewis
University of Reading
School of Biosciences
P. O. Box 226
Reading RG6 6AP
United Kingdom
Tel: +44 (0)1189 318713
Fax: +44 (0) 1189 310080
E-mail: M.J.Lewis@afnovell.reading.ac.uk

Dr Elena Molina
Instituto de Fermentaciones Industriales (CSIC)
Juan de la Cierva 3
28006 Madrid
Spain
Tel: +34 91 5622900 Ext. 243
Fax: +34 91 5644853
E-mail: e.molina@ifi.csic.es

Mr Micheál O'Mahony
Chief Specialist
Veterinary Public Health
Consumer Protection Division
Food Safety Authority of Ireland
Abbey Court
Lower Abbey Street
Dublin 1
Republic of Ireland
Tel: +353 1 8171 328
Fax: +353 1 8171 228
E-mail: momahony@fsai.ie

Ms Ana Quirós
Instituto de Fermentaciones Industriales (CSIC)
Juan de la Cierva 3
28006 Madrid
Spain
Tel: +34-91-5622900 Ext. 220
Fax: +34-91-5644853
E-mail: anaquiros@ifi.csic.es

Dr Isidra Recio
Instituto de Fermentaciones Industriales (CSIC)
Consejo Superior de Investigaciones Científicas
Juan de la Cierva 3
28006 Madrid
Spain
Tel: +34 91 5622900
Fax: +34 91 5644853
E-mail: recio@ifi.csic.es

Dr Allen Sayler
International Dairy Foods Association
Tel: +1 (0)202 220 3544
Fax: +1 (0)202 331 7820
Mobile: +1 (0)202 841 1029
E-mail: asayler@idfa.org

Dr Maarten A.I. Schutyser
Project Manager
NIZO Food Research B.V.
P.O. box 20
6710 BA Ede
The Netherlands
Tel: +31 318 659617
Fax: +31 318 650400
E-mail: maarten.schutyser@nizo.nl

Dr Meike C. te Giffel
Manager Operations
NIZO Food Research B.V.
P.O. Box 20
6710 BA Ede
The Netherlands
Tel: +31 318 659 590
Fax: +31 318 650 400
E-mail: Meike.te.Giffel@nizo.nl

Dr Visalsok Touch
University of Battambang
National Road No 5, Prek Preah Sdach Commune,
Battambang District, Battambang Province, Cambodia
Tel: +855 12 73 54 16
Fax: +855 53 952 905
Email: visalsoktouch@yahoo.com

Dr M. Villamiel
Tenured Scientist
Instituto de Fermentaciones Industriales (CSIC)
c/Juan de la Cierva, 3
28006 Madrid
Spain

Tel: +34 91 5622900 Ext. 397
Fax: +34 91 5644853
E-mail: mvillamiel@ifi.csic.es

Dr Marc M.M. Vissers
NIZO Food Research
Department of Health and Safety
Kerhemseweg 2
P.O. Box 20
6710 BA Ede
The Netherlands
Tel: +31 318 659515
Fax: +31 318 650400
E-mail: Marc.Vissers@nizo.nl

Dr Paul Whyte
Centre for Food Safety
School of Agriculture, Food Science and Veterinary Medicine
College of Life Sciences
University College Dublin
Belfield
Dublin 4
Republic of Ireland
Tel: +353 1 716 6074
Fax: +353 1716 6091
E-mail: paul.whyte@ucd.ie

1 On-Farm Hygienic Milk Production

M.M.M. Vissers and F. Driehuis

1.1 Introduction

Food producers are responsible for the safety of their products, and to guarantee food safety of dairy products, the dairy industry has implemented hazard analysis of critical control points (HACCP) systems. This enables quality assurance of final products via a chain management approach (European Commission, 2004b). The quality and safety of raw milk is essential for the quality and safety of milk and dairy products. The quality and safety of milk is related to the contamination of milk with microorganisms, chemical residues and other contaminants. This chapter focuses on microbial contamination.

Human microbial pathogens that can be found in raw milk include *Listeria monocytogenes*, *Salmonella* spp. and *Campylobacter jejuni* (Jayaroa & Henning, 2001). In addition to their significance for public health, a very good microbial quality of raw milk is also important to prevent production losses and to achieve an optimal shelf life of dairy products. For example, spore formers of butyric acid bacteria in raw milk are responsible for defects in semi-hard cheeses (Klijn *et al*., 1995), and the contamination of raw milk with spores of *Bacillus cereus* limits the shelf life of pasteurised dairy products (Te Giffel *et al*., 1997). To ensure a good microbial quality of bulk tank milk, quality assurance systems for dairy farms are being developed and bacteriological schemes are being implemented in payment systems of farm raw bulk milk (IDF, 2006). In addition, hygienic milk production by dairy farmers is important with respect to animal welfare and the image of the dairy sector. Pathogenic microorganisms can infect cows (e.g. gastrointestinal tract, udder tissue), and result in reduced milk yields and even the death of animals. Thus, in summary, control of the microbial ecology at the dairy farm resulting in on-farm hygienic milk production is important for all elements of the dairy production chain.

In this chapter, on-farm hygienic milk production is defined as the control of the microbial contamination of bulk milk tank. Microbial control includes minimisation of microbial sources in the farm environment, minimisation of microbial transmission, prevention of microbial growth and infection of animals and maximisation of microbial inactivation and removal. Microorganisms are present in all parts of the farm environment. Many aspects of farm management (e.g. feed management, facility hygiene and milking operations) are involved in the control of the microbial contamination of bulk tank milk. However, the total bacterial count will also be affected by factors that are independent of farm management, such as seasonal variations.

This chapter discusses the origin of microorganisms in bulk tank milk (Section 1.2), various aspects of microbial control at the dairy farm (Section 1.3), some future developments (Section 1.4) and draws some general conclusions (Section 1.5).

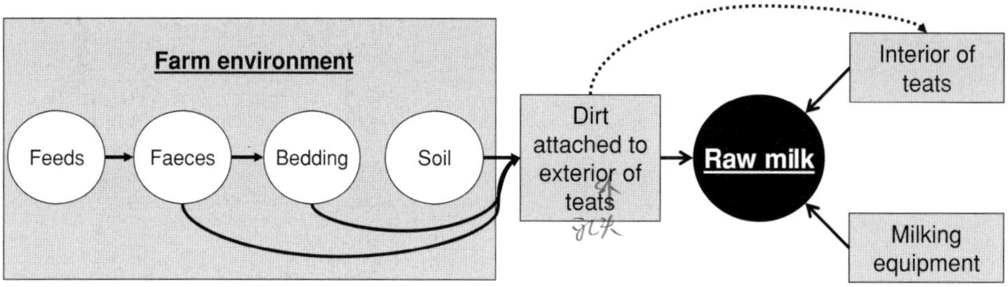

Fig. 1.1 Possible routes for the contamination of raw milk with microorganisms. Adapted from Akam *et al.* (1989).

1.2 Sources of microbial contamination of bulk tank milk

1.2.1 *Background*

Milk is sterile when secreted into the alveoli of the udder. Microbial contamination occurs mainly during and after milking (Figure 1.1). Microorganisms in bulk tank milk originate from the interior of teats, the farm environment and surfaces of the milking equipment (Bramley & McKinnon, 1990; Chambers, 2002). Microorganisms are mainly transferred from the farm environment to milk via dirt (e.g. faeces, bedding and soil) attached to the exterior of teats; in addition, microorganisms attached to the exterior of the teats can enter the teat canal and cause mastitis (Makovec & Ruegg, 2003). Finally, contamination can originate from insufficiently cleaned milking equipment when, during milking, microorganisms adhered to surfaces of the milking equipment are released into the milk (Bramley & McKinnon, 1990; Chambers, 2002). Aerial contamination is insignificant under normal production conditions (Akam *et al.*, 1989). The concentration of microorganisms in bulk tank milk can further increase due to their growth.

The microbial population in bulk tank milk consists of a variety of bacterial species. Most species have a specific origin. For example, the presence of *Staphylococcus aureus* in bulk tank milk will, generally, be traced back to cows suffering from mastitis, and silage is the most likely origin of spores of butyric acid bacteria in bulk tank milk (Stadhouders & Jørgensen, 1990; Haven *et al.*, 1996). Table 1.1 lists the origin and dominant contamination route(s) for various microorganisms found in bulk tank milk.

In the case of high microbial concentrations in bulk tank milk, determination of the composition of the microbial flora can reveal the cause of the elevated concentration. Holm *et al.* (2004) examined 73 samples of bulk tank milk with more than 4.5 $\log_{10}$ colony-forming units (cfu) mL^{-1}. In 48 samples, one microbial species dominated, for example *Lactococcus* spp. or *S. aureus*. In these samples, high microbial concentrations were in 64% of the cases, which were traced back to poor hygiene (dirty teats and insufficiently cleaned milking equipment). Psychrotrophic microorganisms, which could have grown during the storage of cooled bulk tank milk, overshadowed other microorganisms in 28% of the samples. Mastitis bacteria were found in 48% of all samples, and formed the dominant flora in 8% of the samples tested.

Table 1.1 Main source of microorganisms occurring in milk and associated spoilage and safety issues in dairy products.

Microbial species	Associated problem	Contamination source (main pathway[a])	Possible growth in bulk milk tank
Bacillus cereus (spores)	Spoilage of pasteurised dairy products	Environment[a] (feeds → faeces + soil), milking equipment	Yes
Bacillus sporothermodurans (spores)	Spoilage of UHT-treated dairy products	Environment (feeds → faeces)	No
Butyric acid bacteria (spores)	Spoilage of Gouda and Emmenthal cheeses	Environment (feeds → faeces)	No
Campylobacter jejuni	Food safety (products made of raw milk)	Environment (faeces)	No
Escherichia coli	Spoilage and food safety (products made of raw milk)	Environment (faeces and bedding)	Yes
Listeria monocytogenes	Food safety (products made of raw milk and soft or surface ripened cheeses)	Environment[a] (e.g. feeds, faeces)	Yes
Mycobacterium paratuberculosis	Food safety (products made of raw milk)[b]	Environment[a] (faeces)	No
Pseudomonas spp.	Spoilage	Environment[a] (bedding, soil), milking equipment	Yes
Salmonella spp.	Food safety (products made of raw milk)	Environment[a] (faeces)	Yes
Streptococcus thermophilus	Spoilage	Environment[a] (faeces, bedding, soil), milking equipment	Yes
Staphylococcus aureus	Food safety (products made of raw milk)	Interior of teats	Yes

[a] For species having the environment as the major source of contamination and are the main microbial carries indicated between brackets.
[b] Relevance for human health is unclear.
Data compiled from Fenlon (1988), Haven *et al.* (1996), Slaghuis *et al.* (1997), Stadhouders & Jørgensen (1990), Te Giffel *et al.* (1995) and Vaerewijck *et al.* (2001).

1.2.2 *Mastitis*

Mastitis organisms enter the teat canal and infect the interior tissues of the teats. After inflammation, the levels of mastitis organisms within the teat increase significantly. Consequently, during milking, high concentrations of the infectious organisms can be transmitted to milk. The concentration of mastitic-associated microorganisms in bulk tank milk depends on the type of microorganism, infection status within a herd (clinical/sub-clinical), stage of infection and fraction of the herd infected (Bramley & McKinnon, 1990; Chambers, 2002).

A large variety of microorganisms causes mastitis. Table 1.2 presents the frequency of different mastitis organisms as the dominant flora in milk samples of infected cows. In general, contagious and environmental pathogens are distinguished, although a strict classification

Table 1.2 Frequency (%) of different mastitis organisms as the dominant flora in milk samples submitted for microbial analysis in the United States and the Netherlands.

Microorganisms	US 1994–2001 (Makovec & Ruegg, 2003)	The Netherlands 2000 (Sampimon *et al.*, 2004)
Contagious mastitis organisms		
Staphylococcus aureus	9.7	32.2
Streptococcus agalactiae	13.2	5.3
Corynebacterium bovis	2.7	0.0
Environmental mastitis organisms		
Streptococcus uberis	12.2[a]	18.9
Streptococcus dysgalactiae		7.6
Escherichia coli	4.0	12.9
Klebsiella spp.	1.2	0.3

[a] Streptococci not including *Streptococcus agalactiae*.

is not possible for all species. Contagious pathogens are mainly transmitted from cow to cow, with or without an intermediate vector such as teat cup liners. The most important contagious pathogens are *S. aureus*, *Streptococcus agalatiae* and *Corynebacterium bovis* (Makovec & Ruegg, 2003).

Environmental pathogens are a natural part of the farm environment. They are, for example, present in faeces, bedding and mud. After the teats are soiled with (contaminated) faeces and bedding, these pathogens enter the teat canal and cause an infection (Smith, 1983). *Streptococcus uberis*, *Streptococcus dysgalactiae* and gram-negative bacteria, such as *Escherichia coli* and *Klebsiella* spp., are the most important environmental pathogens (Makovec & Ruegg, 2003; Ruegg, 2003b; Sampimon *et al.*, 2004). Unlike contagious pathogens, environmental pathogens cannot be eliminated entirely from the farm environment (Smith & Hogan, 1993). In amongst others in the Netherlands and the USA, the relative incidence of mastitis caused by environmental pathogens has increased in the recent decades, presumably due to the successful implementation of measures that reduce spreading of contagious pathogens (Makovec & Ruegg, 2003; Sampimon *et al.*, 2004).

Mastitis can be classified as clinical or sub-clinical. In the case of the former type, cows show recognisable and apparent symptoms, and their milk generally has a deviant colour. Since cows with clinical mastitis are relatively easy to recognise, they are generally removed from the milking herd and, thus, only accidentally contribute to the concentration of mastitis organisms in bulk tank milk. Cows suffering from sub-clinical mastitis show no apparent symptoms of mastitis and, in general, laboratory testing is necessary for diagnosis. The lack of apparent symptoms makes it difficult to recognise cows suffering from sub-clinical mastitis and, as a consequence, sub-clinical mastitis forms a greater threat for the microbial quality of bulk tank milk than clinical mastitis.

Depending on the stage of infection, a single cow can excrete up to $7 \log_{10}$ mastitis pathogens mL^{-1}. In a herd of 100 milking cows, only 1 cow can thus be responsible for a total bulk tank count of $5 \log_{10}$ cfu mL^{-1} (Bramley & McKinnon, 1990; Chambers, 2002). In theory, all mastitis organisms can increase the microbial contamination of bulk tank milk, and Zadoks *et al.* (2004) found that streptococci species to be responsible for 69% of the bacterial count variability at 48 dairy farms sampled, where *S. aureus* and gram-negative bacteria were responsible only for 3% of the variation. Hayes *et al.* (2001) characterised sudden elevations of the total microbial count in bulk tank milk (i.e. spike values); *S. uberis* was responsible for 55% and *E. coli* for 20% of the spike values. However, both *S. uberis* and *E. coli* are environmental pathogens and, therefore, do not necessarily originate from the interior of infected teats.

1.2.3 *Environment*

As mentioned elsewhere, the most common microbial sources in the farm environment are feeds, faeces, bedding material and soil. Microorganisms from these sources are transferred to milk in a number of steps (see Figure 1.1). The consecutive steps from source to milk are referred to as the contamination pathway. A crucial step in the contamination pathway is the transmission of dirt, composed of, for example, faeces, bedding and/or soil, to milk. Microorganisms from transmitted dirt dilute in the milk and pass the filter of the milking system (Akam *et al.*, 1989). Dirt is mainly transmitted to milk when it is attached to the exterior of teats and rinses off during the milking operations (Stadhouders & Jørgensen, 1990; Murphy & Boor, 2000). Additional dirt and microorganisms can be transmitted from the farm environment to bulk tank milk when the teat cups (that fall on the ground or are kicked off the teats) get contaminated or even suck up dirt from the milking parlour floor (Stadhouders & Jørgensen, 1990). The mass of transmitted dirt per unit of volume can be calculated using a marker method (Stadhouders & Jørgensen, 1990). At eleven farms, Vissers *et al.* (2007c) found between 3 and 300 mg of dirt per litre of milk with an average of 59 mg L^{-1}.

The strains and concentrations of microorganisms transmitted from the farm environment to milk via the exterior of teats depends on the composition of the attached dirt and microbial concentration in the dirt. When cows are at pasture, the teats are predominantly contaminated with soil, whereas teats of cows housed in the barn are mainly contaminated with faeces and bedding material (Christiansson *et al.*, 1999; Magnusson *et al.*, 2007). The contamination of teats with soil during the grazing period is considered to be the main cause of elevated concentrations of spores of *B. cereus* in bulk tank milk (Slaghuis *et al.*, 1997; Vissers *et al.*, 2007a,d).

Table 1.3 lists concentrations of important microbial groups observed in feeds, faeces, bedding and soil. Microorganisms in faeces include natural inhabitants, infectious microorganisms and microorganisms or their spores that originate directly from the feeds. Spore concentrations in faeces are between 2 and 10 times as high as the concentration in the ration of the cows (Hengeveld, 1983). This increase is explained by digestion of feed components while spores pass the gastrointestinal tract unaffected.

Different materials are used for bedding in barns, for example, straw, sawdust, wood shavings and shredded paper. Fresh bedding contains a large variety of microorganisms. Microbial concentrations in fresh bedding are usually much lower than concentrations in used bedding (Hogan *et al.*, 1990; Te Giffel *et al.*, 1995; Hogan & Smith, 1997; Slaghuis *et al.*, 1997). Especially, during the first day when the bedding is laid down, the concentrations in bedding material seem to increase significantly due to contamination with faeces and microbial growth (Hogan *et al.*, 1990, 1999; Hogan & Smith, 1997). However, high coliforms counts (7–9 $\log_{10}$ cfu g^{-1}) have also been measured in unused bedding material (Knappstein *et al.*, 2004b).

Table 1.3 Concentration ($\log_{10}$ g^{-1}) of aerobic microorganisms, spores of aerobic microorganisms and spores of gas-forming anaerobic microorganisms in feeds, faeces, bedding and soil.

Source		Aerobic microorganisms	Spores of *Bacillus* spp.	Spores of (gas-forming) clostridia[a]
Feed	Roughage	4.5 to more than 9.0	2.5–8.7	<1.7–6.2
	Concentrate	2.3–7.5	<1.0–6.7	<1.7–2.9
Faeces		5.6–8.0	3.3–6.8	<1.7–6.6
Bedding	Fresh	3.1–5.7	2.1–6.0	2.2–5.8[b]
	Used	7.4–9.7	3.9–7.2	
Soil		6.0–7.9	4.8–6.6	3.4–4.4

[a] Enumerated using most probable number (MPN) method.
[b] No separate data for unused and used bedding material.
After NIZO Food Research (unpublished data).

Feeds introduce a large variety of microorganisms to the farm environment, and subsequently to milk. The impact of feed as a hazard of microbial contaminants of raw milk is twofold: *firstly*, feed can be a source or transmission vehicle of pathogens causing infection in cattle, and *secondly*, feed is an important source of bacterial spores in raw milk.

Basically, the diet of high-yielding dairy cows consists of two categories of feedstuffs, roughages and concentrate. The former feed provides the animal with dietary fibre, which is essential for the normal functioning of the cow's rumen. The most important roughage crops are grass, maize and lucerne (Wilkinson & Toivonen, 2003). Ensiling and haymaking are the two most common methods to conserve the nutritional value after harvesting. A special situation exists for grass, for example during the growing season, it is usually fed fresh, and outside the growing season, it usually fed as silage or hay. To meet the high nutritional requirements of high-yielding dairy cows, roughage-based diets are supplemented with concentrate feeds, which are high in energy and/or protein. Some examples include cereal grains, bran of cereals and pulses and by-products of the processing of soybeans, rapeseed and other oilseeds. These feeds have low moisture content and may be fed as individual ingredients or blended into particular formulations by compound feed manufacturers. In addition, concentrate feeds with high moisture content are also utilised (e.g. sugar beet pulp, brewers' grains and other co-products of crop-processing industries). These products are usually supplied directly by the processor to the farmer and, subsequently, conserved as silage.

Animal pathogens associated with feed include *L. monocytogenes*, *E. coli* O157:H7 and *Salmonella enterica*. Outbreaks of listeriosis in cattle herds have been associated with the feeding of poorly conserved silages contaminated with *L. monocytogenes* (Fenlon, 1988; Wiedmann *et al.*, 1996). Furthermore, there is evidence supporting a role of silage in the contamination of raw milk with *L. monocytogenes* (Sanaa *et al.*, 1993). In addition, recent studies suggest that cattle feed can be a vehicle for transmission of *E. coli* O157:H7 and *S. enterica* (Fenlon & Wilson, 2000; Davis *et al.*, 2003; Dodd *et al.*, 2003; Dargatz *et al.*, 2005). However, the significance of feed in the ecology of the bacterium in the farm environment and colonisation of cattle remains to be quantified.

Spore-forming bacteria isolated from feeds belong to the genera *Clostridium* and *Bacillus*. In contrast to vegetative cells, spores can survive the passage through the alimentary tract of the dairy cow, and are excreted with the faeces. *Clostridium* species, with particular dairy relevance, are *C. tyrobutyricum*, *C. butyricum*, *C. beijerinckii* and *C. sporogenes*. In cheeses such as Gouda and Emmenthal, the growth of these species, *C. tyrobutyricum* in particular, can cause off-flavours and excessive gas formation; a defect called late-blowing (Klijn *et al.*, 1995; Cocolin *et al.*, 2004; Le Bourhis *et al.*, 2005). Species of *Bacillus* organisms are associated with spoilage of heat-treated dairy products (Te Giffel *et al.*, 1997; Huemer *et al.*, 1998). Spores of *Clostridium* and *Bacillus* species are ubiquitous, and can be isolated from a wide variety of sources in the dairy farm environment, including soil, plants, bedding materials, concentrate feeds, roughages and cattle faeces (Te Giffel *et al.*, 1995; Vaerewijck *et al.*, 2001; Pahlow *et al.*, 2003). Silage is generally recognised as the most important source of *C. tyrobutyricum* spores in raw milk (Stadhouders & Jørgensen, 1990; Vissers *et al.*, 2006). Several studies indicate that silage is also a significant source of contaminating the milk with *Bacillus* spores (Slaghuis *et al.*, 1997; Te Giffel *et al.*, 2002; Vissers *et al.*, 2007b), which is due to growth of spore-forming bacteria in poorly conserved silages. This topic is further discussed in Section 3.3.

1.2.4 *Milking equipment*

Contamination of milk via the milking equipment occurs when (a) microorganisms adhere to surfaces of the milking equipment and (b) milk residues that remain in the equipment after the cleaning cycle (Figure 1.2). Under these conditions, growth of adhered microorganisms may occur, especially in cracked and decayed rubber parts, that are sensitive to accumulation of microorganisms (Akam *et al.*, 1989). During the next milking, adhered microorganisms can be released into the milk.

The level and type of contamination of milk via the milking equipment depends largely on the cleaning procedure applied. The milking machine is cleaned after each milking or in the case of automatic milking systems at regular intervals, to remove residues and prevent contamination during milking. In general, microorganisms originating from the farm environment (e.g. soil, faeces, bedding and feeds) are found on equipment surfaces, but also *S. aureus* has been recovered from surface of milking equipment (Bramley & McKinnon, 1990; Zadoks *et al.*, 2002). Cleaning the milking equipment at low temperatures or cleaning without sanitisers gives rise to fast growing gram-negative rods like coliforms and *Pseudomonas* (Murphy & Boor, 2000). Increasing the times between two milking

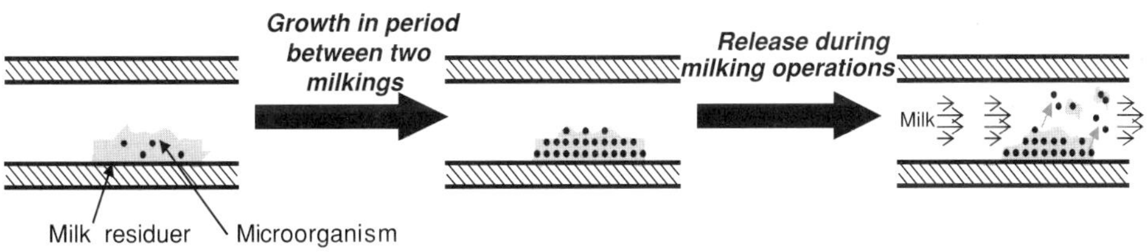

Fig. 1.2 Contamination of milk via the surfaces of the milking equipment.

intervals (i.e. more time available for growth) and higher temperatures during this period (i.e. increased growth rate) increase the number of microorganisms present in the equipment prior to milking and, thus, the level of contamination of milk.

1.2.5 *Microbial growth during milk storage*

It is common practice to collect milk of several milkings in a bulk tank before transportation of the milk to the dairy (e.g. milk of about four milkings in the UK and milk of six milkings in the Netherlands). To prevent microbial growth in the farm bulk tank, milk has to be cooled during storage. The European Union requires cooling of bulk tank milk to a temperature below 8°C when milk is collected daily, and to a temperature below 6°C when milk is not collected daily (European Commission, 2004b). However, cooling of milk does not prevent the growth of microorganisms completely. Some psychrotrophic organisms, such as *Pseudomonas* spp. and *L. monocytogenes*, still grow at temperatures below 6°C, although at reduced rates (Ratkowsky *et al.*, 1982; Te Giffel & Zwietering, 1999). Modelled studies showed that, in adequately functioning bulk tanks, the concentrations of psychrotrophic *L. monocytogenes* and *B. cereus* will not increase significantly (Albert *et al.*, 2005; Vissers *et al.*, 2007a).

1.3 Control of microbial contamination of bulk tank milk

1.3.1 *Good farming practice*

The HACCP approach has been implemented throughout the food and dairy industry, and it is a science-based quality management system developed to ensure the production of safe foods. Guidelines for the application of HACCP can be found in the Codex Alimentarius Code of Practice (FAO, 2003). Application of HACCP principles to dairy farms is discussed, but considered to be not yet generally feasible. The necessity for critical multidisciplinary review of management processes, difficulties in establishing limits via the identification of critical control points, the use of routine surveillance procedures and effective record keeping and documentation of standard processes restrict the widespread adoption of HACCP programme to dairy farms (Ruegg, 2003a). Furthermore, adequate monitoring is an essential principle in the HACCP methodology. Application of HACCP programmes for dairy farms is limited by the absence of adequate and low-cost monitoring tests (Gardner, 1997).

As an alternative to HACCP, the formulation of guides to good farming practices has been proposed (European Commission, 2004a). These guides should encourage the use of appropriate hygiene practices at farm level; however, the International Dairy Federation (IDF) and the Food and Agriculture Organisation (FAO) of the United Nations have developed such a guide (Morgan, 2004). The central objective is that the milk should be produced from healthy animals under generally accepted conditions. Good dairy farming practices require that people working and supervising at the farm are skilled in animal husbandry, hygienic milking of animals and administration of veterinary drugs. The guide contains guidelines with respect to different aspects of farm management.

In the next sections of this chapter, different aspects of farm management and their relations to hygienic milk production are discussed (Sections 1.3.2–1.3.7), whilst in Section 1.3.8, the use of mathematical models to identify effective control measures is briefly discussed.

1.3.2 *Animal health management*

Animal health management is extremely important for hygienic milk production. Mastitis infections lead to contamination of milk via the interior of teats, and gastrointestinal infections will increase the contamination via the exterior of teats. Furthermore, regulations of the European Union require that raw milk comes from animals that do not show any symptoms of infectious diseases that are communicable to humans via milk, and are in a good state of health and do not have udder wounds likely to affect milk; separation of milk of animals treated with authorised treatment products is also required (European Commission, 2004b).

Basically animal health management is aimed at achieving and sustaining a disease-free herd (Hillerton, 2004). This can be achieved when infected animals are cured or removed (e.g. culling) from the herd, and new infections are prevented. A closed herd, i.e. no import of animals from other farms, is an important measure to sustain a disease-free herd. Treatment and separation of infected animals from the rest of the herd prevents transmission of pathogens from cow to cow (Hillerton, 2004). In addition, a high feed quality, facility hygiene and hygienic milking operations are important to prevent infection of healthy cows with pathogens present in the farm environment.

As an example, mastitis control is an important issue for the dairy sector. In many countries, mastitis control programmes have been developed and implemented (Ekman *et al.*, 2005; Olde Riekerink *et al.*, 2005; Van der Zwaag, 2005). These programmes are usually based on five basic principles:

- Post-milking teat disinfection (see Section 3.5)
- Dry cow antibiotic therapy
- Appropriate treatment of clinical cases
- Culling of chronically infected cows
- Regular milking machine maintenance (Akam *et al.*, 1989).

In Norway, a successful udder health programme was implemented in 1982, and the main focus in this programme was on milking operations and correction of milking machines; however, less emphasis was put on dry cow therapy and teat dipping. In combination with changed farming attitudes and breeding programmes, this has led to a 50% reduction of treatments of clinical mastitis, a reduction of somatic cell counts (an indicator of subclinical mastitis) from 250 000 to 114 000 mL^{-1}, and a significant reduction in mastitis costs between 1994 and 2004 (Østeras & Sølverød, 2005).

1.3.3 *Control of feed*

Control of microbiological contaminants in feed is a critical factor, in particular for the contamination of raw milk with microbial spores. Because of the fundamental differences in microbiological hazard properties and control measures between concentrate feeds and roughages, these feed categories will be reviewed separately.

Factors of importance for the microbiological quality of concentrate feeds are the initial contamination level, of (cross-)contamination during processing, and contamination during storage (ICMSF, 2005). Commonly applied processing methods used in feed manufacturing, such as solvent extraction, extrusion and pelleting, reduce the concentration of vegetative bacteria, but generally do not inactivate spores. The low moisture content of concentrate raw materials and compound feeds prevents microbial growth. However, unintentional rehydration during storage can create conditions permitting microbial growth. In many countries, feed manufacturers have developed quality assurance systems, either individually or on a national level, aiming to control chemical and microbiological safety hazards in feed (Den Hartog, 2003).

The microbiological quality of roughages depends strongly on the effectiveness of the conservation, and the conservation principle of hay is low in moisture content. High moisture conditions in hay cause deterioration, especially by moulds and bacilli. Rapid drying of the crop in the field to at least 85 g dry matter 100 g^{-1} is important to achieve a high-quality product. The main principles of conservation by ensilage are a rapid achievement of a low pH by lactic acid fermentation involving lactic acid bacteria and the maintenance of anaerobic conditions. The pH after fermentation is determined mainly by the concentration of water-soluble sugars, buffer capacity and dry matter content of the crop and the activity of the lactic acid bacteria.

Undesirable microorganisms in silage are involved in anaerobic spoilage (primarily *Clostridium* species, especially *C. tyrobutyricum*) and aerobic/facultative anaerobic spoilage (e.g. acid-tolerant yeasts, moulds, *Bacillus* and/or *Listeria* species). The final concentration of spores of anaerobic microorganisms in silage is determined by the ensiling conditions, permitting or inhibiting growth of clostridia. Growth of clostridia can be prevented when a sufficiently low pH is achieved by fermentation. The pH needed for an anaerobically stable silage decreases with decreasing dry matter content (which is inversely related to water activity), and ranges from pH 4.1 to pH 5.0 for silage with dry matter content of 150–500 g kg^{-1} (Pahlow *et al.*, 2003). Silage additives are available that aim to control the fermentation process (Driehuis & Oude-Elferink, 2000). Wilting is commonly used to increase the dry matter content of grass and lucerne prior to ensiling. Another important factor, which affects the survival and growth of clostridia, is the nitrate concentration of the crop. Crops low in nitrate are more susceptible to spoilage by clostridia than crops high in nitrate (Kaiser *et al.*, 1999). The initial concentration of clostridia spores in the crop at ensiling is of minor importance (Hengeveld, 1983; Rammer, 1996).

Aerobic spoilage of silage is associated with penetration of air into the silage during storage or feeding. Lactate-oxidising yeasts are generally responsible for the initiation of aerobic spoilage, and the growth of these microorganisms causes an increase in pH, which subsequently permits the growth of other organisms. This secondary spoilage flora consists of moulds, bacilli, *Enterobacteriaceae*, *Listeria* and even clostridia (Woolford, 1990; Pahlow *et al.*, 2003; Vissers *et al.*, 2007b). Under practical farming conditions, exposure of silages to air is inevitable after opening a silo for feeding. The extent of penetration of air into the silage mass mainly depends on the porosity and density of the material, pressure gradients in the silo and the rate of silage removal (Honig, 1991). However, aerobic spoilage often starts during the storage period, for instance because in practice the silo construction and sealing materials, usually plastics, do not prevent air infiltration completely, and there may be damage to the sealing materials.

To a great extent, the success of silage conservation is dependent on management decisions. Silage management factors that are under control by the farmer are (a) time of harvest, (b) dry matter content at ensiling, (c) harvesting and ensiling machinery, (d) use of silage additives, (e) silo construction, (f) compaction and sealing and (g) rate of silage removal and face management. Not all farmers are aware of the effects that these factors can have on the quality and safety of silage and, ultimately, on milk quality.

Communicating relevant information on these topics to farmers is of great importance for hygienic milk production. In this context, the recent implementation in Europe of the Feed Hygiene Regulation is of importance too (European Commission, 2005). This regulation establishes requirements for feed hygiene on the basis of HACCP principles and applies to feed manufacturers and other operators in the feed production chain, including individual farmers.

1.3.4 *Facility hygiene*

Facility hygiene comprises amongst others the cleanliness of the barn, access alleys and milking parlour, and is an integral part of hygienic milk production and quality control programme. The cleanliness of cows (e.g. udder and teats) and, thus, microbial contamination of milk via the exterior of teats and the incidence of mastitis are affected by measures related to facility hygiene (Ruegg, 2003b). They include, for example, regular removal of dung from the barn, regular refreshment of bedding materials, clean entries to the milking parlour, one or more cubicles per cow and non-crossing walking paths (Haven *et al.*, 1996; Ouweltjes *et al.*, 2003; Ruegg, 2003b).

Surveys have been performed to establish relations between measures related to facility hygiene and microbial counts in bulk tank milk. A correlation between the dirtiness of the access alley to the milking parlour and the concentration of spores of *B. cereus* in bulk tank milk has been reported by Christiansson *et al.* (1999). Relations between cleanliness of housing areas and the rate of clinical mastitis and high somatic cell counts in bulk tank milk have also been established (Barkema *et al.*, 1998). Herlin & Christannson (1993) compared tied and loose housing systems; remarkably, lower concentrations of clostridial spores in milk were detected at farms with tied housing systems despite the fact that cubicles, parlour floors and teats were considered less clean than at farms with loose housing systems. A more intensive care and management of the barn and thorough cleaning of dirty teats (prior to milking) in tied housing systems were considered to be the cause of these opposing observations. In their survey, Hutchinson *et al.* (2005) did not find any significant relation between various microbial parameters (total viable counts, coliforms, *Bacillus* spp., *Bifidobacteria* spp. and *Pseudomonas* spp.) and hygienic practices, such as milking parlour cleaning regime and barn cleanliness.

The lack of clear and significant relation between measured facilities and microbial counts in bulk tank milk may be due to various reasons. First, microbial concentrations in faeces, bedding and soil vary more than the amount of dirt transferred to milk (see Section 1.2.3 and Table 1.3). This means that the microbial concentration in bulk tank milk depends more on the concentration of microorganisms in the dirt rather than the mass of transferred dirt. With respect to the contamination of bulk tank milk with spores of butyric acid bacteria and *B. cereus*, it has been concluded on the basis of mathematical models that the concentration of these spores in silage and soil are more important than hygienic measures associated with the cleanliness of teats and milking operations (Vissers *et al.*, 2006, 2007a). Second, hygienic practices comprise a large number of measures, and negligence of one or more of these measures can diminish the positive effect of other measures on microbial counts in bulk tank milk. The study of Herlin & Christiansson (1993) is an example of the complexity of the effect of facility hygiene on the microbiological quality of milk. Third, seasonal variations and outbreaks of mastitis can affect bulk tank milk counts independent of facility hygiene (Slaghuis *et al.*, 1997; Zadoks *et al.*, 2001).

1.3.5 *Milking operations*

Hygienic milking operations start with a clean and stress-free milking environment, teat cleaning, pre-dipping, fore-stripping, careful attachment of the teats cups and post-milking teat disinfection. Teat cleaning is performed to reduce the microbial load on the teats prior to milking. Pre-dip agents are often used to disinfect the teats prior to milking and reduce the risk of environmental mastitis. Pre-dipping should be applied with care since residues may contaminate milk. Fore-stripping is expressing two or three streams of milk before attachment of the milk liners in order to visibly check the milk quality and to stimulate milk let down. Post-milking teat disinfection is important to increase the hygienic defence against infection of the teats after milking is completed. An overview of literature and methods that were proved to be useful for improving milking performance were reported by Reinemann *et al.* (2005).

Table 1.4 summarises experimentally determined efficiencies of four commonly used manual cleaning methods. In general, cleaning with a moist plus a dry towel has the highest efficiency, whilst cleaning with a dry towel the lowest, although for the microbial contamination of bulk tank milk differences are marginal. The difference between the least and most effective method on bulk tank milk microbial counts is $<0.5 \log_{10}$ cfu mL^{-1}; however, in the light of mastitis, this difference can be of importance.

In automatic milking systems (for a description of automatic milking systems see Section 1.3.6), teats are cleaned automatically. Teat cleaning efficiencies of different brands and

Table 1.4 Mean and range (%) of bacteria removal reported for different teat cleaning methods.

	Total viable count		Coliforms		Clostridia (i.e. spores)	
Teat cleaning method	Mean	Range	Mean	Range	Mean	Range
Dry towel	41	3–90	51	4–99	60	45–79
Moist towel	42[a]	40–43	35[b]	75	50–91	
Moist towel + dry towel	77	67–97	71	55–84	95	92–96
Pre-dip + dry towel	63	44–80	63	48–74	88[a]	85–91

[a] Data based on two observations.
[b] Result of one observation.
After Galton *et al.* (1982, 1986), Pankey (1989), Rasmussen *et al.* (1991) and Magnusson *et al.* (2006).

types of automatic milking systems differ, but significant differences were also observed between farms using the same system (Jepsen *et al.*, 2004). Cleaning efficiencies range from 50 to 98% (Melin *et al.*, 2002; Knappstein *et al.*, 2004a), and are comparable with efficiencies of manual methods (Table 1.4). Teat cleaning intervals (2 or 3 times a day) do not affect microbial counts in bulk tank milk significantly (Benfalk & Gustafsson, 2004).

Finally, it should be realised that teat cleaning efficiencies determined under experimental conditions may not be translated directly to the practical situation on farms. The farmer's perception of hygiene may be more important; this is supported by the results of a survey in the UK demonstrating that teat cleaning efficiencies achieved in practice were lower than the theoretical efficiencies (Gibson *et al.*, 2005). Moreover, in 28% of the cases, teat cleaning resulted in an increase of the microbial concentration in milk. The use of the same towel for more than one cow, not using an effective disinfectant, insufficient general parlour hygiene, microbiologically dirty hands and contamination of cleaning materials and solutions were all considered to have the potential to increase microbial concentrations in milk.

1.3.6 *Milking machine design and operations*

In former days, hand milking was applied, but at present in developed countries, milking machines are more widely used. In these machines, milk is extracted from the udder and conveyed to the cooled bulk tank automatically using air–vacuum pulses to extract milk (Figure 1.3) (Akam *et al.*, 1989; Haven *et al.*, 1996). ISO standard 5707 (ISO, 2001) covers the construction and performance of milking machines, and important aspects of the design are the possibility to drain the different elements and the cleanability of the different parts. With respect to maintenance, it is important to regularly check and replace rubber parts; cracked and decayed rubber parts are very sensitive to accumulation of microorganisms.

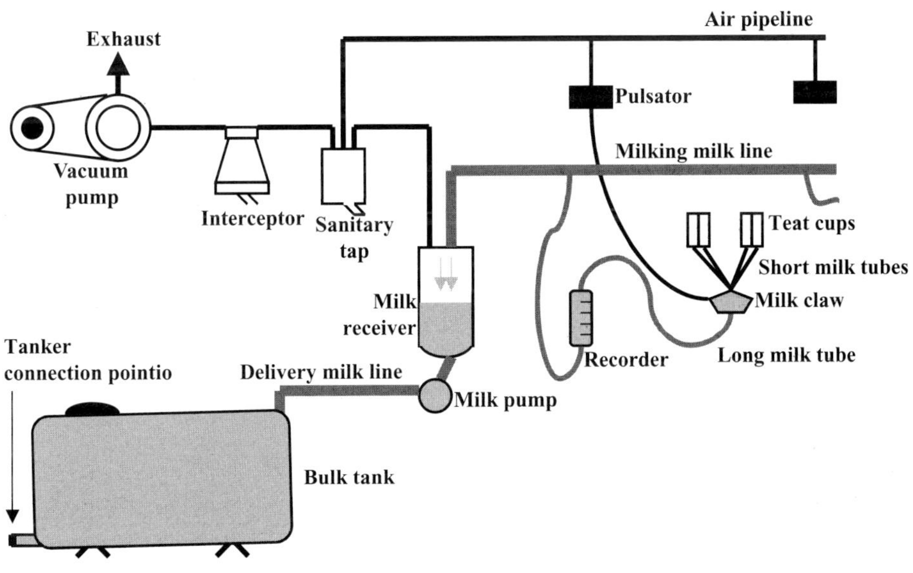

Fig. 1.3 Schematic representation of milking machines.

More and more, larger farms are implementing automatic milking systems. Cows can enter these systems freely and are milked without interference of the farmer. To assure a high hygienic level of the milk, automatic milking systems are equipped with animal recognition functions, online monitoring systems for pre-defined milk quality indicators and teat cleaning functions. Online monitoring systems are used to divert milk from pre-defined animals (e.g. cows treated with antibiotics), off-quality milk and foremilk from the main milk flow. More information can be obtained from the 'Code of Good Hygienic Practices for Milking with Automatic Milking Systems', which was published by the IDF (Jepsen *et al.*, 2004).

Contamination of milk via surfaces of milking equipment depends largely on the effectiveness of cleaning procedures. The choice of a cleaning regime depends on national regulations, the energy costs for heating water and chemicals and local habits. In most European countries, the standard cleaning regime starts with a pre-rinse with hot water (35–45°C), followed by 8 to 10 min cleaning with an alkaline detergent and disinfectant, and finally, a cold water rinse. Additionally, the equipment can be rinsed with an acid solution to remove milk stone, and in Denmark, an acid rinse is generally implemented directly after the cold water rinse, and an extra cold water rinse is performed before the start of each milking. In the USA, the most common routine consists of a pre-rinse, alkaline detergent, acid rinse and pre-milking disinfection. More information about cleaning and sanitation of milking machines have been reported by Reinemann *et al.* (2003).

1.3.7 *Bulk tank design and operations*

Milk has to be cooled during storage in bulk tank at the farm, and in the ISO standard 5708 (ISO, 1983), requirements for design and operation of refrigerated bulk tanks milk are described together with methods for testing the performance. The materials used and design of bulk tank should allow proper cleaning of the tank using an automatic (or semi-automatic) system and fast drainage by gravity. The cooling system of the tank must be able to cool a full tank from 35 to 4°C within 2.5–3.5 h, and milk of the second milking from 10 to 4°C within 0.8–1.75 h. Installation of a plate cooler in the milk line could further reduce cooling time; however, plate coolers increase the risk of contamination of the milk via the surfaces of the equipment and extra attention should be paid when cleaning it. Isolation of the tank should prevent heating of the milk with more than 0.25°C h^{-1} when the cooling system does not work. Tanks should be equipped with a system to monitor the milk temperature, and important parameters of bulk tank cleaning procedures involve the disinfectants used, temperature at which cleaning is performed and the rinsing procedure to remove disinfectants (Reinemann *et al.*, 2003).

1.3.8 *Identification of effective control measures*

In the field of food safety, there has been a shift from qualitative approaches like good manufacturing practices (comparable with guidelines for good farming practices) and HACCP to more quantitative risk analyses using mathematical models (Cassin *et al.*, 1998;

Van Gerwen, 2000). Quantitative risk analyses are used to assess the risks throughout the food chain, and to identify effective measure to improve food safety.

The use of mathematical models at the farm level is also useful to identify effective measures to control animal welfare and the microbial contamination of bulk tank milk. Over the years, various models describing microbial process at the farm have been developed starting with models to simulate and identify measure to control animal diseases. The development of epidemiological models to simulate outbreaks of mastitis and to identify measures to control spreading of mastitis started more than 25 years ago, and are still in progress (Oltenacu & Natzke, 1975; Allore & Erb, 1999; Grinberg *et al.*, 2005). Groenendaal *et al.* (2002) developed a simulation model to improve control of Johne's disease, an infection of the gastrointestinal tract of cows. These models, amongst other, increase the insight into the dynamics of the spread of infectious pathogens within a herd, and reveal whether it is, for example, more effective to cull or treat infected animals. Also, microbial processes in silage have been modelled extensively (Leibensperger & Pitt, 1987; Ruxton & Gibson, 1993, 1995; Kelly *et al.*, 2000). An interesting example is a model that links aerobic deterioration of big-bale silage with growth of *L. monocytogenes* in silage (Ruxton & Gibson, 1995). Strategies to control the contamination of bulk tank milk with spores of butyric acid bacteria and *B. cereus* have been identified using models based on the contamination pathway of these microorganisms (Vissers *et al.*, 2006, 2007a). Controlling the silage quality is most important to prevent concentrations of butyric acid bacteria above 1 spore mL^{-1}. To control the contamination of bulk tank milk with spores of *B. cereus*, it is most important to keep teats clean during the grazing period, and to assure a high feed quality during the housing period.

1.4 Future developments in handling of the milk on the farm

Since prehistoric times, humans have kept animals for the production of milk for human consumption. Economical, social and technological developments have forced dairy farmers and dairy producers to continuously change and improve their production processes. Nowadays globalisation puts milk prices under pressure; also, the quality and safety of raw milk receives more and more attention from consumers and governmental bodies. On the other hand, the increasing sizes of dairy farms and technological developments, such as the introduction of automatic milking systems, generate new opportunities for improving the production processes at the farm. In the future, three relevant technological trends for milk handling on the farm will be:

1.4.1 *Concentration of milk*

Milk consists for more than 90% of water. This means that transportation of milk from the dairy farm to the processing dairy, for a large part, is transportation of water. Concentration of milk at the dairy farm, using membrane filtration, is a way to decrease costs of transportation and energy use. Especially, with increasing herd sizes, this option becomes more and more attractive to dairy farmers. In New Zealand, concentration of the milk at the farm is already applied, and in many countries application of this technology is under discussion.

1.4.2 *Heat treatment of the milk*

In general, raw milk undergoes a preliminary heat treatment (± 10 s, 65°C) directly after arriving at the dairy processing plant and before storage in a silo. This treatment is known as thermisation, and it is applied to inactivate psychrotrophic microorganisms. These organisms produce heat-stable enzymes like lipases and proteinases at low temperatures, and are responsible of spoilage of dairy products. Thermisation or pasteurisation at the dairy farm decreases the time for psychrotrophic microorganisms to multiply and produce heat-stable enzyme and, thus, decreases risks of spoilage.

1.4.3 *In/online monitoring of bulk tank milk quality*

Sensors and measuring systems to automatically analyse quality indicators are getting faster and more accurate. Especially, automatic milking systems offer possibilities for the implementation of online monitoring tools. Online application of these automatic measuring systems combined with modern information and communication technology solutions enables the farmer and the industry to gain more information about raw milk quality during milking operations and storage of the milk in the bulk tank. Based on these data, the farmer can react; for example, milk from cows under medical treatment can be separated, and batches of the bulk tank milk can be selected and sent to dedicated dairy plants. In addition, if the concentration of spores of butyric acid bacteria can be measured at the dairy farm and the count is high, the milk of this farm can be sent to a processing plant where no cheese is produced.

1.5 Conclusions

On-farm hygienic milk production is important for farmers, the dairy industry and consumers. For farmers, hygienic milk production is not only important with respect to the quality of the bulk tank milk, but also for animal welfare. Microorganisms in bulk tank milk at the farm originate from the interior of teats, the farm environment and surfaces of the milking equipment. Different microorganisms have different origins and, hence, require different control measures. Therefore, hygienic milk production involves many aspects of farm management, varying from animal welfare and feed management to bulk tank design. Mathematical models are useful means to identify effective measures for control of microbial contamination. It should be kept in mind that complete control is not possible. The contamination of bulk tank milk is also affected by uncontrollable aspects, such as seasonal variations in microbial concentrations in, for example, the soil and periodic stress, such as calving. The awareness of farmers of the impact of hygiene in various aspects of farm management on milk quality and their attitude towards hygiene in everyday practices are key factors in hygienic milk production. Therefore, to make progress in this field, more attention to education of and communication with farms is needed.

References

Akam, F.D., Dodd, F.H. & Quick, A.J. (1989) *Milking, Milk Production Hygiene and Udder Health*, Report No. 78, Food and Agriculture Organization of the United Nation, Rome, pp. 56–95.

Albert, I., Pouillot, R. & Denis, J.-B. (2005) Stochastically modeling *Listeria monoytogenes* growth in farm tank milk. *Risk Analysis*, **25**, 1171–1185.

Allore, H. & H. Erb, H. (1999) Approaches to modeling intramammary infections in dairy cattle. *Preventive Veterinary Medicine*, **39**, 279–293.

Barkema, H.W., Schukken, Y.H., Lam, T.J.G.M., Beoboer, M.L., Benedictus, G. & Brand, A. (1998) Management practices associated with low, medium and high somatic cell counts in milk. *Journal of Dairy Science*, **81**, 1917–1927.

Benfalk, C. & Gustafsson, M. (2004) Investigations of different cleaning frequencies in automatic milking systems. In: *Farm Hygiene and Teat cleaning Requirements in Automatic Milking* (eds A. Meijering, H. Hogeveen & C.J.A.M. de Koning), pp. 101–105, Wageningen Academic Publishers, Wageningen, The Netherlands.

Bramley, A.J. & McKinnon, C.H. (1990) The microbiology of raw milk. In: *Dairy Microbiology*, 2nd edn, Vol. 1, (ed. R.K. Robinson), pp. 163–208, Elsevier Science Publishers, London.

Cassin, M., Paoli, G. & Lammerding, A. (1998) Simulation modeling for microbial risk assessment. *Journal of Food Protection*, **61**, 1560–1566.

Chambers, J.V. (2002) The microbiology of raw milk. In: *Dairy Microbiology Handbook*, 3rd edn (ed. R.K. Robinson), pp. 39–90, John Wiley & Sons, New York.

Christiansson, A., Bertilsson, J. & Svensson, B. (1999) *Bacillus cereus* spores in raw milk: factors affecting the contamination of milk during the grazing period. *Journal of Dairy Science*, **82**, 305–314.

Cocolin, L., Innocente, N., Biasutti, M. & Comi, G. (2004) The late blowing in cheese: a new molecular approach based on PCR and DGGE to study the microbial ecology of the alteration process. *International Journal of Food Microbiology*, **90**, 83–91.

Dargatz, D.A., Strohmeyer, R.A., Morley, P.S., Hyatt, D.R. & Salman, M.D. (2005) Characterization of *Escherichia coli* and *Salmonella enterica* from cattle feed ingredients. *Foodborne Pathogens and Disease*, **2**, 341–347.

Davis, M.A., Hancock, D.D., Rice, D.H., Call, D.H., di Giacomo, R., Samadpour, M. & Besser, T.E. (2003) Feedstuffs as a vehicle of cattle exposure to *Escherichia coli* O157:H7and *Salmonella enterica*. *Veterinary Microbiology*, **95**, 199–210.

Den Hartog, J. (2003) Feed for food: HACCP in the animal feed industry. *Food Control*, **14**, 95–99.

Dodd, C.C., Sanderson, M.W., Sargeant, J.M., Nagaraja, T.G., Oberst, R.D., Smith, R.A. & Griffin, D.D. (2003). Prevalence of *Escherichia coli* O157 in cattle feeds in Midwestern feedlots. *Applied Environmental Microbiology*, **69**, 5243–5247.

Driehuis, F. & Oude-Elferink, S.J.W.H. (2000) The impact of the quality of silage on animal health and food safety a review. *Veterinary Quarterly*, **22**, 212–216.

Ekman, T., Landin, H.L., Waldner, J., Gyllenswärd, M. & Hallén Sandgren, C. (2005) 'Friskko–Juver' ('Healthycow–Udder') A systematic veterinary approach to udder disease in dairy herds. *4th IDF International Mastitis Conference*, (ed. H. Hogeveen), pp. 605–610, Wageningen Academic Publishers, Wageningen, The Netherlands.

European Commission (2004a) Regulation (EC) No 852/2004 on the hygiene of foodstuffs. *Official Journal of the European Union*, **L139**, 1–54.

European Commission (2004b) Regulation (EC) No 853/2004 laying down specific hygiene rules for food of animal origin. *Official Journal of the European Union*, **L139**, 55–205.

European Commission (2005) Regulation (EC) No 183/2005 laying down requirements for feed hygiene. *Official Journal of the European Union*, **L35**, 1–22.

FAO (2003) *General Principles of Food Hygiene*, Vol. 1, 4th revision of the 1969 edition, pp. 31–32, Codex Alimentarius of the Food and Agriculture Organisation of the United Nation Rome.

Fenlon, D.R. (1988) Listeriosis. In: *Silage and Health* (eds B.A. Stark & J.M. Wilkinson), pp. 7–18, Chalcombe Publications, Marlow, UK.

Fenlon, D.R. & Wilson, J. (2000) Growth of *Escherichia coli* O157 in poorly fermented laboratory silage: a possible environmental dimension in the epidemiology of *E. coli* O157. *Letters in Applied Microbiology*, **30**, 118–121.

Galton, D.M., Adkinson, R.W., Thomas, C.V. & Smith, T.W. (1982) Effect of premilking udder preparation on environmental bacterial contamination of milk. *Journal of Dairy Science*, **65**, 1540–1543.

Galton, D.M., Petersson, L.G. & Merill, W.G. (1986) Effects of premilking preparation practices on bacterial counts in milk and on teats. *Journal of Dairy Science*, **69**, 260–266.

Gardner, I.A. 1997. Testing to fulfil HACCP Requirements: principles and examples. *Journal of Dairy Science*, **80**, 3453–3457.

Gibson, H., Protheroe, R.G., Sinclair, L.A., Brizuela, C.M. & Worton, H. (2005) *Investigation of the Effectiveness of Pre-Milking Teat Cleaning Regimes*, FSA: Code – B12003, pp 1–79, University of Wolverhampton and Harper Adams University College (Newport), Wolverhampton, UK.

Grinberg, A., Lopez-Villalobos, N., Lawrence, K. & Nulsen, M. (2005) Prediction of penicillin resistance in *Staphylococcus aureus* isolates from dairy cows with mastitis, based on prior test results. *New Zealand Veterinary Journal*, **53**, 332–335.

Groenendaal, H., Nielen, M., Jalvingh, A.W., Horst, S.H., Galligan, D.T. & Hesselink, J.W. (2002). A simulation of Johne's disease control. *Preventive Veterinary Medicine*, **54**, 225–245.

Haven, M.C., De Koning, C.J.A.M., Wemmenhove, H. & Westerbeek, R. (1996) Kwaliteit van Melk. In: *Handboek Melkwinning* (eds K. de Ridder & H. Wolbers), pp. 47–70, Praktijkonderzoek Rundvee Schapen en Paarden, Lelystad, The Netherlands.

Hayes, M.C., Ralyea, R.D., Murphy, S.C., Carey, N.R., Scarlett, J.M. & Boor, K.J. (2001) Identification and characterization of elevated microbial counts in bulk tank raw milk. *Journal of Dairy Science*, **84**, 292–298.

Hengeveld, A.G. (1983) Spores of butyric acid bacteria in silage. *Rapport-Proefstation-voor-de-Rundveehouderij,-Schapenhouderij-en-Paardenhouderij*, **88**, 1–54.

Herlin, A.H. & Christiansson, A. (1993) Cheese-blowing anaerobic spores in bulk milk from loose-housed and tied dairy cows. *Milchwissenschaft*, **48**, 686–690.

Hillerton, J.E. (2004) Control of MAP in milk. In: *New Applications of Mid-Infra-Red Spectrometry for the Analysis of Milk and Milk Products*, Proceedings of IDF Symposium of Advancement in Analytical Techniques (Holstebro, 21 May 2003), pp. 17–19, Document No. 383, International Dairy Federation, Brussels.

Hogan, J.S., Bogacz, V.L., Thompson, L.M., Romig, S., Schoenberger, P.S., Weiss, W.P. & Smith, K.L. (1999) Bacterial counts associated with sawdust and recycled manure bedding treated with commercial conditions. *Journal of Dairy Science*, **82**, 1690–1695.

Hogan, J.S. & Smith, K.L. (1997) Bacterial counts in sawdust bedding. *Journal of Dairy Science*, **80**, 1600–1605.

Hogan, J.S., Smith, K.L., Todhunter, D.A. & Schoenberger, P.S. (1990) Bacterial counts associated with recycled newspaper bedding. *Journal of Dairy Science*, **73**, 1756–1761.

Holm, C., Jepsen, J., Larsen, M. & Jespersen, L. (2004) Predominant microflow of downgraded Danish bulk tank milk. *Journal of Dairy Science*, **87**, 1151–1157.

Honig, H. (1991) Reducing losses during storage and unloading of silage. In: *Silage Conservation Towards 2000* (eds G. Pahlow & H. Honig), pp. 116–128, Landbauforschung Völkenrode Sonderheft 123, Bundesforschungsanstalt für Landwirtschaft Braunschweig-Völkenrode, Braunschweig, Germany.

Huemer, I.A., Klijn, N., Vogelsang, H.W.J. & Langeveld, L.P.M. (1998) Thermal death kinetics of spores of Bacillus sporothermodurans isolated from UHT milk. *International Dairy Journal*, **8**, 851–855.

Hutchinson, M.L., Thomas, D.J.I., Moore, A., Jackson, D.R. & Ohnstad, I. (2005) An evaluation of raw milk microorganisms as markers of on-farm hygiene practices related to milking. *Journal of Food Protection*, **68**, 764–772.

ICMSF (International Commission on Microbiological Specifications of Foods, ed.) (2005) *Microbial Ecology of Food Commodities*, Vol. 6, Kluwer Academic & Plenum Publishers, New York.

IDF (2006) *Payment Systems for Ex-Farm Milk*, Document No. 403, International Dairy Federation, Brussels.

ISO (1983) *Refrigerated Bulk Milk Tanks*, Document No. 5708, International Standard Organization, Geneva.

ISO (2001) *Milking Machine Installations – Construction and Performance*, Document No. 5707, International Standard Organization, Geneva.

Jayaroa, B.M. & Henning, D.R. (2001) Prevalence of foodborne pathogens in bulk tank milk. *Journal of Dairy Science*, **84**, 2157–2162.

Jepsen, L., Heggum, D., Zecconi, A., Bouchard, R. & Knappstein, K. (2004) *Code of Good Hygienic Practice for Milking with Automatic Milking Systems*, Document No. 386, pp. 44–48, International Dairy Federation, Brussels.

Kaiser, E., Weiss, K. & Milimonka, A. (1999) Investigations on the fermentation quality of silage from low-nitrate green forage. *Archiv Tierernahrung*, **52**,75–93.

Kelly, L.A., Gibson, G., Gettingby, G., Donachie, W. & Low, J.C. (2000) A Predictive model of the extend of listerial contamination within damaged silage bales. *Quantitative microbiology*, **2**, 171–188.

Klijn, N., Nieuwenhof, F.F.J., Hoolwerf, J.D., Van Der Waals, C.B. & Weerkamp, A.H. (1995) Identification of *Clostridium tyrobutyricum* as the causative agent of late blowing in cheese by species-specific PCR amplification. *Applied and Environmental Microbiology*, **61**, 2919–2924.

Knappstein, K., Roth, N., Slaghuis, B.A., Ferweda-van Zonneveld, R.T., Walte, H.G. & Reichmuth, J. (2004a) Farm hygiene and teat cleaning requirements in automatic milking. In: *Automatic Milking – A Better Understanding*, (eds A. Meijering, H. Hogeveen & C.J.A.M. de Koning), pp. 83–93, Wageningen Academic Publishers, Wageningen, The Netherlands.

Knappstein, K., Roth, N., Walte, H.G. & Reichmuth, J. (2004b) *Hygiene Measures Resulting in Adequate Teat Cleaning, Deliverable D15*, pp. 4–26, Institute for Hygiene and Food Safety, Federal Research Centre for Nutrition and Food, Kiel, Germany.

Le Bourhis, A.G., Saunier, K., Dore, J., Carlier, J.P., Chamba, J.F., Popoff, M.R. & Tholozan, J.L. (2005) Development and validation of PCR primers to assess the diversity of *Clostridium* spp. in cheese by temporal temperature gradient gel electrophoresis. *Applied and Environmental Microbiology*, **71**, 29–38.

Leibensperger, R.Y. & Pitt, R.E. (1987) A model of clostridial dominance in ensilage. *Grass and Forage Science,* **42**, 297–314.

Magnusson, M., Christiansson, A. & Svensson, B. (2007). *Bacillus cereus* spores during housing of dairy cows: factors affecting contamination of raw milk. *Journal of Dairy Science*, **90**, 2745–2754.

Magnusson, M., Christiansson, A. & Svensson, B. & Kolstrup, C. (2006) Effect of different premilking manual teat-cleaning methods on bacterial spores in milk. *Journal of Dairy Science*, **89**, 3866–3875.

Makovec, J.A. & Ruegg, P.L. (2003) Results of milk samples submitted for microbiological examination in Wisconsin from 1994 to 2001. *Journal of Dairy Science*, **86**, 3466–3472.

Melin, M., Wiktorsson, H. & Christiansson, A. (2002) Teat cleaning efficiency before milking in Delaval VMS^{TM} versus conventional manual cleaning using *Clostridium tyrobutyricum* spores as mark. *1st North American Conference on Robotic Milking*, pp. 60–63, Toronto.

Morgan, T.G. (2004) Good agricultural practice. In: *Guide to Good Dairy Farming Practice*, pp. 5–8, Food and Agriculture Organization of the United Nations Rome and International Dairy Federation, Brussels.

Murphy, S.C. & Boor, K.J. (2000) Sources and causes of high bacteria counts in raw milk: an abbreviated review. *Dairy, Food and Environmental Sanitation*, **20**(8), 1–4.

Olde Riekerink, R.G.M., Barkema, H.W., Scholl, D.T., Kelton, D.F., Keefe, G.P. & Sandgren, C.H. (2005) Implementation of recommended mastitis prevention management practices and the herd level prevalence of contagious mastitis pathogens on Canadian dairy farms. *4th IDF International Mastitis Conference* (ed. H. Hogeveen), pp. 587–592, Wageningen Academic Publishers, Wageningen, The Netherlands.

Oltenacu, P.A. & Natzke, R.P. (1975) Mathematical modelling of the mastitis infection process. *Journal of Dairy Science*, **59**, 515–521.

Østeras, O. & Sølverød, L. (2005) Mastitis control system: the Norwegian experience. *4th IDF International Mastitis Conference* (ed. H. Hogeveen), pp. 91–101, Wageningen Academic Publishers, Wageningen, The Netherlands.

Ouweltjes, W., Van Dooren, H.J.C., Ruis-Heutinck, L.F.M., Dijk, G.J. & Meijering, A. (2003) *Huisvesting van Melkvee: Knelpunten uit Oogpunt van Welzijn*, Praktijkonderzoek Veehouderij, Wageningen, The Netherlands.

Pahlow, G., Muck, R.E., Driehuis, F., Oude Elferink, S.J.W.H. & Spoelstra, S.F. (2003) Microbiology of ensiling. In: *Silage Science and Technology* (eds D.R. Buxton, R.E. Muck & J.H. Harrison), pp. 31–93, Agronomy Monograph 42, American Society of Agronomy Inc., Crop Science Society of America Inc., Soil Science Society of America Inc. Publishers, Madison, WI.

Pankey, J.W. (1989) Premilking udder hygiene. *Journal of Dairy Science*, **72**, 1308–1312.

Rammer, C. (1996) Quality of grass silage infected with spores of *Clostridium tyrobutyricum*. *Grass and Forage Science*, **51**, 88–95.

Rasmussen, M.D., Galton, D.M. & Petersson, L.G. (1991) Effects of premilking teat preparation on spores of anaerobes, bacteria and iodine residues in milk. *Journal of Dairy Science*, **74**, 2471–2478.

Ratkowsky, D.A., Olley, J., McMeekin, T.A. & Ball, A. (1982) Relationships between temperature and growth rate of bacterial cultures. *Journal of Bacteriology*, **149**, 1–5.

Reinemann, D.J., Mein, G.A., Rasmussen, M.D. & Ruegg, P.L. (2005) Simple machine checks and milking-time observations, Basic test of pulsator function and of vacuum during milking, and Complete professional machine evaluation. In: *Evaluating Milking Performance*, Document No. 396, pp. 3–9, 12–14 and 14–16, respectively, International Dairy Federation, Brussels.

Reinemann, D.J., Wolters, G., Billon, P., Lind, O. & Rasmussen, M.D. (2003) *Review of Practices for Cleaning and Sanitation of Milking Machines*, Document No. 381, pp.1–12, International Dairy Federation, Brussels.

Ruegg, P.L. (2003a) Practical food safety intervention for dairy production. *Journal of Dairy Science*, **86**(E. Suppl.), E1–E9.

Ruegg, P.L. (2003b) The role of hygiene in efficient milking. *Milking and Milk Quality*, **406**, 1–6.

Ruxton, G.D. & Gibson, G. (1993) A critical introduction to mathematical modelling of the aerobic deterioration of silage. *Journal Science of Food and Agriculture*, **62**, 209–217.

Ruxton, G.D. & Gibson, G. (1995) A mathematical model of the aerobic deterioration of big-bale silage and its implications for the growth of *Listeria monocytogenes*. *Grass and Forage Science*, **50**, 331–344.

Sampimon, O.C., Vernooij, J.C.A. & Sol, J. (2004) Praktische aspecten rond droogzetten. *Tijdschrift Diergeneeskunde*, **129**, 823–833.

Sanaa, M., Poultrel, B., Menard, J.L. & Seieys, F. (1993) Risk factors associated with contamination of raw milk by *Listeria monocytogenes* in dairy farms. *Journal of Dairy Science*, **76**, 2891–2898.

Slaghuis, B.A., Te Giffel, M.C., Beumer, R.R. & Andre, G. (1997) Effect of pasturing on the incidence of *Bacillus cereus* spores in raw milk. *International Dairy Journal*, **7**, 201–205.

Smith, K.L. (1983) Mastitis control: a discussion. *Journal of Dairy Science*, **66**, 1790–1794.

Smith, K.L. & Hogan, J.S. (1993) Environmental mastitis. *Veterinary Clinical North America*, **9**, 489–498.

Stadhouders, J. & Jørgensen, K. (1990) Prevention of the contamination of raw milk by a hygienic milk production. *Production by Microorganisms and its Use for Cheesemaking – Detection and Prevention of Anaerobic Spore Formers and Cheese Quality*, Document No. 251, pp. 32–36, International Dairy Federation, Brussels.

Te Giffel, M.C., Beumer, R.R., Granum, P.E. & Rombouts, F.M. (1997) Isolation and characterization of *Bacillus cereus* from pasteurized milk in household refrigerators in the Netherlands. *International Journal of Food Microbiology*, **34**, 207–318.

Te Giffel, M.C., Beumer, R.R., Slaghuis, B.A. & Rombouts, F.M. (1995) Occurrence and characterization of (psychrotrophic) *Bacillus cereus* on farms in the Netherlands. *Netherlands Milk and Dairy Journal*, **49**, 125–138.

Te Giffel, M.C., Wagendorp, A., Herrewegh, A. & Driehuis, F. (2002) Bacterial spores in silage and raw milk. *Antonie van Leeuwenhoek*, **81**, 625–630.

Te Giffel, M.C. & Zwietering, M.H. (1999) Validation of predictive models describing the growth of *L. monocytogenes*. *International Journal of Food Microbiology*, **46**, 135–149.

Vaerewijck, M., De Vos, P., Lebbe, L., Scheldeman, P., Hoste, B. & Heyndrickx, M. (2001) Occurrence of *Bacillus sporothermodurans* and other aerobic sporeforming species in feed concentrate for dairy cattle. *Journal of Applied Microbiology*, **91**, 1074–1084.

Van Der Zwaag, H.G. (2005) Mastitis control program in the Netherlands: goals, tools and conditions. *4th IDF International Mastitis Conference* (ed. H. Hogeveen), pp. 599–604, Wageningen Academic Publishers, Wageningen, The Netherlands.

Van Gerwen, S.J.C. (2000) *Microbiological Risk Assessment of Food*. Wageningen University, The Netherlands, pages 158.

Vissers, M.M.M., Driehuis, F., Te Giffel, M.C., De Jong, P. & Lankveld, J.M.G. (2006) Improving farm management by modeling the contamination of farm tank milk with butyric acid bacteria. *Journal of Dairy Science*, **89**, 850–858.

Vissers, M.M.M., Driehuis, F., Te Giffel, M.C., De Jong, P. & Lankveld, J.M.G. (2007b) Concentrations of butyric acid bacteria spores in silage and relationships with aerobic deterioration. *Journal of Dairy Science*, **90**, 928–936.

Vissers, M.M.M., Driehuis, F., Te Giffel, M.C., De Jong, P. & Lankveld, J.M.G. (2007c) Quantification of the transmission of microorganisms to milk via dirt attached to the exterior of teats. *Journal of Dairy Science*, **90**, 3579–3582.

Vissers, M.M.M., Te Giffel, M.C., Driehuis, F., De Jong, P. & Lankveld, J.M.G. (2007d) Minimizing the level of *Bacillus cereus* spores in farm tank milk. *Journal of Dairy Science*, **90**, 3286–3293.

Vissers, M.M.M., Te Giffel, M.C., Driehuis, F., De Jong, P. & Lankveld, J.M.G. (2007a) Predictive modeling of *Bacillus cereus* spores in farm tank milk during grazing and housing periods. *Journal of Dairy Science*, **90**, 281–292.

Wiedmann, M., Bruce, J.L., Knorr, R., Bodis, M., Cole, E.M., McDowell, C.I., McDonough, P.L. & Batt, C.A. (1996) Ribotype diversity of *Listeria monocytogenes* strains associated with outbreaks of listeriosis in ruminants. *Journal of Clinical Microbiology*, **34**, 1086–1090.

Wilkinson, J.M. & Toivonen, M.I. (2003) *World Silage—A Survey of Forage Conservation Around the World*. Chalcombe Publications, Marlow, Buckinghamshire, UK.

Woolford, M.K. (1990) The detrimental effects of air in silage. *Journal of Applied Bacteriology*, **68**, 101–116.

Zadoks, R.N., Allore, H.G., Barkema, H.W., Sampimon, O.C., Wellenberg, G.J., Grohn, Y.T. & Schukken, Y.H. (2001) Cow- and quarter-level risk factors for *Streptococcus uberis* and *Staphlococcus aureus* mastitis. *Journal of Dairy Science*, **84**, 2649–2663.

Zadoks, R.N., Gonzalez, R.N., Boor, K.J. & Schukken, Y.H. (2004) Mastitis causing streptococci are important contributors to bacterial counts in bulk tank milk. *Journal of Food Protection*, **67**, 2644–2650.

Zadoks, R.N., Van Leeuwen, W.B., Kreft, D., Fox, L.K., Barkema, H.W., Schukken, Y.H. & Van Belkum, A. (2002) Comparison of *Staphylococcus aureus* isolates from bovine and human skin, milking equipment and bovine milk by phage typing, pulsed-field gel electrophoresis and binary typing. *Journal of Clinical Microbiology*, **40**, 3894–3902.

2 Properties and Constituents of Cow's Milk

T. Huppertz and A.L. Kelly

2.1 Introduction

Milk contains all the nutrients required by the neonate and has thus long been recognised as perhaps nature's ultimate food; furthermore, it is also a rich source of protective agents, enzymes and growth factors. For millennia, humans have domesticated mammals for the purpose of milk supply. The first animals domesticated for this purpose were probably sheep and goats, ~8000–10 000 years ago, followed by cattle, which are now the dominant species for dairying (Fox, 2003a). The annual total world milk production has since then risen to ~600 million tonnes at present (IDF, 2003, 2005; Table 2.1). Cow's milk represents by far the vast majority of milk produced worldwide (>80% of total milk production); predominant other types of milk include buffalo, goat and sheep milk (Table 2.1).

The relative ease with which milk can be converted into a wide variety of products makes it an extremely useful base material. In some cases, milk undergoes relatively limited processing, consisting of heat treatment to increase the bacterial shelf life of the product and homogenisation to increase the physical shelf life through retarding fat separation. Other well-known processes involve the acid-induced coagulation of milk to produce yoghurt, or the enzymatic coagulation of milk to manufacture cheese. In addition, milk may be spray-dried or be used as a base from which constituents, e.g. proteins or fats, are isolated. As a result of the widespread applications and use of milk, and products derived there from, in human nutrition, it has been the subject of scientific study for over a century.

In this chapter, a concise overview of the constituents and properties of milk is provided, in particular in relation to heat-induced changes in milk. The focus of this chapter is primarily cow's milk; milk from the other major dairying species (buffalo, goat and sheep) differs from cow's milk in some respects, but is outside the scope of this chapter. For a comparison on constituents and properties of cow, buffalo, goat and sheep milk, the reader is referred to Huppertz *et al.* (2006). Section 2.2 focuses on milk composition, whereas Section 2.3 discusses the primary constituents of milk in more detail, with specific attention to heat-induced changes therein. Section 2.4 focuses on physico-chemical properties of milk and heat-induced changes therein, whereas Section 2.5 briefly discusses heat-induced changes in processing characteristics of milk. Finally, the influence of milk quality on the processing characteristics of milk is described in Section 2.6.

Table 2.1 World production figures (in million tonnes) for cow, sheep, goat or buffalo milk.

Type of milk	Year						
	1993	1995	1997	1999	2001	2003	2005
Cow	460.1	465.2	472.1	483.0	492.0	505.3	527.0
Buffalo	50.0	54.5	59.7	64.9	68.9	72.7	77.5
Goat	9.9	11.8	12.1	12.1	12.5	12.4	12.3
Sheep	7.8	8.0	8.2	8.0	8.2	7.8	8.2
Other	1.2	1.3	1.3	1.3	1.3	1.3	1.3
Total	**529.0**	**540.8**	**553.4**	**569.4**	**582.9**	**599.6**	**626.3**

Data compiled from IDF (2003, 2005).

2.2 Milk composition

Milk is often described as a colloidal suspension, containing emulsified globules of fat, a heterogeneous family of major and minor proteins, the carbohydrate lactose, minerals, vitamins and enzymes. While the classes of constituents are similar for milk from most species, there are considerable inter-species differences, both qualitatively (i.e. the exact nature of constituents) and quantitatively (i.e. the amount of each constituent per litre). An overview of the composition of milk from a wide variety of species is given in Table 2.2, from which it is clear that milk composition is highly dependent on the producing species. The composition and components of bovine milk are discussed in more detail in this section.

The composition and properties of fresh cow's milk shows considerable variability. The main factors from which such variability arises are: (a) genetic factors (e.g. breed and individual), (b) stage of lactation, (c) health status of the cow and (d) environmental factors (e.g. feed, climate or method of milking).

Table 2.2 Approximate composition (g 100 g^{-1}) of milk from dairying species.

Species	Total solids	Fat	Protein	Lactose	Ash
Human	12.2	3.8	1.0	7.0	0.2
Cow	12.7	3.7	3.4	4.8	0.7
Goat	12.3	4.5	2.9	4.1	0.8
Sheep	19.3	7.4	4.5	4.8	1.0
Pig	18.8	6.8	4.8	5.2	—
Horse	11.2	1.9	2.5	6.2	0.5
Donkey	11.7	1.4	2.0	7.4	0.5
Reindeer	33.1	16.9	11.5	2.8	—
Domestic rabbit	32.8	18.3	11.9	2.1	1.8
Bison	14.6	3.5	4.5	5.1	0.8
Indian elephant	31.9	11.6	4.9	4.7	0.7
Polar bear	47.6	33.1	10.9	0.3	1.4
Grey seal	67.7	53.1	11.2	0.7	—

After Fox & McSweeney (1998).

Table 2.3 Composition (g 100 g^{-1}) of cow's milk from various breeds.

Breed	Fat	Protein	Lactose	Ash	Total solids
Ayrshire	4.0	3.3	4.6	0.7	12.7
Brown Swiss	3.8	3.2	4.8	0.7	12.7
Guernsey	4.6	3.5	4.8	0.8	13.7
Holstein	3.6	3.0	4.6	0.7	11.9
Jersey	5.0	3.7	4.7	0.8	14.2

After Nickerson (1995).

Various cow breeds have been selected according to intended use (milk or meat) and environmental conditions (e.g. climate, feed and terrain), which have led to a wide variability in cow's milk composition (see Table 2.3). However, the strong selection of cow breeds, particularly over the last 50 years, has reduced this type of variability considerably. The changes in milk composition during a lactation cycle of a cow are well described. Colostrum, the initial mammary secretion after parturition, contains considerably more protein, particularly serum protein. High levels of immunoglobulin are primarily responsible for this high protein level. Seasonal variation of other milk components is also observed.

The health status of the cow also has a considerable effect on milk composition and yield. Oestrus and gestation affect milk yield primarily, but mastitis, for example, a severe inflammation of the mammary gland as a result of the entering of pathogenic bacteria, affects milk composition as well as milk yield. Mastitis is characterised by increased levels of blood components in milk. The influence of mastitis on the quality of milk and dairy products is discussed in more detail in Section 2.6. Finally, several environmental factors also influence milk yield and composition, such as extreme climates, stress, exhaustion, housing, milking technique and milking frequency.

2.3 Milk constituents

2.3.1 *Lactose*

Introduction

Lactose is a disaccharide (4-*O*-β-D-galactopyranosyl-D-glucopyranose) which is present in milk of most mammalian species, with the exception of milk from sea lions, some seals and opossums, where it is absent or present only in very low concentrations (Jenness & Holt, 1987). In milk from most mammals, lactose is the major carbohydrate, but small amounts of other carbohydrates also occur; cow's milk contains ~10 mg L^{-1} monosaccharides (glucose and galactose) and ~100 mg L^{-1} oligosaccharides (Renner, 1983). Cow's milk contains ~4.8 g lactose 100 g^{-1} (Harper, 1992), which is responsible for ~50% of the osmotic pressure of milk. The concentration of lactose decreases progressively and significantly with lactation stage as well as with increasing somatic cell count of the milk (Walstra & Jenness, 1984); in both cases, this is due to the influx of NaCl from the blood and the resultant need for a reduction in lactose concentration to maintain the osmotic equilibrium (Fox, 2003b).

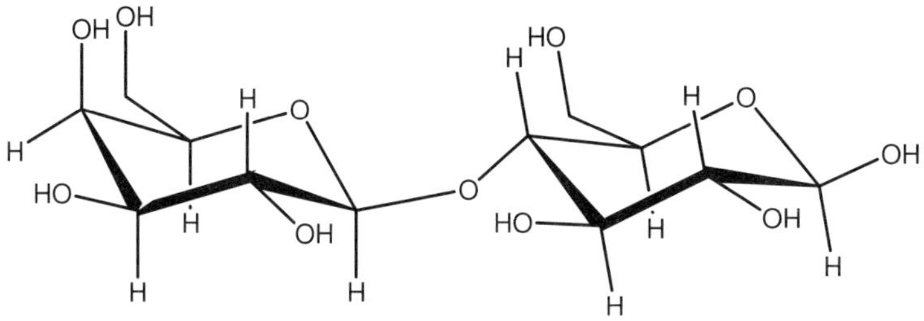

Fig. 2.1 Primary structure of lactose.

Lactose is a disaccharide composed of D-glucose and D-galactose; the aldehyde group of galactose is linked to glucose, through a β-1,4-glycosylic linkage (see Figure 2.1). Lactose is synthesised from glucose in the Golgi apparatus of the mammary secretory cells (Larsen, 1985). The whey protein α-lactalbumin, which is described in more detail in Section 2.3.4, plays an important role in the synthesis of lactose, because in its presence, the non-specific galactosyltransferase becomes highly specific for glucose. Thus, α-lactalbumin has the ability to, if necessary, terminate the synthesis of lactose and regulate and control osmotic pressure.

Through the open-chain aldehyde form, the C_1 atom of glucose can easily change from the α- to the β-form and vice versa; this process is called mutarotation and will eventually lead to equilibrium (mutarotation). α-Lactose and β-lactose differ in their specific rotation to polarised light, i.e., +89.4° or +35.0°, respectively, in water at 20°C. An aqueous solution of lactose in equilibrium at 20°C contains 37.3 g 100 g^{-1}α-lactose and 62.7 g 100 g^{-1} β-lactose and thus has an optical rotation of +55.7°.

Heat-induced changes in lactose

Lactose is one of the many constituents in milk that undergo a variety of changes on heat treatment. Heat-induced changes in lactose, which include isomerisation and degradation as well as its participation in the Maillard reaction, were reviewed in detail by O'Brien (1996, 1997), and are described briefly below.

If heating of milk is not too intense, lactose will be decomposed primarily via the isomerisation reactions, whereas only a small proportion will be degraded via the Maillard reaction; at intense heating, however, degradation of lactose is primarily via the Maillard reaction. Heat-induced isomerisation of lactose leads to the formation of lactulose via a 1,2-enediol intermediate, which may subsequently degrade via β-elimination to give galactose, tagatose and saccharinic acids. Alternatively, lactulose may epimerise via a 2,3-endidiol to form epilactose, a C2 epimer of lactose (Figure 2.2). Lactulose concentrations have been of interest as a possible indicator for the thermal history of milk and can stimulate the growth of bifidobacteria. Isomerisation via the enolisation reaction is often accompanied by the degradation to carboxylic acids. Lactose and lactulose may be degraded into galactose and another C_6 compound, whereas lactulose may also be degraded into galactose, a C_5 components and formic acid (O'Brien, 1996, 1997).

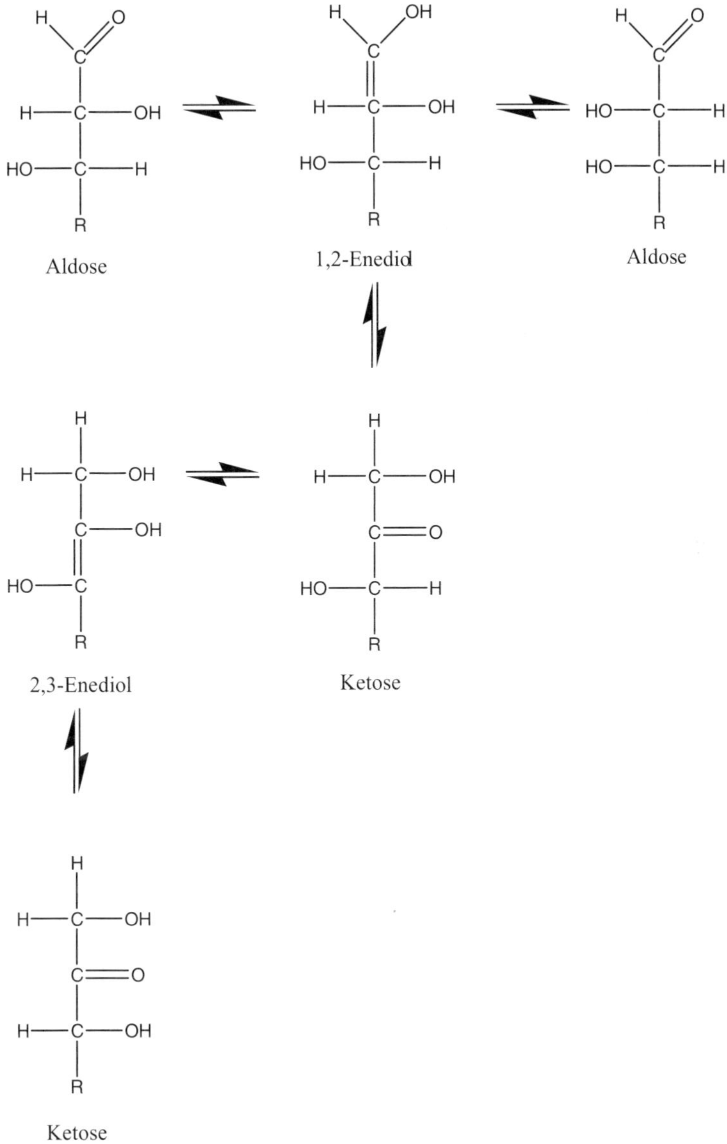

Fig. 2.2 Heat-induced isomerisation of sugars.

The Maillard reaction is a chemical reaction between an amino acid and a reducing sugar, usually requiring the addition of heat. Like caramelisation, it is a form of non-enzymatic browning. The reactive carbonyl group of the sugar interacts with the nucleophilic amino group of the amino acid (in milk mainly lysine) and interesting, but poorly characterised, odour and flavour molecules result. The complete pathway of the Maillard reaction is outside the scope of this chapter, but a schematic overview is presented in Figure 2.3. It is only

Maillard Reactions:

Initial
Lactose + lysine-R $\rightarrow$ Lactulosyl-lysine-R

Galactose + lysine-R $\rightarrow$ Tagatosyl-lysine-R

Intermediate
Lactulosyl-lysine-R $\rightarrow$ Lysine-R + Galactose + C_6
$\rightarrow$ Lysine-R + Galactose + C_5 + formic acid

Tagatosyl-lysine-R $\rightarrow$ Lysine-R + C_n (n = 1–6)

Advanced
C_n + Lysine-R $\rightarrow$ Melanoidins
C_n + Arginine-R $\rightarrow$ Melanoidins

Fig. 2.3 Schematic overview of the Maillard reaction.

in the later stage of the heating of milk that the Maillard reaction manifests itself in the changes in flavour and colour, as a result of the formation of melanoidins.

2.3.2 *Milk salts*

Background

The primary salts in milk are phosphates, citrates, chlorides, sulphates, carbonates and bicarbonates of sodium, potassium, calcium and magnesium. The average mineral composition of bovine milk is given in Table 2.4. Since milk contains organic and inorganic salts, the level of salts is not equivalent to the level of mineral substances, and the level of salts is by no means equivalent to the ash content. Milk salt composition is influenced by a number

Table 2.4 Mineral content of mature cow's milk.

Constituent	Molar mass (Da)	Range (mmol kg^{-1})	Average (mg 100 g^{-1})	Fraction present (in serum)
Cations				
Na	23	17–28	48	0.95
K	39.1	31–43	143	0.94
Ca	40.1	26–32	117	0.32
Mg	24.3	4–6	11	0.66
Amines		~1.3		~1
Anions				
Cl	35.5	22–34	110	1
CO_3	60	~2	10	~1
SO_4	96.1	~1	10	1
$PO_{4(\text{inorganic only})}$	95	19–23	203	0.53
Citrate	189	7–11	175	0.92
Carboxylic acids		1–4		~1
Phosphoric esters (soluble)		2–4		1

After Walstra *et al.* (2006).

of factors, including species, stage of lactation and feed, as well as by the breed of species. For example, milk from Jersey cows usually contains more calcium and phosphorus, but less sodium and chloride, than milk from other breeds.

The solubility of some milk salts, e.g. chlorides and sodium and potassium salts, is sufficiently high for them to be present almost entirely in soluble the milk serum. However, the concentration of several other salts, particularly calcium phosphate, exceeds the solubility at the normal pH of milk (~6.6). Thus, these salts exist partially in soluble form and partially in a colloidal form, i.e. associated with the casein micelles (for review see Holt, 1985, 1997). The colloidal salts are commonly referred to as micellar calcium phosphate (MCP) or colloidal calcium phosphate (CCP), since calcium and phosphate predominate, although some magnesium and citrate are also present. MCP plays an important role in the structure and stability of the casein micelle, as discussed in more detail in Section 2.3.4.

Heat-induced changes in the salt balance

The salt balance in milk, e.g. the distribution of milk salts between the serum and micellar phase, is influenced by various treatments (for review see Holt, 1985; De la Fuente, 1998), including heat treatment (for review see Holt, 1995). The solubility of calcium phosphate is strongly temperature-dependent and, unlike for most other compounds, decreases with increasing temperature. Comparisons of levels of micellar and serum minerals between unheated and heated milks are widely reported, but the influence of heat treatment on the mineral balance of milk is perhaps best described by data from the in situ measurements of Pouliot *et al.* (1989a), who used ultra-filtration at high temperature (for a description of the technique, see Pouliot *et al.* 1989b). As shown in Figure 2.4, levels of calcium and phosphate in the milk serum decrease progressively with increasing temperature (in the range 4–90°C); heat-induced decreases in levels of serum magnesium and citrate are also observed, but to a considerably smaller extent (Figure 2.4). The level of sodium and potassium in the milk serum does not change on heat treatment (Pouliot *et al.*, 1989a). Heat-induced (85°C) reductions in levels of serum calcium and phosphate are rapidly, and almost completely, reversible on subsequent cooling (Pouliot *et al.*, 1989c). More severe heat treatments (>90°C) may result in irreversible changes in the mineral balance in milk (Holt, 1995).

2.3.3 *Lipids*

Introduction

Lipids are esters of fatty acids and related components that are soluble in apolar solvents (e.g. ethyl/petroleum ether or chloroform/methanol). The lipid content of cow's milk ranges from 33 to 47 g L^{-1} (Christie, 1995). The concentration of lipids for a species varies with breed, individual animal, stage of lactation, mastitic infection and plane of nutrition. The lipids of cow's milk are composed of 98% triglycerides and ~1% phospholipids, plus small amounts of diglycerides, monoglycerides, cholesterol, cholesteryl esters and traces of fat-soluble vitamins and other lipids. Over 400 different fatty acids have been detected in cow's milk fat, although most occur only in trace amounts (Christie, 1995). The concentration of

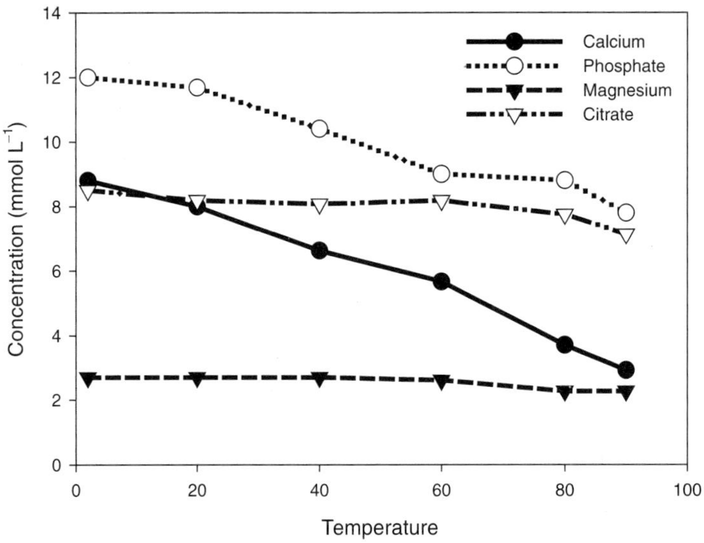

Fig. 2.4 Influence of temperature on the concentrations of calcium, phosphorus, magnesium and citrate in the milk serum. Data from Pouliot *et al.* (1989a).

the principal fatty acids in cow's milk fat is shown in Table 2.5. Crystallisation properties and the melting point of the fat are markedly influenced by fatty acid composition (i.e. the melting temperature increases with fatty acid chain length and with the level of saturation); hence, fatty acid composition is important for products such as butter or spreads. Lactation stage significantly affects the fatty acid profile, so large seasonal variations in the hardness of butter can be observed in countries where milk production is seasonal.

Table 2.5 Principal fatty acids (% weight of total) in triacylglycerols in cow's milk.

Fatty acid	% weight of total
$C_{4:0}$	3.1
$C_{6:0}$	1.6
$C_{8:0}$	1.3
$C_{10:0}$	3.0
$C_{12:0}$	3.1
$C_{14:0}$	9.5
$C_{16:0}$	26.3
$C_{16:1}$	2.3
$C_{18:0}$	14.6
$C_{18:1}$	29.8
$C_{18:2}$	2.4
$C_{18:3}$	0.8
C_{20}–C_{22}	Trace

After Christie (1995).

Milk fat globules

Since almost all the lipids in milk are found in milk fat globules, milk can be considered as an oil-in-water emulsion, the physico-chemical aspects of which are essential when considering changes that occur on storage or processing of milk. Cow's milk typically contains $>10^{10}$ milk fat globules per millilitre. These globules are spherical and range in diameter from <0.2 to >15 μm. These globules are naturally emulsified by a surface layer, the milk fat globule membrane (MFGM), which comprises $\sim$2.6 g 100 g^{-1} of the globule mass. The MFGM resembles the mammary cell membrane, from which it is largely derived, quite closely, and consists primarily of protein ($\sim$75 g 100 g^{-1}; including enzymes such as alkaline phosphatase and xanthine oxidase), phospholipids ($\sim$20 g 100 g^{-1}), cerebrosides ($\sim$3 g 100 g^{-1}) and cholesterol ($\sim$2 g 100 g^{-1}); other substances (neutral glycerides, water, caretenoids, vitamin A, iron and copper) are also present at lower levels (Walstra *et al.*, 2006). The MFGM consists of an innermost layer, which existed on the surface prior to secretion, plus a bilayer membrane that has a dense, proteinaceous coat 10- to 50-nm thickness oriented on the inner membrane face (Keenan & Mather, 2006).

Heat-induced changes in milk fat and fat globules

Thermal degradation of lipids in milk is generally not observed, because the temperature required for non-oxidative decomposition of fatty acids (>200°C) is well outside the range in which milk products are heated. However, heat treatment of milk can alter the properties of the MFGM. Denaturation of MFGM proteins occurs at temperatures >70°C, leading to the exposure of reactive groups. A pronounced effect is the release of sulphydryl compounds, most notably H_2S. The cysteine residues of the membrane components become very active, thus facilitating sulphydryl–disulphide interchange reactions. This may result in the association of whey proteins and κ-casein with the MFGM. Other heat-induced changes in the MFGM include the removal of phospholipids at elevated temperature (Van Boekel & Walstra, 1995).

2.3.4 Proteins

Background

Cow's milk generally contains 30–35 g L^{-1} protein which is commonly divided into two classes on the basis of the solubility at pH 4.6: the insoluble caseins, which represent $\sim$80% of total milk protein, and the soluble whey (or serum) proteins, which represent $\sim$20% of total milk protein. Milk proteins have been the subject of study for almost 200 years and form an important constituent of an extremely wide range of food and non-food products (Fox, 2003c).

Whey proteins

The whey, or serum, proteins represent $\sim$20% of total milk protein in cow's milk (Walstra *et al.*, 2006). Whey proteins in their native form are soluble at pH 4.6 or in saturated NaCl, remain soluble after rennet-induced coagulation of casein micelles and cannot be sedimented by ultracentrifugation. The class of whey proteins consists of a number of proteins, primarily β-lactoglobulin (β-lg), α-lactalbumin (α-la), blood serum albumin, immunoglobulins and proteose peptones.

In milk from most species, β-lg is the most abundant individual whey protein. β-Lg is synthesised in the epithelial cells of the mammary gland. Monomeric cow's β-lg consists of 162 residues per monomer, with a molecular mass of 18.3 kDa and an iso-electric point at pH 5.1. Cow's β-lg is generally present in the A or B genetic variant, although the C and D variant have also been identified (Sawyer, 2003). At natural milk pH, β-lg in cow's milk is present in the form of dimers, formed through hydrophobic interaction (Sawyer, 2003). Native β-lg is a globular protein with a well-defined three-dimensional structure, consisting of an α-helix along one side of a β-barrel (Creamer & Sawyer, 2003). β-Lg contains two intra-molecular disulphide bridges plus one free sulphydryl group which, in the proteins native state, is unavailable for interaction.

α-La is the second most abundant whey protein in cow's milk. The polypeptide chain α-la consists of 123 amino acid residues, and has a molecular mass of 14.2 kDa and an iso-electric point at pH 4.80 (Brew, 2003). α-La is synthesised in the rough endoplasmic reticulum; from there, it is transported to the Golgi apparatus, where it has an important function in the synthesis of lactose. α-La contains eight cysteine residues, which form four intra-molecular disulphide bonds, and all known α-la contain a tightly bound calcium ion. The three-dimensional structure of α-la consists of three α-helices and a 3_{10}-helix (Brew, 2003).

Serum albumin (SA) is the most abundant protein in the circulatory system of the cow, consisting of ~50% of the protein in bovine blood serum, but is present in only small quantities in milk. Bovine SA has a molecular weight of ~66 kDa and consists of 582 amino acids; it contains seventeen disulphide bonds and one free sulphydryl group (Fox, 2003c). The SA in milk is present at concentrations of 0.1–0.4 g L^{-1} (Fox, 2003c). Owing to its low concentration, the SA probably has little influence on the properties of cow's milk.

Immnoglobulins (Ig) are present in the colostrum and, milk of all lactating species, with the biological function of providing immunological protection to the offspring against microbial pathogens and toxins. Levels of immunoglobulins are very high in colostrum, but decline rapidly with advancing stage of lactation. The Ig classes of cow's milk are IgG, IgM and IgA (Butler, 1999; Marnila & Korhonen, 2003). IgG has a molecular mass of ~146 kDa, and occurs predominantly in two subclasses, IgG_1 and IgG_2. IgA or IgM have a molecular mass of ~160 or ~970 kDa, respectively (Marnila & Korhonen, 2003). IgM plays an important role in the creaming of cow's milk.

The proteose peptones are often classified as the pH 4.6-soluble proteins that are not denatured by heat treatment, but are insoluble in 12 g 100 mL^{-1} trichloroacetic acid. The proteose peptone fractions of milk appears to consist of two groups of proteins/peptides, i.e. those that are indigenous to milk (e.g. osteopontin, proteose peptone 3 – PP3), and those that are derived from the action of proteolytic enzymes, primarily plasmin, on caseins (Fox, 2003c).

Heat-induced changes in whey proteins

Like most proteins containing a significant degree of tertiary structure, the whey proteins are susceptible to heat-induced denaturation. Denaturation can be described as any alteration in the native state of the protein which does not affect the primary structure. As such, denaturation of whey proteins is most accurately measured by differential scanning calorimetry (DSC). DSC analysis of individual whey proteins indicates denaturation temperatures of 63, 74, 79 and 87°C for α-la, β-lg, immunoglobulins and bovine serum albumin, respectively. A more routine method for measurement of denaturation of whey proteins is the loss in solubility at a pH ~4.6. Heat-induced denaturation of whey proteins in dairy systems has been described in detail by Mulvihill & Donovan (1987) and Jelen & Rattray (1995).

Denaturation of the whey proteins leads to unfolding which, in the case of β-lg, leads to the exposure of a free sulphydryl group. The exposure of the sulphydryl group initiates aggregation of β-lg with other proteins though sulphydryl–disulphide interchange reactions. Proteins available for such interactions with β-lg include other β-lg molecules, α-la, bovine serum albumin, α_{s2}-casein, κ-casein and some of the proteins of the MFGM. Because β-lg is the primary whey protein, its denaturation and precipitation properties tend to govern that of the whole whey protein system in milk.

Heat-induced denaturation of whey proteins in milk is affected by a wide range of environmental conditions. The extent of heat-induced denaturation increases with increasing milk pH, an effect which is at least partially explained by increased reactivity of the sulphydryl-group of β-lg at higher pH. Furthermore, the extent of heat-induced denaturation increases with increasing level of ionic calcium in the medium. Heat-induced denaturation of whey proteins in milk affects many processing characteristics of milk, as described in more detail in Section 2.6.

Caseins

The caseins represent ~80% of total protein in cow's milk, and are thus the most abundant class of milk proteins. The caseins are a class of phosphoproteins whose properties differ considerably from most other proteins; they are hydrophobic, have a relatively high charge and contain many proline and only few cysteine residues. Caseins have little tertiary structure, with only small α-helical regions present. Cow's milk contain four types of caseins, denoted α_{s1}-, α_{s2}-, β- and κ-casein, which occur at a ratio of ~4:1:4:1.6, respectively.

Of the caseins in cow's milk α_{s1}-casein has the highest charge; it consists of 199 amino acids, has a molecular mass of ~23.6 kDa and contains eight phosphoserine residues per molecule (Martin *et al.*, 2003). α_{s1}-Casein exhibits progressive self-association to dimers, tetramers, hexamers, etc. The degree of association is strongly dependent on pH and ionic strength. α_{s1}-Casein is easily precipitated by addition of calcium (Swaisgood, 2003). α_{s2}-Casein, the least abundant of the caseins in cow's milk, is the least hydrophobic and most highly and variably phosphorylated of the caseins (Creamer, 2003). α_{s2}-Casein consist of 207 amino acid residues (Martin *et al.*, 2003), resulting in a molecular mass of ~25.2 kDa (Ng-Kwai-Hang, 2003). The structure of α_{s2}-casein is characterised by distinct domains of high charge (Swaisgood, 2003). Association studies have shown that α_{s2}-casein behaves very similarly to α_{s1}-casein (Creamer, 2003).

β-Casein is the most hydrophobic of the caseins and contains a large number of proline residues, has a hydrophilic C-terminal end and a very hydrophobic N-terminal end (Creamer, 2003). β-Casein consists of 209 amino acid residues (Martin *et al.*, 2003), and has a molecular mass of ~24.0 kDa (Creamer, 2003); it contains five phosphoseryl groups.

β-Casein is readily cleaved by the indigenous milk proteinase, plasmin, leading to the formation of γ-caseins and proteose peptones, as described in more detail in Section 2.3.5. β-Casein is precipitated in the presence of calcium and, at a temperature >5°C, the β-casein molecules undergo self-association, leading to the formation of micelles (O'Connell *et al.*, 2003).

κ-Casein differs greatly from the other caseins, primarily because it is the only one of the caseins which is glycosylated. In cow's milk, approximately two-thirds of κ-casein molecules are glycosylated; carbohydrate groups include galactosamine, galactose and *N*-acetylneuraminic acid residues. κ-Casein is amphiphatic, with a very hydrophobic N-terminal end and a rather hydrophilic C-terminal end, which plays an important role in stabilising the casein micelles. The κ-casein consists of 169 amino acid residues, and has a molecular weight of ~19.0 kDa (Creamer, 2003). Unlike the other caseins, κ-casein is not sensitive to calcium, but it does, like β-casein, tend to form micelles in solution (Swaisgood, 2003).

Casein micelles

The majority of caseins in cow's milk exist not in solution but in the form of casein micelles. These micelles are highly hydrated (~3 g H_2O g^{-1} protein) and also contain inorganic mineral constituents (primarily calcium and phosphate but also some magnesium and citrate, collectively referred to as micellar calcium phosphate – MCP). Casein micelles have an average molecular mass of ~10^8 Da, with a radius ranging from 50 to 300 nm (average 100 nm). The microstructure of caseins micelles has been subject to considerable research and discussion over the last five decades, but there is still a lack of general consensus about this topic. Numerous models have been proposed for the cow casein micelle; a detailed discussion of casein micelle substructure is outside the scope of this chapter, but a brief overview will follow.

Perhaps the initially most widely supported model is the sub-micelle model, several variations of which have been published, as reviewed by Rollema (1992). These sub-micelle models assume that the casein micelle consist of sub-micelles with a molecular weight of ~10^6–10^7 Da; κ-casein is located on the micellar surface and its hydrophilic C-terminal end protrudes from the micelles, creating a hairy layer around the micelle and providing electrostatic and steric stabilisation. Differences between the various sub-micelle models arise mainly from the manner in which the sub-micelles are bound in the micelle, i.e. through interactions between the calcium-sensitive caseins or through MCP. Some variations of this model claim the existence of two types of sub-micelles; those devoid of κ-casein, which are located in the core of the micelle, and those containing a relatively high level of κ-casein, which can be found primarily in the outside of the micelles. In the last decade or so, reservations about the sub-micelle model have been expressed and alternatives have been proposed.

Holt (1992) proposed a model of the casein micelle in which no sub-units are present. In this model, the calcium-sensitive caseins are linked by nanoclusters (micro-crystals) of MCP, leading to a depiction of the micelle as a tangled web of casein polypeptide chains cross-linked by calcium phosphate interactions. This model was recently refined by De Kruif & Holt (2003), who proposed a more or less homogenous distribution of protein, with the periphery partially drained by solvent. It is assumed that no calcium phosphate is present in this drained layer, but that calcium phosphate nanoclusters are otherwise randomly distributed throughout the micelle. Casein micelles grow to colloidal dimensions because of a balance between cross-linking of casein and the formation of loops in the protein chains. Growth terminates because the formation of nanoclusters terminates or because the weak attractive interactions between the calcium-sensitive caseins are prevented from propagating by a competitive equilibrium interaction with κ-casein (De Kruif & Holt, 2003).

A somewhat alternative view on the microstructure of the casein micelle is presented in the dual-binding model proposed by Horne (1998, 2003), who assumes that micellar assembly and growth takes place through a polymerisation process involving two distinct forms of bonding – cross-linking through hydrophobic bonding or bridging across calcium phosphate nanoclusters. Micellar integrity is maintained through a localised excess of hydrophobic interaction over electrostatic repulsion. In this model, α_{s1}-, α_{s2} and β-caseins can interact through both hydrophobic bonding and cross-linking between a negatively charged phosphoserine cluster and a positively charged calcium phosphate nanocluster. κ-Casein can only link through hydrophobic bonding at its N-terminal end; thus, chain and network growth are terminated at this point, leaving the casein micelle network with a surface layer of primarily κ-casein molecules (Horne, 1998, 2003).

To date, however, there has been no conclusive evidence presented for the complete accuracy of one of the models for the microstructure of the casein micelles, and this topic will undoubtedly remain an active and interesting issue of considerable research, discussion and debate in the future.

Heat-induced changes in caseins and casein micelles

Although caseins are extremely heat-stable, primarily due to their lack of tertiary structure, casein micelles are susceptible to heat-induced changes. In extreme cases, this may lead to heat-induced coagulation, which is described in Section 2.5.3. In this section, a brief overview of heat-induced changes in caseins and casein micelles is given; for extensive reviews on this subject, the reader is referred to Singh (1995), Van Boekel (1999) and O'Connell & Fox (2003).

Heating at relatively low temperature (<70°C) causes some reversible changes in the association of micellar caseins, and may result in the interaction of denatured whey protein with the casein micelles. Heat treatment of milk at >70°C leads to increased levels of non-micellar casein in milk, suggesting heat-induced dissociation of caseins; particularly, levels of non-micellar κ-casein are high in heated milk. The extent of heat-induced dissociation of caseins increases with increasing temperature and pH, but the mechanism for this phenomenon has thus far not been explained adequately. Furthermore, heating at 110–150°C causes extensive dephosphorylation of caseins, which may affect micelle structure; however, the true significance of thermal dephoshorylation on heat treatment of milk remains to be elucidated. Treatment of milk at temperatures >100°C may also lead to the hydrolysis of caseins, leading to the formation of peptides. Of the caseins, α_{s1}-, α_{s2}- and κ-casein are particularly susceptible to thermal hydrolysis and peptide bonds involving aspartic acid residues are hydrolysed rapidly. Finally, heating milk at a high temperature can lead to inter- and intra-molecular cross-linking of caseins as a result of the high reactivity of nucleophilic amino acid residues (e.g. lysine or cysteine) in the presence of lactose. The nature, factors affecting and significance of these covalent bonds remains to be elucidated and warrants further study (Singh, 1995; Van Boekel, 1999; O'Connell & Fox, 2003).

2.3.5 *Indigenous milk enzymes*

In cow's milk, ~20 enzymes have been characterised. Furthermore, the presence of another 40 enzymes has been demonstrated via their activity. Indigenous milk enzymes are found in, or associated with various, casein micelles, milk fat globule membrane, milk serum or somatic cells and may originate from blood, somatic cells, the MFGM or the cell cytoplasm. These milk enzymes can be used as indices of animal health or thermal history of the milk, they can result in quality deterioration or induce desirable changes in milk and dairy products or they may also offer protective effects (Fox, 2003d). In this section, the principal, technologically important indigenous milk enzymes, i.e. plasmin, lipoprotein lipase, alkaline phosphatase and lactoperoxidase are described.

Plasmin (EC 3.4.21.7) is the predominant indigenous proteinase in milk. It is part of a complex protease system in milk, consisting of its inactive precursor, plasminogen, plasminogen activators, which catalyse the conversion of plasminogen to plasmin, plus inhibitors of plasmin and plasminogen activators. Plasmin and plasminogen originate from the mammal's blood and are predominantly associated with the casein micelle in milk. Plasmin is a serine protease, which is optimally active at pH ~7.5 and ~37°C (Bastian & Brown, 1996; Kelly & McSweeney, 2003; Nielsen, 2003). Plasmin is active on all caseins, but particularly β-casein and α_{s2}-casein; primary β-casein cleavage sites are Lys_{28}–Lys_{29}, Lys_{105}–His_{106} and Lys_{107}–Glu_{108}, which leads to the formation of γ-caseins and proteose peptones. Plasmin cleaves α_{s2}-casein at eight sites (Lys_{21}–Gln_{22}, Lys_{24}–Asn_{25}, Arg_{114}–Asn_{115}, Lys_{149}–Lys_{150}, Lys_{150}–Thr_{151}, Lys_{181}–Thr_{182}, Lys_{188}–Ala_{189} and Lys_{197}–Thr_{198}) (Fox, 1992; Bastian & Brown, 1996; Kelly & McSweeney, 2003). The other caseins (i.e. α_{s1}- and κ-casein) are hydrolysed at a considerably lower rate, whereas α-la and β-lg are not hydrolysed by plasmin. Plasmin is of significant interest for several dairy products. The cheesemaking properties of milk deteriorate as a result of plasmin activity (e.g. increased rennet coagulation time and curd moisture content), whereas in the cheese itself, plasmin contributes to primary proteolysis, i.e. the conversion of caseins into polypeptides. Furthermore, the role of plasmin and plasminogen in the physical instability or age gelation of ultra-high temperature (UHT) milk has also received considerable attention, although the exact mechanism remains to be elucidated (Fox, 1992; Bastian & Brown, 1996; Kelly & McSweeney, 2003; Nielsen, 2003). Most plasmin in milk survives pasteurisation, although considerable inactivation is observed at higher temperatures. Thermal inactivation of plasmin depends also on the presence of β-lg, which, though sulphydryl–disulphide interchange reactions, promotes inactivation of plasmin (Farkye & Imafidon, 1995).

Lipoprotein lipase (LPL; EC 3.1.1.34) is a milk lipase which is synthesised in the mammary gland secretory cells. LPL is a glycoprotein which consists of 450 amino acid residues (M_W 100 kDa) with optimal activity at pH 9.2 and 37°C (Olivecrona *et al.*, 2003; Shakeel-Ur-Rehman & Farkye, 2003c). LPL liberates fatty acids from the 1 and 3 position in tri-, di- and monoglycerides in two steps: *firstly*, the enzyme absorbs at the lipid–water interface, and *secondly*, the enzyme aligns its active site at the substrate molecule and hydrolyses it. Lipolysis leads to the release of free fatty acids, which can result in the development of hydrolytic rancidity in milk (Deeth, 2003). In cow's milk, most LPL is associated with the casein micelles, some is in the serum phase and only a very small amount is associated with the MFGM. LPL is a relatively heat-labile enzyme; very little LPL activity survives pasteurisation and complete thermal inactivation occurs when the severity of heat treatment exceeds 75°C for 15 min (Farkye & Imafidon, 1995).

Alkaline phosphatase (ALP; EC. 3.1.3.1) is a phosphomonoesterase with optimum activity at pH 9.0–10.5 and ~37°C which originates from the mammary gland. ALP is a dimer of two identical 85 kDa subunits, and contains four zinc atoms per molecule, which are required for activity. ALP is predominantly associated closely with phospholipid particles in the MFGM (Andrews, 1992; Shakeel-Ur-Rehman *et al.*, 2003; Shakeel-Ur-Rehman and Farkye, 2003a). This enzyme is active against a wide range of substrates, hydrolyses most phosphate ester bonds and can dephosphorylate caseins under suitable conditions. The technological significance of ALP for most milk products, however, is related to its thermal inactivation, and the relative ease of determination thereof (Shakeel-Ur-Rehman *et al.*, 2003; Shakeel-Ur-Rehman & Farkye, 2003a). ALP is relatively heat-sensitive and its thermal stability is only slightly higher than that of non-spore forming pathogenic bacteria present in milk. Hence, thermal inactivation of ALP has been effectively used as a sensitive indicator for adequate pasteurisation of milk; however, care needs to be applied as reactivation on subsequent storage occurs, leading to false ALP-positive test results (Farkye & Imafidon, 1995).

Lactoperoxidase (LPO; EC 1.11.1.7) is synthesised in the mammary gland and is a glycoprotein with a molecular weight of ~78 kDa and contains one heme-group. LPO has a pH optimum of ~8.0, and exists primarily in the milk serum (Bjorck, 1992; Pruitt, 2003; Shakeel-Ur-Rehman & Farkye, 2003b) and catalyses the oxidation of a donor compound according to:

$$\text{Donor} + H_2O_2 \rightarrow \text{Oxidised donor} + 2H_2O$$

In milk, LPO has antibacterial activity in the presence of H_2O_2 and thiocyanate (SCN^-), though the catalysis of the oxidation of thiocyanate to hypothiocyanite ($OSCN^-$):

$$\text{Donor} + H_2O_2 \rightarrow \text{Oxidised donor} + H_2O$$

$$\text{Oxidised donor} + SCN^- \rightarrow \text{Donor} + OSCN^-$$

Hypothiocyanite and its conjugate acid (HOSCN), with which it is at equilibrium, are responsible for the antibacterial activity of this system (Pruitt, 2003). LPO is a relatively heat-stable milk enzyme; heating up to 80°C appears required to ensure thermal inactivation (Farkye & Imafidon, 1995).

2.4 Heat-induced changes in physico-chemical properties of milk

2.4.1 *pH*

The pH of normal milk from a healthy cow is in the range 6.6–6.7; lower pH values would suggest growth of microorganisms that ferment lactose to lactic acid or extensive lipolysis. However, higher pH values in milk may be encountered in times of physiological stress for the producing animal, when the mineral balance of milk is altered by changes in the permeability of the blood–milk barrier (e.g. in very late stages of the lactation cycle or

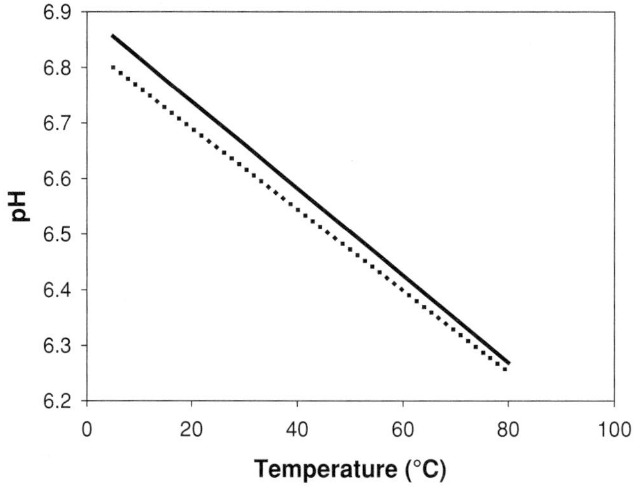

Fig. 2.5 Correlation between temperature and milk pH according to data of Ma & Barbano (2003) (—) and Chapman & Lyster (1988) (· · · · ·).

during mastitic infection). Milk pH is affected by temperature, generally decreasing with increasing temperature, due to changes in dissociation of ionisable groups. Lipolysis can also decrease the pH of milk, due to hydrolysis of esters, especially phosphoric esters.

Heating milk causes a decrease in its pH, and at temperatures <80°C, milk pH decreases in linear fashion with increasing temperature (Figure 2.5) (Chaplin & Lyster, 1988; Ma & Barbano, 2003). Heat-induced reductions in pH at this temperature range are primarily due to shifts in the mineral balance of milk, and are largely reversible on subsequent cooling. The heat-induced loss of CO_2 provides a counteracting increase in pH, but generally this is of lesser magnitude. Severe heat treatment decreases the pH of milk further, due to thermolytic breakdown of lactose to organic acids, especially formic acid.

2.4.2 *Buffering capacity*

The buffering capacity of milk is a measure of its ability to resist changes in pH in either alkaline or acidic directions. Many constituents of milk contribute to its buffering capacity, including small molecules (e.g. salts and organic acids) and proteins. The caseins have maximal buffering capacity around pH 5–5.5, while the maximum for whey proteins is measured around pH 3–4; the exact maximum depends on the presence of acidic amino acids and phosphoserine and histidine residues in individual proteins (for reviews, see Salaün *et al.*, 2005).

Heat treatment of milk affects its buffering capacity to an extent dependent on the severity of treatment applied. Heating milk to <100°C slightly increases the buffering capacity at around pH 5, while heating to 120°C for 10 min increases the buffering capacity, and shifts the pH of maximum buffering capacity from 5.0–5.2 (for unheated milk) to 4.3–4.5. These changes are largely due to the factors discussed above for milk pH (i.e. insolubilisation of calcium phosphate, acidification of lactose). Thermal degradation of urea to products such as carbonic acid may also influence the buffering capacity.

2.4.3 *Creaming*

Creaming or gravity-driven separation of milk into cream and skim milk phases is due to the density difference between milk fat globules and milk serum. However, the rate of creaming of raw milk is much faster than would be predicted on the basis of density difference alone, considering the relatively small size of milk fat globules. Rapid creaming occurs because the fat globules form clusters via a complex system including immunoglobulins (or cryoglobulins) and lipoproteins; these clusters rise rapidly due to a greatly increased effective diameter (for review, see Huppertz & Kelly, 2006).

Cold agglutination, and hence creaming, is influenced by processing conditions. For example, agitation of milk during cold storage impairs its creaming capacity, but heating milk to 40–50°C restores this property and heating milk at higher temperatures, up to ~62°C, can improve the creaming capacity compared to fresh milk. The exact mechanism, however, for heat-induced increases in creaming of milk is yet to be identified. Heating milk at higher temperatures (Orla-Jensen *et al.*, 1929; Rowland, 1937) impairs the rate of creaming of milk fat globules as a result of denaturation of key proteins involved in the cold agglutination pathway, i.e. IgM. Furthermore, heat-induced interactions of caseins or whey proteins with the MFGM may also prevent cold agglutination (Van Boekel & Walstra, 1995).

2.5 Heat-induced changes in processing characteristics of milk

2.5.1 *Rennet-induced coagulation*

Heating milk can result in significant changes in rennet coagulation properties of milk, to an extent dependent on the severity of heating applied. For example, a mild heat treatment, such as high-temperature short-time (HTST) pasteurisation, has no negative impact on rennet coagulation properties, and indeed is preferred to restore the mineral equilibrium in milk which has been cold-stored, during which time solubilisation of MCP will occur. However, heating milk to temperatures above those used for HTST pasteurisation can result, as discussed in Section 'heat-induced changes in whey proteins', particularly β-lg, and transfer of soluble forms of calcium to the colloidal phase (i.e. association with micellar casein). For example, UHT treatment of milk (i.e. heating to temperatures in the range 135–140°C for 3–4 s) reduces ionic calcium levels by 10–20%. These changes have sufficiently negative effects on the rennet coagulation properties of milk that such treatments are undesirable before the manufacture of most types of cheese (e.g. hard cheese varieties, such as Cheddar and Gouda).

Heat-induced denaturation of β-lg does not affect the primary stage of rennet coagulation, i.e. the enzymatic hydrolysis of κ-casein by chymosin. However, heat-induced denaturation of β-lg does interfere with the subsequent aggregation of para-casein micelles. This has a negative effect on the formation of the rennet gel, which is weaker and has a considerably higher moisture content; this is caused by the water-holding capacity of denatured β-lg and by impaired syneresis of rennet gels. Denatured β-lg can also interfere with the subsequent ripening of cheese.

2.5.2 *Acid-induced coagulation*

Whereas, as discussed above, severe heat treatment has negative effects on the rennet coagulation of milk, and is hence avoided for many cheese varieties, it is desirable for many acid-coagulated dairy products, such as yoghurt. For such products, the key processing objective is to avoid syneresis during storage, or contraction of the gel with resulting undesirable expulsion of whey. Heat-induced denaturation of β-lg and its interaction with casein micelles have a number of significant effects on acidified milk gels, including a contribution to the water-binding capacity of the gel, increasing the complexity of the network and gel strength, increasing the pH at which coagulation occurs and ultimately reducing syneresis (O'Kennedy & Kelly, 2000; Vasbiner *et al.*, 2003; Robinson *et al.*, 2006). In addition, for products such as yoghurt, the milk (typically fortified by concentration or addition of milk powder, frequently as well as hydrolloid stabilisers, sweeteners and other ingredients) is also homogenised. This results in incorporation of milk fat globules into the gel network structure, as the much smaller fat globules after homogenisation are partially stabilised by adsorbed casein micelles or fragments thereof and thereby become structural elements (i.e. pseudocasein particles) within the gel.

Hence, overall, the combination of heat treatment and homogenisation, along with increasing the solids content of the mix, are intended to maximise the quantity of gel-forming elements (i.e. casein micelles, casein/fat globule complexes and denatured whey proteins) with the overall objective of avoiding syneresis.

2.5.3 *Heat-induced coagulation*

Compared to most other food systems, milk is extremely stable on heating. However, on prolonged heating at very high temperatures (>120°C) milk becomes unstable, which leads to coagulation. As a result of the implications of heat-induced coagulation for retort-sterilised dairy products, the mechanisms behind such instability have been the subject of much investigation. Classically, the heat stability of milk is quantified in terms of the time required for coagulation (i.e. heat coagulation time – HCT) to occur at a specified temperature (typically at 140°C or 120°C for unconcentrated or concentrated milk, respectively); heat stability is usually evaluated in a mechanically rocking unit in an oil bath maintained at the required temperature. The heat stability of milk was reviewed in detail by O'Connell & Fox (2003).

The heat stability of milk is influenced by a myriad of factors, of which pH is the most significant. When HCT is determined as a function of pH, all bulked milk samples and most individual milk samples display a so-called type A pH-HCT profile, in which HCT increases with pH to a maximum at pH ~6.6; at higher pH values, HCT decreases to minimum at pH ~7.0, followed by a progressive increase in HCT on the alkaline side of the minimum (Figure 2.6). Some individual milk samples display a so-called type B profile, in which HCT increases with pH (Figure 2.6). The main significance of the pH dependence of heat

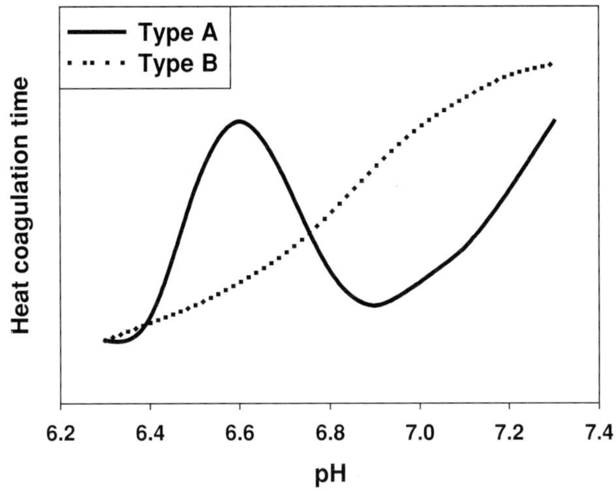

Fig. 2.6 Schematic pH-heat coagulation time profile for type A (—) and type B (· · · · ·) milk.

stability is that the likelihood of milk coagulation, e.g. during sterilisation, changes quite dramatically over a relatively small pH range.

As well as pH, a number of other factors affect the heat stability of milk, including:

- levels of Ca^{2+} or Mg^{2+}, reduction of which increases stability in the pH range 6.5–7.5;
- removal of lactose, which increases heat stability throughout the pH range;
- addition of κ-casein, which eliminates the minimum in the type A profile;
- addition of β-lg to type B milk, which converts it to a type A profile;
- addition of phosphates, which increase heat stability;
- addition of oxidising agents, which convert a type A to a type B profile, or reducing agents, which decrease heat stability and
- presence of alcohols and sulphydryl-blocking agents, which reduce the heat stability of milk.

Concentrated or evaporated milk is much less thermally stable than unconcentrated milk, and its heat stability profile, normally assayed at 120°C for that reason, is quite different. It shows a maximum at ~pH 6.5, with decreasing stability at higher and lower pH values. The possibility that concentrated milk will coagulate on sterilisation may be reduced by preheating the milk, adjustment of the pH of the milk, or addition of stabilising salts, such as polyphosphates.

2.6 Relationship between the quality of raw milk and that of products

Milk is a surprisingly variable raw material for the production of dairy products; its composition and processing characteristics can vary in a complex manner, due to factors including stage of lactation, seasonality, breed, diet, milking technology and health. Perhaps most significantly, in later stages of the cow's lactation cycle, the levels of many milk constituents and enzymes change, including alterations in the mineral balance, reduced casein, increased whey protein content and elevated plasmin activity (Auldist *et al.*, 1998). Broadly, similar changes in milk composition and processability occur in mastitis, and changes in milk composition are generally correlated with milk somatic cell count (SCC) (Auldist & Hubble, 1998; Hortet & Seggers, 1998). There are also interactive effects of factors, such as SCC and stage of lactation (Auldist *et al.*, 1995, 1996a, 1996b, 1996c; O'Brien *et al.*, 2001), and diet and stage of lactation (Kefford *et al.*, 1995; Mackle *et al.*, 1999) on milk and dairy product quality.

These changes can have significant implications for the quality of dairy products. For example, cheese quality is quite sensitive to changes in milk quality, with negative changes in rennet coagulation properties, the cheese yield, composition (e.g. moisture content), texture and flavour all being affected by milk quality, such as stage of lactation and SCC (Coulon *et al.*, 1996, 2004; Rownie & Christian, 1996; Klei *et al.*, 1998). In a recent study, it was demonstrated that the rate of proteolysis was higher in Cheddar cheese made from late-lactation milk than that made from early- or mid-lactation milk from a spring-calving herd, and that volatile profiles differed between cheeses made from milk at different stages of lactation (Hickey *et al.*, 2006). The composition and quality of milk are also dependent on the frequency of milking; for example, once-daily milking reduces milk yield, increases fat and protein content, reduces lactose content and alters the mineral balance relative to twice-daily milking (Davis *et al.*, 1999; Remond & Pomies, 2005). The cheesemaking characteristics of milk from Friesian and Jersey cows (Auldist *et al.*, 2004) and cows of different genetic variants of β-lg (Ng-Kwai-Hang *et al.*, 2002a,b) have also been compared. It has been proposed that genotypic selection can be used to identify cows that suit specific milking frequencies or extension of lactation, and those specific cows within herds may be grouped to supply milk for specific applications (Davis, 2005).

In addition, the quality of dairy products is obviously dependent on the microbiological quality of milk. Many key spoilage or pathogenic bacteria of concern in dairy products originate from milk. Key spoilage species include:

- *Bacillus* species (especially *Bacillus cereus*), spore-forming thermophilic bacteria which may act both as a spoilage agent (e.g. resulting in 'bitty' cream) and, at very high numbers, as a pathogen.
- *Pseudomonas* spp., which are of particular significance due to their psychrotrophic nature, and their ability to produce heat-stable proteinases and lipases, which can impair quality of cheese or UHT milk.

Also, coliform and lactic acid bacteria, which may be contaminants in raw milk, may cause spoilage of milk or dairy products. Many steps in dairy processes (e.g. refrigeration, pasteurisation) are intended specifically to inactivate or control these microorganisms. However, it is nonetheless recognised that a key priority is hygienic production of high microbiological quality raw milk.

2.7 Conclusions

Milk is an exceedingly complex biological material, containing a multi-phase system of several groups of constituents of nutritional and technological significance, the levels of which are variable due to a range of factors. It is also a dynamic system, which is exceedingly sensitive to changes in environmental conditions, e.g. temperature and pH. Changes due to manipulation of such parameters have been exploited for centuries to produce a range of dairy products, and to ensure the safety of such products for consumers.

References

Andrews, A.T. (1992) Phosphatases in milk. In: *Advanced Dairy Chemistry Volume 1: Proteins*, 2nd edn (ed. P.F. Fox), pp. 322–331, Elsevier Applied Science, London.

Auldist, M.J., Coats, S.T., Rogers, G.L. & McDowell, G.H. (1995) Changes in the composition of milk from healthy and mastitic dairy cows during the lactation cycle. *Australian Journal of Experimental Medicine*, **35**, 427–436.

Auldist, M.J., Coats, S.T., Sutherland, B.J., Clarke, P.T., McDowell, G.H. & Rogers, G.L. (1996a) Effects of somatic cell count and stage of lactation on the quality of full cream milk powder. *Australian Journal of Dairy Technology*, **51**, 94–98.

Auldist, M.J., Coats, S.T., Sutherland, B.J., Hardham, J.F., McDowell, G.H. & Rogers, G.L. (1996b) Effects of somatic cell count and stage of lactation on the quality and storage life of ultra high temperature milk. *Journal of Dairy Research*, **63**, 377–386.

Auldist, M.J., Coats, S.T., Sutherland & Mayes, J.J. (1996c) Effects of somatic cell count and stage of lactation on raw milk composition and the yield and quality of Cheddar cheese. *Journal of Dairy Research*, **63**, 269–280.

Auldist, M.J. & Hubble, I.B. (1998) Effects of mastitis on raw milk and dairy products. *Australian Journal of Dairy Technology*, **53**, 28–36.

Auldist, M.J., Johnston, K.A., White, N.J., Fitzsimons, W.P. & Boland, M.J. (2004) A comparison of the composition, coagulation characteristics and cheesemaking capacity of milk from Friesian and Jersey dairy cows. *Journal of Dairy Research*, **71**, 51–57.

Auldist, M.J., Walsh, B.J. & Thomson, N.A. (1998) Seasonal and lactational influences on bovine milk composition in New Zealand. *Journal of Dairy Research*, **65**, 401–411.

Bastian, E.D. & Brown, R.J. (1996) Plasmin in milk and dairy products: an update. *International Dairy Journal*, **6**, 435–457.

Bjorck, L. (1992) Lactoperoxidase. In: *Advanced Dairy Chemistry, Volume 1: Proteins*, 2nd edn (ed. P.F. Fox), pp. 332–338, Elsevier Applied Science, London.

Brew, K. (2003) α-Lactalbumin. In: *Advanced Dairy Chemistry, Volume 1: Proteins*, 3rd edn (eds P.F. Fox & P.L.H. McSweeney), pp. 387–419, Kluwer Academic/Plenum Publishers, New York.

Butler, J.E. (1999) Immunoglobulins and immunocytes in animal milks. In: *Mucosal Immunology* (eds P.L Ogra, J. Mestecky, M.E. Lamm, W. Strober, J. Bienenstock & J.R. McGhee), pp. 1531–1554, Academic Press, San Diego, CA.

Chaplin, L.C. & Lyster, R.L.J. (1988) Effect of temperature on the pH of skim milk. *Journal of Dairy Research*, **55**, 277–280.

Christie, W.W. (1995) Composition and structure of milk lipids. In: *Advanced Dairy Chemistry, Volume 2: Lipids*, 2nd edn (ed. P.F. Fox), pp. 1–36, Chapman & Hall, London.

Coulon, J.B., Delacroxi-Buchet, A., Martin, B. & Pirisi, A. (2004) Relationships between ruminant management and sensory characteristics of cheeses: a review. *Lait*, **84**, 221–241.

Coulon, J.B., Verdier, I. & Pradel, P. (1996) Effect of forage type (hay or grazing) on milk cheesemaking ability. *Lait*, **76**, 479–486.

Creamer, L.K. (2003) Casein nomenclature, structure and association properties. In: *Encyclopedia of Dairy Sciences* (eds H. Roginski, J.W. Fuquay & P.F. Fox), pp. 1895–1902, Academic Press, Amsterdam.

Creamer, L.K. & Sawyer, L. (2003) β-Lactoglobulin. In: *Encyclopedia of Dairy Sciences* (eds H. Roginski, J.W. Fuquay & P.F. Fox), pp. 1932–1939, Academic Press, Amsterdam.

Davis, S.R. (2005) Lactational trends of importance in dairy cows and applications for emerging technologies. *New Zealand Veterinary Journal*, **53**, 400–405.

Davis, S.R., Farr, V.C. & Sterwagen, K. (1999) Regulation of yield loss and milk composition during once-daily milking: a review. *Livestock Production Science*, **59**, 77–94.

Deeth, H.C. (2003) Lipolysis. In: *Encyclopedia of Dairy Sciences* (eds H. Roginski, J.W. Fuquay & P.F. Fox), pp. 1595–1600, Academic Press, Amsterdam.

De Kruif, C.G. & Holt, C. (2003) Casein micelle structure, functions and interactions. In: *Advanced Dairy Chemistry, Volume 1: Proteins*, 3rd edn (eds P.F. Fox & P.L.H. McSweeney), pp. 233–276, Kluwer Academic/Plenum Publishers, New York.

De la Fuente, M.A. (1998) Changes in the mineral balance of milk submitted to technological treatments. *Trends in Food Science and Technology*, **9**, 281–288.

Farkye, N.Y. & Imafidon, G.I. (1995) Thermal denaturation of indigenous milk enzymes. In: *Heat-Induced Changes in Milk* (ed. P.F. Fox), Special Issue No. 9501, pp. 331–348, International Dairy Federation, Brussels.

Fox, P.F. (1992) Proteinases. In: *Advanced Dairy Chemistry, Volume 1: Proteins*, 2nd edn (ed. P.F. Fox), pp. 310–321, Elsevier Applied Science, London.

Fox, P.F. (2003a) Milk. In: *Encyclopedia of Dairy Sciences* (eds H. Roginski, J.W. Fuquay & P.F. Fox), pp. 1805–1812, Academic Press, Amsterdam.

Fox, P.F. (2003b) The major constituents of milk. In: *Dairy Processing, Improving Quality* (ed. G. Smit), pp. 5–41, CRC Press LLC, Boca Raton, FL.

Fox, P.F. (2003c) Milk proteins: general and historical aspects. In: *Advanced Dairy Chemistry, Volume 1: Proteins*, 3rd edn (eds P.F. Fox & P.L.H. McSweeney), pp. 1–48, Kluwer Academic/Plenum Publishers, New York.

Fox, P.F. (2003d) Indigenous enzymes in milk. In: *Advanced Dairy Chemistry, Volume 1: Proteins*, 3rd edn (eds P.F. Fox & P.L.H. McSweeney), pp. 467–471, Kluwer Academic/Plenum Publishers, New York.

Fox, P.F. & McSweeney, P.L.H. (1998) *Dairy Chemistry and Biochemistry*, Kluwer Academic/Plenum Publishers, New York.

Harper, W.J. (1992) Lactose and lactose derivatives. In: *Whey and Lactose Processing* (ed. J.G. Zadow), pp. 317–360, Elsevier Applied Science. London.

Hickey, D.K., Kilcawley, K.N., Beresford, T.P., Sheehan, E.M. & Wilkinson, M.G. (2006) The influence of a seasonal milk supply on the biochemical and sensory properties of Cheddar cheese. *International Dairy Journal*, **16**, 697–690.

Holt, C. (1985) The milk salts: their secretion, concentrations and physical chemistry. In: *Developments in Dairy Chemistry, 3: Lactose and Minor Constituents* (ed. P.F. Fox), pp. 143–181, Elsevier Applied Science, London.

Holt, C. (1992) Structure and stability of bovine casein micelles. *Advances in Protein Chemistry*, **43**, 63–151.

Holt, C. (1995) Effect of heating and cooling on the milk salts and their interaction with casein. In: *Heat-induced Changes in Milk* (ed. P.F. Fox), Special Issue No. 9501, pp. 105–133, International Dairy Federation, Brussels.

Holt, C. (1997) The milk salts and their interaction with casein. In: *Advanced Dairy Chemistry, Volume 3: Lactose, Water, Salts and Vitamins*, 2nd edn (ed. P.F. Fox), pp. 233–256, Chapman & Hall, London.

Horne, D.S. (1998) Casein interactions: casting light on the *Black Boxes*, the structure in dairy products. *International Dairy Journal*, **8**, 171–177.
Horne, D.S. (2003) Caseins, micellar structure. In: *Encyclopedia of Dairy Science* (eds. H. Roginski, J.W. Fuquay & P.F. Fox), pp. 1902–1909, Academic Press, Amsterdam.
Hortet, P. & Seggers, H. (1998) Loss in milk yield and related compositional changes resulting from clinical mastitis in dairy cows. *Preventative Veterinary Medicine*, **37**, 1–20.
Huppertz, T. & Kelly, A.L. (2006) Physical chemistry of milk fat globules. In: *Advanced Dairy Chemistry 2: Lipids*, 3rd edn (eds P.F. Fox & P.L.H. McSweeney), pp. 173–212, Springer-Verlag, New York.
Huppertz, T., Upadhyay, V.K., Kelly A.L. & Tamime, A.Y. (2006) Constituents and properties of milk from different species. In: *Brined Cheeses* (ed. A.Y. Tamime), pp. 1–42, Blackwell Publishing, Oxford.
IDF (2003) *World Dairy Situation*, Bulletin No. 384, International Dairy Federation, Brussels.
IDF (2005) *World Dairy Situation*, Bulletin No. 399, International Dairy Federation, Brussels.
Jelen, P. & Rattray, W. (1995) Thermal denaturation of whey protein. In: *Heat-induced Changes in Milk* (ed. P.F. Fox), Special Issue No. 9501, pp. 66–85, International Dairy Federation, Brussels.
Jenness, R. & Holt, C. (1987) Casein and lactose concentrations in milk of 31species are negatively correlated. *Experentia*, **43**, 1015–1017.
Keenan, T.M. & Mather, I.H. (2006) Intracellular origin of milk fat globules and the nature of the milk fat globule membrane. In: *Advanced Dairy Chemistry 2: Lipids*, 3rd edn (eds P.F. Fox & P.L.H. McSweeney), pp. 137–171, Springer Verlag, New York.
Kefford, B., Christian, M.P., Sutherland, B.J., Mayes, J.J. & Grainger, C. (1995) Seasonal influences on Cheddar cheese manufacture – influence of diet quality and stage of lactation. *Journal of Dairy Research*, **62**, 529–537.
Kelly, A.L. & McSweeney, P.L.H. (2003) Indigenous proteinases in milk. In: *Advanced Dairy Chemistry, Volume 1: Proteins*, 3rd edn (eds P.F. Fox & P.L.H. McSweeney), pp. 495–521, Kluwer Academic/Plenum Publishers, New York.
Klei, L., Yun, J., Sapru, A., Lynch, J., Barbano, D., Sears, P. & Galton, D. (1998) Effects of somatic cell count on cottage cheese yield and quality. *Journal of Dairy Science*, **81**, 1205–1213.
Larsen B.C. (1985) Biosynthesis and cellular secretion of milk. In: *Lactation* (ed. B.L. Larsen), pp. 1063–1069, Iowa State University Press, Ames, IA.
Ma, Y. & Barbano, D.M. (2003) Milk pH as a function of CO_2 concentration, temperature, and pressure in a heat exchanger. *Journal of Dairy Science*, **86**, 3822–3830.
Mackle, T.R., Bryant, A.M., Petch, S.F., Hooper, R.J. & Auldist, M.J. (1999) Variation in the composition of milk protein from pasture-fed dairy cows in late lactation and the effect of grain and silage supplementation. *New Zealand Journal of Agricultural Research*, **42**, 147–154.
Marnila, P. & Korhonen, H. (2003) Immunoglobulins. In: *Encyclopedia of Dairy Sciences* (eds H. Roginski, J.W. Fuquay & P.F. Fox), pp. 1950–1956, Academic Press, Amsterdam.
Martin, P., Ferranti, P., Leroux, C & Addeo, F. (2003) Non-bovine caseins: qualitative variability and molecular diversity. In: *Advanced Dairy Chemistry, Volume 1: Proteins*, 3rd edn (eds P.F. Fox & P.L.H. McSweeney), pp. 277–318, Kluwer Academic/Plenum Publishers, New York.
Mulvihill, D.M. & Donovan, M. (1987) Whey proteins and their thermal denaturation – a review. *Irish Journal of Food Science and Technology*, **11**, 43–75.
Ng-Kwai-Hang, K.F. (2003) Milk proteins: heterogeneity, fractionation and isolation. In: *Encyclopedia of Dairy Sciences* (eds H. Roginski, J.W. Fuquay & P.F. Fox), pp. 1881–1894, Academic Press, Amsterdam.
Ng-Kwai-Hang, K.F., Johnston, K., Boland, M.J. & Auldist, M.J. (2002a) Cheddar cheesemaking characteristics of Jersey milk containing genetic variants A, B and C of beta-lactoglobulin. *Milchwissenschaft*, **57**, 383–387.

Ng-Kwai-Hang, K.F., Otter, D.E., Lowe, E., Boland, M.J. & Auldist, M.J. (2002b) Influence of genetic variants of beta-lactoglobulin on milk composition and size of casein micelles. *Milchwissenschaft*, **57**, 303–306.

Nickerson, S.C. (1995) Milk production: factors affecting milk composition. In: *Milk Quality* (ed. F. Harding), pp. 3–24, Blackie Academic and Professional, London.

Nielsen, S.S. (2003) Plasmin system in milk. In: *Encyclopedia of Dairy Sciences* (eds H. Roginski, J.W. Fuquay & P.F. Fox), pp. 929–934, Academic Press, Amsterdam.

O'Brien, J. (1996) Heat-induced changes in lactose: isomerization, degradation, Maillard Browning. In: *Heat-induced Changes in Milk* (ed. P.F. Fox), Special Issue No. 9602, pp. 134–170, International Dairy Federation, Brussels.

O'Brien, J. (1997) Reaction chemistry of lactose: non enzymatic degradation pathways and their significance in dairy products. In: *Advanced Dairy Chemistry, Volume 3: Lactose, Water, Salts and Vitamins*, 2nd edn (ed. P.F. Fox), pp. 155–231, Chapman & Hall, London.

O'Brien, B., Meaney, W.J., McDonagh, D. & Kelly, A. (2001) Influence of somatic cell count and storage interval on composition and processing characteristics of milk from cows in late lactation. *Australian Journal of Dairy Technology*, **56**, 213–218.

O'Connell, J.E. & Fox, P.F. (2003) Heat-induced coagulation of milk. In: *Advanced Dairy Chemistry, Volume 1: Proteins* (eds P.F. Fox & P.L.H. McSweeney), pp. 879–945, Kluwer Academic/Plenum Publishers, New York.

O'Connell, J.E., Grinberg, V. Ya. & De Kruif, C.G. (2003) Association behaviour of β-casein. *Journal of Colloid and Interface Science*, **258**, 33–39.

O'Kennedy, B.T. & Kelly, P.M. (2000) Evaluation of milk protein interactions during acid gelation using a simulated yoghurt model. *Milchwissenschaft*, **55**, 187–190.

Olivecrona, T., Vilaro, S. & Olivecrona, G. (2003) Lipases in milk. In: *Advanced Dairy Chemistry, Volume 1: Proteins*, 3rd edn (eds P.F. Fox & P.L.H. McSweeney), pp. 473–494, Kluwer Academic/Plenum Publishers, New York.

Orla-Jensen, S., Le Dous, C., Jacobsen, J. & Otte, N.C. (1929) Recherches sur la pasteurisation basse du lait. Une nouvelle reaction pour controller la temperature de chauffage. *Le Lait*, **9**, 622–648, 724–740, 816–831, 914–927 & 1032–1046.

Pouliot, Y., Bouler, M. & Paquin, P. (1989a) Observations on the heat-induced salt balance changes in milk I. Effect of heating time between 4 and 90°C. *Journal of Dairy Research*, **56**, 185–192.

Pouliot, Y., Boulet, M. & Paquin, P. (1989b) An experimental technique for the study of milk salt balance. *Journal of Dairy Science*, **72**, 36–40.

Pouliot, Y., Boulet, M. & Paquin, P. (1989c) Observations on the heat-induced salt balance changes in milk, II. Reversibility on cooling. *Journal of Dairy Research*, **56**, 193–199.

Pruitt, K. (2003) Lactoperoxidase. In: *Advanced Dairy Chemistry, Volume 1: Proteins*, 3rd edn (eds P.F. Fox & P.L.H. McSweeney), pp. 563–570, Kluwer Academic/Plenum Publishers, New York.

Remond, B. & Pomies, D. (2005) Once-daily milking of dairy cows: a review of recent French experiments. *Animal Research*, **54**, 427–442.

Renner, E. (1983) *Milk and Dairy Products in Human Nutrition*, N-GmbH., Volkswirtschaftlicher Verlag, Munich.

Robinson, R.K., Lucey, J.A. & Tamimne, A.Y. (2006) Manufacture of yoghurt. In: *Fermented Milks* (ed. A.Y. Tamime), pp. 53–75, Blackwell Publishing, Oxford.

Rollema, H.S. (1992) Casein association and micelle formation. In: *Advanced Dairy Chemistry, Volume 1: Proteins*, 2nd edn (ed. P.F. Fox), pp. 111–140, Elsevier Applied Science, London.

Rowland, S.J. (1937) The creaming power of milk. The relationship between the denaturation of albumin and globulin and the reduction in creaming power. *Journal of Dairy Research*, **8**, 195–202.

Rownie, M. & Christian, M. (1996) Effect of cow diet and stage of lactation on the composition of milk fat for cheese manufacture. *Australian Journal of Dairy Technology*, **51**, 118–122.

Salaün, F., Mietton, B. & Gaucheron, F. (2005) Review: buffering capacity of dairy products. *International Dairy Journal*, **15**, 95–109.

Sawyer, L. (2003) β-Lactoglobulin. In: *Advanced Dairy Chemistry, Volume 1: Proteins*, 3rd edn (eds P.F. Fox & P.L.H. McSweeney), pp. 319–386, Kluwer Academic/Plenum Publishers, New York.

Shakeel-Ur-Rehman & Farkye, N.Y. (2003a) Phosphatases. In: *Encyclopedia of Dairy Sciences* (eds H. Roginski, J.W. Fuquay & P.F. Fox), pp. 934–938, Academic Press, Amsterdam.

Shakeel-Ur-Rehman & Farkye, N.Y. (2003b) Lactoperoxidase. In: *Encyclopedia of Dairy Sciences* (eds H. Roginski, J.W. Fuquay & P.F. Fox), pp. 938–941, Academic Press, Amsterdam.

Shakeel-Ur-Rehman & Farkye, N.C. (2003c) Lipases and esterases. In: *Encyclopedia of Dairy Sciences* (eds H. Roginski, J.W. Fuquay & P.F. Fox), pp. 926–929, Academic Press, Amsterdam.

Shakeel-Ur-Rehman, Flemming, C.M., Farkye, N.Y. & Fox, P.F. (2003) Indigenous phosphatases in milk. In: *Advanced Dairy Chemistry, Volume 1: Proteins*, 3rd edn (eds P.F. Fox & P.L.H. McSweeney), pp. 523–543, Kluwer Academic/Plenum Publishers, New York.

Singh, H. (1995) Heat-induced changes in casein, including interactions with whey proteins. In: *Heat-induced Changes in Milk* (ed. P.F. Fox), Special Issue No. 9501, pp. 86–104, International Dairy Federation, Brussels.

Swaisgood, H.E. (2003) Chemistry of caseins. In: *Advanced Dairy Chemistry, Volume 1: Proteins*, 3rd edn (eds P.F. Fox & P.L.H. McSweeney), pp. 139–202, Kluwer Academic/Plenum Publishers, New York.

Van Boekel, M.A.J.S. (1999) Heat-induced deamidation, dephosphorylation and breakdown of caseinate. *International Dairy Journal*, **9**, 237–241.

Van Boekel, M.A.J.S. & Walstra, P. (1995) Effect of heat treatment on chemical and physical changes to milk fat globules. In: *Heat-induced Changes in Milk* (ed. P.F. Fox), Special Issue No. 9501, pp. 51–65, International Dairy Federation, Brussels.

Vasbiner, A.J., Alting, A.C., Visschers, R.W. & De Kruif, C.G. (2003) Texture of acid milk gels: formation of disulphide cross-links during acidification. *International Dairy Journal*, **13**, 29–38.

Walstra, P. & Jenness, R. (1984) *Dairy Chemistry and Physics*, John Wiley and Son, New York.

Walstra, P., Wouters, J.T.M. & Geurts, T.J. (2006), *Dairy Science and Technology*, 2nd edn, CRC Press, Boca Raton, FL.

3 Microbiology of Raw and Market Milks

V. Touch and H.C. Deeth

3.1 Introduction

Milk is an ideal medium for the growth of many organisms, having a high water content and abundant nutrients, and being nearly neutral pH (6.4–6.8). A plentiful supply of food for energy is available in the form of milk sugar (lactose), milk fat, citrate, and nitrogenous compounds (proteins, amino acids, ammonia, urea and other non-protein nitrogenous compounds) (Frank & Hassan, 2003). In addition, the Eh of milk is above 0.3 volts, thus enabling aerobes to grow readily (Fields, 1979).

Microorganisms present in milk can be classified into two main groups: pathogenic and spoilage organisms, although some may play a dual role (e.g. *Bacillus cereus*). Pathogenic organisms are those capable of inducing food poisoning, thus posing a threat to public health. By virtue of their elaborated enzymes (e.g. protease, peptidase, lipase, esterase, oxidase, polymerase, β-galactosidase), spoilage organisms are capable of hydrolysing milk components such as protein, fat and lactose in order to yield compounds suitable for their growth. Such reactions can lead to spoilage of milk, manifested as off-flavours and odours, and changes in texture and appearance (Frank & Hassan, 2003).

This chapter reviews the spoilage and pathogenic microorganisms of raw, pasteurised, extended shelf life (ESL) and ultra-high temperature (UHT) milk, with particular emphasis on spoilage organisms.

3.2 Microflora of raw milk

The types of organisms present in raw milk are influenced by temperatures and time of storage as well as methods of handling during and after milking (Varnam & Sutherland, 2001).

3.2.1 *Spoilage organisms*

Gram-negative psychrotrophic bacteria

The growth of psychrotrophic bacteria is of major concern when raw milk is kept at low temperature, which is the normal practice in the modern dairy industry (Muir, 1996a). During growth of these bacteria, heat-stable enzymes such as proteases and lipases are formed and consequently cause protein and lipid breakdown and related defects

(Muir, 1996b). Therefore, the metabolic processes of psychrotrophs are considered more important than the total numbers of organisms (Cousin, 1980).

Pseudomonas *spp.*

Species of the *Pseudomonas* genus are the most important because of their ability to produce heat-stable enzymes (particularly proteases and lipases) during growth under refrigerated storage (Champagne *et al.*, 1994; Muir, 1996a). They are motile, gram-negative rods, with the ability to grow at temperatures just above freezing, despite their optimum growth temperature being between 25 and 30°C (Muir, 1996a). This genus is represented by species with short generation times (growth rate) at 0–7°C (Sorhaug, 1992); their generation times can be even shorter in the presence of air (Vasavada & Cousin, 1993). The generation times of the most rapidly growing psychrotrophic *Pseudomonas* spp. isolated from raw milk are 8–12 h at 3°C and 5.5–10.5 h at 3–5°C (Suhren, 1989). These growth rates are sufficient to cause spoilage within 5 days at these temperatures if the milk initially contains only 1 colony count unit (cfu) mL^{-1} (Frank, 1997).

About 50% of *Pseudomonas* spp. are the fluorescent type, characterised by the production of a diffusible pigment (pyoverdin) during growth (Muir, 1996a). *Pseudomonas fluorescens* and *Pseudomonas fragi*, common milk contaminants, are notorious for producing heat-stable protease and lipase (Shelley *et al.*, 1987).

Enterobacteriaceae

Enterobacteriaceae account for 5–33% of psychrotrophic microflora present in raw milk. These organisms are small, motile, gram-negative rods. Their optimum growth temperature (>30°C) tends to be higher than that of pseudomonads, but they adapt well to growth at refrigeration temperature (Muir, 1996a). Coliforms belonging to this group are able to ferment lactose with production of acid and gas at 32°C within 48 h (Vasavada & Cousin, 1993). Members of the genera *Enterobacter* and *Klebsiella* are most often associated with coliform spoilage, while *Escherichia* spp. only occasionally exhibit sufficient growth to produce a defect (Frank, 1997).

Other psychrotrophic bacteria

Other types of psychrotrophs commonly found in raw milk include *Flavobacterium*, *Achromobacter*, *Aeromonas*, *Alcaligenes* and *Chromobacterium*. They, like pseudomonads, are gram-negative rods capable of growing at low temperature (Muir, 1996a). *Acinetobacter* and *Psychrobacter* are of limited spoilage potential, because their growth at low temperatures is slower than that of other psychrotrophs and so they are often overgrown by *Pseudomonas* spp. (Varnam & Sutherland, 2001).

Gram-positive bacteria

Spore-forming bacteria

Spore-forming bacteria in raw milk are predominantly *Bacillus* spp. (Jay, 1996). *Bacillus* contamination levels, although variable, are found up to 10^5 cfu mL^{-1}. The optimum growth temperature for most *Bacillus* spp. is 20–40°C (Vasavada & Cousin, 1993), although for some, such as *Bacillus stearothermophilus* (now *Geobacillus stearothermophilus*), it is higher. Therefore, the generation and lag times of psychrotrophic *Bacillus* spp. at 2–7°C are considerably longer than those of *Pseudomonas* spp. (McKellar, 1989), although psychrotrophic spore-formers may become the dominating microflora in spoiled milk at 10°C (Meer *et al.*, 1991). Of the *Bacillus* spp. found in milk, *B. licheniformis*, *B. cereus*, *B. subtilis* and *B. megaterium* are most commonly isolated (Shehata *et al.*, 1983). These gram-positive motile, spore-forming, rod-shaped organisms have also been implicated as the cause of a variety of proteolytic defects (Vasavada & Cousin, 1993). *B. cereus* is a common contaminant of raw milks (Champagne *et al.*, 1994), being present in over 80% of raw milk samples (Frank, 1997). There is a distinct seasonality in the occurrence of the organism in raw milk supplies with the highest levels being in late summer and early fall (Champagne *et al.*, 1994), indicating that contamination occurs at the farm. *B. cereus* usually gives rise to milk spoilage defects such as bitty cream, sweet curdling and various off-flavours.

Clostridium spp. are present in raw milk at such low levels that enrichment and most probable number techniques must be used for quantification (Rosen *et al.*, 1989). Populations in raw milk vary seasonally. In temperate climates, clostridia are at higher levels in raw milk collected in the winter than that collected in the summer, because in the winter, in many countries, cows are housed and lie on spore-contaminated bedding materials and are more likely to consume spore-laden silage (Bramley & McKinnon, 1990).

Lactic acid bacteria

Spoilage of raw milk resulting from growth of acid-producing fermentative lactic acid bacteria occurs when storage temperatures are sufficiently high for these microorganisms to outgrow psychrotrophic bacteria or when product composition is inhibitory to gram-negative aerobic organisms (Frank, 1997). This particularly occurs in countries where milk is still stored on farm and transported in unrefrigerated containers. Spoilage can occur, especially in hot weather, before milk is delivered to the factory. Species of *Streptococcus*, *Enterococcus*, *Lactobacillus*, *Leuconostoc*, *Lactococcus* and *Pediococcus* are involved. *Lactococcus lactis* subsp. *lactis* is the main species responsible for spoilage of raw milk at 10–37°C, being able to produce acid (about 0.25%, mostly lactic acid but also small amounts of acetic and propionic acids) to cause milk to sour (Lim, 1998). Some *Enterococcus* isolates can grow at 7°C and have detectable proteolytic activity. These microorganisms constitute only a minor population of the microflora in raw milk, but their numbers may be proportionally higher in pasteurised milk because of their resistance to pasteurisation.

3.2.2 *Pathogenic organisms*

Numerous milkborne pathogens have been isolated from raw milk. The prevalence of these varies considerably, depending on geographical area, season, farm size, number of animals on farm, hygiene and farm management practices (Rohrbach *et al.*, 1992). Although the growth of these pathogens in milk is known to be inhibited by cooling and competing non-pathogenic microorganisms (Frank & Hassan, 2003), outbreaks of illness caused by *Campylobacter jejuni*, Shiga toxin-producing *Escherichia coli*, *Staphylococcus aureus*, *Listeria monocytogenes*, *Salmonella* spp. and *Yersinia enterocolitica* have been reported following consumption of raw milk (Rohrbach *et al.*, 1992; Steele *et al.*, 1997).

Pathogenic organisms in raw milk are of two types: those that are involved in bovine mastitis and those that externally contaminate milk. The bacteria causing mastitis, which vary geographically and with different animal husbandry practices, include *S. aureus*, *Streptococcus agalactiae*, *Str. dysgalactiae*, *Str. uberis*, *Listeria* spp. and *E. coli* (Hillerton & Berry, 2005). The staphylococci are of concern to the dairy industry and public health as some of them, particularly *S. aureus*, produce heat-stable enterotoxins that can cause food poisoning. *Str. agalactiae* causes bacteraemia and meningitis, which are potentially fatal to infected infants (Chambers, 2002).

Salmonellae and thermoduric *Campylobacter* strains are the most common pathogenic bacteria from sources external to the udder (Chambers, 2002). Salmonellae, particularly, continue to be a major concern for the dairy industry because they have been implicated in numerous outbreaks of illness. The emergence of multiple antibiotic-resistant strains of *Salmonella* ser. typhimurium DT104 should be of greater concern to public health authorities since this organism is reportedly resistant to antibiotics commonly used in medical and veterinary practice (El-Gazzar & Marth, 1992).

3.3 Microflora of pasteurised milk

3.3.1 *Spoilage organisms*

The spoilage microflora of pasteurised milk are of two types: (a) post-process contaminants, which enter the milk after heating, and (b) heat-resistant bacteria, which survive heating. Post-process contaminants are of greatest spoilage significance (Sorhaug & Stepaniak, 1997; Varnam & Sutherland, 2001).

Post-process contamination

Gram-negative psychrotrophic bacteria

The most important spoilage of pasteurised milk distributed through a good cold chain is due to recontamination after the heat treatment with gram-negative psychrotrophic bacteria (Fredsted *et al.*, 1996). This post-pasteurisation recontamination takes place most often during the filling process, which is an open process that allows the milk to come in contact with the surrounding air, and with its aerosols. On the other hand, these bacteria often have a degree of resistance to commonly used sanitisers and are able to colonise milk contact surfaces including stainless steel and Buna rubber gaskets (Varnam & Sutherland, 2001).

The shelf life of pasteurised milk contaminated with psychrotrophs is dependent upon the initial contamination, storage temperature, storage time and generation times of the contaminating microorganisms (Vasavada & Cousin, 1993). Pasteurised milk which has high psychrotrophic counts before pasteurisation generally spoils faster at refrigerated temperatures that pasteurised milk produced from low-count milk. Thus Muir (1996b) suggested that raw milk with a count of $\geq 5 \times 10^6$ cfu mL^{-1} should be rejected. Examination made directly after processing indicates that members of the Enterobacteriaceae family including *Serratia*, *Enterobacter*, *Citrobacter* and *Hafnia* are dominant (Varnam & Sutherland, 2001). Nevertheless, the ultimate spoilage microflora usually consist of psychrotrophic gram-negative rods, primarily *Pseudomonas*, *Alcaligens* and, to a lesser extent, *Flavobacterium* (Adams & Moss, 1995). During storage at temperatures below 8°C, the competitive advantage of these species over the Enterobacteriaceae permits relatively rapid development from an initially small population (Varnam & Sutherland, 2001). Even if they are present in low numbers, they will eventually cause spoilage through their ability to multiply and degrade milk protein and fat at refrigeration temperatures (Smithwell & Kailasapathy, 1995; Deeth *et al.*, 2002). The Enterobacteriaceae, however, may remain dominant at storage temperatures above 8°C, and occasionally in other circumstances (Varnam & Sutherland, 2001).

Craven & Macauley (1992a) found *Pseudomonas* to be the only bacterial type to cause defects in pasteurised milk at the required storage temperature (4–7°C). Pasteurisation kills virtually all thermolabile psychrotrophs; however, even if good manufacturing practice is followed, post-pasteurisation contamination by *Pseudomonas* spp. at levels of 10^{-3} cfu mL^{-1} frequently occurs (Stepaniak & Abrahamsen, 1995). The bacterium most often associated with flavour defects in refrigerated milk is *P. fluorescens*, with *P. fragi*, *P. putida*, and *P. aeruginosa* being also encountered. Deeth *et al.* (2002) found the bacteria in pasteurised milk at spoilage comprised five *P.* species: *P. fluorescens*, *P. putida*, *P. aurofaciens*, *P. cepacia* and *P. pseudomallei*, with *P. fluorescens* being the dominant organism at ~50% of isolates tested. The flavour defects in the spoiled milks were described as stale, cheesy, sour, bitter and sour- or bitter-rotten.

Post-process contamination can be monitored using a sensitive test method such as the 'Psychro fast' test (Craven *et al.*, 1994). This test is based on the principle of selective enrichment. A selective agent, benzalkonium chloride, is added to pasteurised milk to inhibit gram-positive bacteria and the milk is incubated at 30°C. The time taken for the (gram-negative) bacterial count to reach approximately 10^7 cfu mL^{-1}, as shown by a colour change in a tetrazonium salt indicator, is indicative of the level of contamination. One gram-negative bacterium in 1 mL of milk can be detected by this method in about 24 h (Craven *et al.*, 1994).

Other post-process contaminants

Other organisms such as *Lactobacillus* and *Lactococcus* are occasionally present as post-process contaminants. Acidification due to these bacteria can occur if milk is held at ambient temperatures (Varnam & Sutherland, 2001).

Heat-resistant bacteria

At the temperatures commonly used for pasteurisation of milk, 72–75°C, most pathogenic and gram-negative psychrotrophic bacteria are eliminated (Fredsted *et al.*, 1996). However, there are a number of survivors from the natural flora which, given suitable conditions, have the ability to promote spoilage (Muir, 1996a). The thermophilic organisms surviving the heating process can attach to the surface of plate heat exchangers with high heat recovery. Growth of these microorganisms preferentially occurs in the temperature range 45–60°C in the regeneration section. Bacterial numbers in pasteurised milk have been found to increase slowly over the initial 8–9 h and then more rapidly over the remaining period of operation. As a result, the already heated milk is recontaminated before it leaves the pasteuriser. The total colony counts of the pasteurised milk may reach levels of 10–100 times higher than the counts of incoming milk (Rademacher *et al.*, 1995). At storage temperatures of >10°C, thermoduric microflora including spore-formers may dominate at the time of spoilage. *B. licheniformis* and *Streptococcus thermophilus* have been implicated (Lehmann *et al.*, 1992, Lehmann, 1996). It has been estimated that 25% of all shelf life problems associated with conventionally pasteurised milk and cream products in the USA are linked to thermoduric psychrotrophs (Meer *et al.*, 1991).

Heat-resistant bacteria present in pasteurised milk are of two types: endospore-forming genera and vegetative genera with high heat resistance. Endospore-forming genera, which are readily isolated in small numbers from pasteurised milk, are of greatest importance (Varnam & Sutherland, 2001).

Endospore-forming genera

Although *Clostridium* endospores are also commonly present, spores of *Bacillus* are of greatest importance because of their ability to grow under refrigeration conditions. *Bacillus circulans* is able to grow at 2°C and *B. cereus* and similar species at 4–5°C. These psychrotrophic species of *Bacillus* become the dominant spoilage organisms at storage temperatures below 5°C when competitive gram-negative microflora are present only in low numbers, possibly as a result of the imposition of severe precautions against post-process contamination in an attempt to extend storage life (Cromie *et al.* 1989; Adams & Moss, 1995; Frank, 1997); otherwise they are overgrown by gram-negative post-process contaminants (Varnam & Sutherland, 2001). This occurs when milk is manufactured under conditions of good hygiene and for which a long storage period is expected.

Of the *Bacillus* spp, *B. cereus*, *B. licheniformis*, *B. mycoides*, *B. circulans* and *B. coagulans* are frequently isolated from pasteurised milk at a level $\leq 10^2$ cfu mL^{-1}. Their vegetative cells are destroyed by pasteurisation and it is the spore form of the organism which is heat-stable. These residual spores may, given the correct conditions, germinate after heat treatment and subsequently grow in the pasteurised milk (Meer *et al.*, 1991; Muir, 1996a). Moreover, psychrotrophic *Bacillus* spp. secrete heat-resistant extracellular proteinases, lipases and phospholipases (lecithinases) that are of comparable heat resistance to those of pseudomonads (Sorhaug & Stepaniak, 1997). Almost 40% of isolates could degrade both milk fat and protein, while 80% of isolates exhibited phospholipase activity, which can destroy the native milk fat globule membrane, resulting in destabilisation of the fat emulsion in milk (Muir, 1996a). Frank (1997) considered psychrotrophic *B. cereus* to be the main spoilage contributor among *Bacillus* spp; it can grow rapidly at temperatures above ca. 8°C (Varnam & Sutherland, 2001) and can reach populations exceeding 10^6 mL^{-1} in milk held for 14 days at 7°C, although slower growth is more common. At 10^4 cfu mL^{-1}, the milk is considered unfit for sale (Christiansson *et al.*, 1996).

Vegetative bacteria

A number of vegetative bacteria are able to survive high-temperature, short-time (HTST) pasteurisation (~72–74°C for 15 s). Most of the genera involved are gram-positive and have only a minimal role in spoilage, especially during refrigerated storage. Common isolates include species of *Microbacterium*, *Micrococcus*, *Enterococcus*, *Streptococcus*, *Lactobacillus* and *Corynebacterium*. Survival rates vary from 100% in the case of *Microbacterium lacticum* to less than 1% in the case of some strains of *Enterococcus*, *Streptococcus*, *Lactobacillus* and coryneform bacteria (Varnam & Sutherland, 2001). The coryneforms, microccoci and streptococci are usually incapable of further growth in pasteurised milk provided the temperature is held below 6°C (Muir, 1996a). According to Varnam and Sutherland (2001), only one gram-negative bacterium, *Alcaligenes tolerans*, is able to survive pasteurisation, although at a level of only 1–10%; its spoilage significance is not known.

3.3.2 *Pathogenic organisms*

Raw milk

There have been numerous reports of disease outbreaks associated with consumption of raw or unpasteurised milk. The most commonly implicated pathogenic organisms are *Campylobacter* spp., *Salmonella* spp., *Listeria* spp., *E. coli* O157, *E. coli* O26, *Yersinia* spp. and *Cryptosporidium* spp. (Stafford, 2005).

Thermised milk

Thermisation (62–65°C for 10–20 s) is sometimes used to delay spoilage of milk during storage in the factory before pasteurisation, usually for manufacture into cheese or other products (see Chapter 7). It is generally insufficient to destroy bacterial pathogens in raw milk with a reasonable margin of safety (Juffs & Deeth, 2007).

Pasteurised milk

Pasteurisation is highly effective in destroying most, if not all, pathogenic vegetative organisms (Juffs & Deeth, 2007). However, outbreaks of food poisoning have occurred through the consumption of pasteurised milk contaminated with *Salmonella* spp., *Brucella* spp., *Listeria* spp. and *E. coli* (Altekruse *et al.*, 1998). Burton (1986) observed that there had been no reported cases of food poisoning from microorganisms surviving the pasteurisation process provided it had been properly carried out in accordance with the recommendations of the International Dairy Federation, that is heating at 63°C for 30 min or 72°C for 15 s. Therefore, the presence of pathogens in the pasteurised product must be the result of post-pasteurisation contamination.

The minimum pasteurisation conditions for milk (63°C for 30 min or 72°C for 15 s) are based on an effective kill of the rickettsia, *Coxiella burnetii*, the causative organism of Q fever. These conditions are slightly more severe than those originally set to destroy *Mycobacterium tuberculosis*, the causative organism of tuberculosis.

Another member of the mycobacteria group, *Mycobacterium avium* subsp. *paratuberculosis* (also known as *M. paratuberculosis* or MAP) is the cause of Johne's disease in ruminants and may be present in the milk of animals infected with the organism. It has been associated with Crohn's disease in humans, but there has been insufficient evidence to prove or disprove whether it has a causal role in this disease and whether it should be treated as a foodborne pathogen (Lund *et al.*, 2002; Sutherland, 2003). However, it is still of concern to dairy processors as some reports have suggested that it may survive pasteurisation at 72°C for 15 s (Grant *et al.*, 2001; Grant, 2004). Several trials on its heat resistance have yielded conflicting results (Lund *et al.*, 2002), although an extensive trial in New Zealand using a pilot-scale HTST pasteuriser with turbulent flow led to the conclusion that pasteurisation at 72°C for 15 s using properly maintained and operated commercial equipment should destroy the organism (Pearce *et al.*, 2004).

As mentioned earlier, spores of the pathogens *Clostridium perfringens* and *B. cereus* are heat-resistant and can occasionally be isolated from pasteurised milk. However, *Cl. perfringens* is unable to germinate and multiply under refrigerated storage conditions (Chambers, 2002). Furthermore, the redox potential is almost invariably too high to permit germination and growth (Varnam & Sutherland, 2001). Although it is believed that the growth of *B. cereus* will lead to the development of a strong unacceptable off-flavour in milk and make it unacceptable to consumers (Lewis, 1999), outbreaks of infection due to consumption of pasteurised milk with *B. cereus* have been reported (Christiansson, 1993). *B. cereus* is often implicated in foodborne gastroenteritis.

3.4 Microflora of UHT milk

The aim of UHT processing is to produce a product which is commercially sterile (see Chapter 7). This means that the product should not contain microorganisms which can grow under the normal conditions of storage. Thus, it is possible for UHT milk to contain thermophilic spore-forming bacteria which will not grow below ~30°C.

Estimates of the spoilage rate of UHT packs vary 1–4 per 100 000 (Cerf & Davey, 2001), 1 per 10 000 (von Boekelman & von Boekelman, 1998) and 2 per 10 000 (Muir, 1990). A common commercial standard is 1 in 10 000 (Robertson, 2003). However, the incidence of non-sterility may be much higher than the spoilage rate if enrichment procedures are used in the analysis (von Boekelman & von Boekelman, 1998; Coelho *et al.*, 2001). The contaminants result from two sources: the survival of heat-resistant spore-formers and post-process contamination. In general, the microbiological failure in UHT milk is due to post-sterilisation contamination, although Cerf & Davey (2001) suggested that it could be explained statistically on the basis of residence time distribution in the UHT plant whereby a very small percentage of spores pass through the holding tube too fast to be destroyed.

3.4.1 *Heat-resistant spore-formers*

There have been several reports of surviving spore-forming bacteria in UHT milk (Forschino *et al.* 1990; Meier *et al.*, 1995). Problems may arise if there is a relatively high population of spore-formers in the raw milk. In this case, a small number may survive UHT treatment, and cause spoilage through the action of their proteolytic and lipolytic enzymes (Rothwell, 1985). They can also cause the 'flat sour' defect characterised by acid production but no gas (Kalogridou-Vassiliadou, 1992).

The spores capable of surviving the UHT process are mainly *G. stearothermophilus*, *B. subtilis* (Muir, 1990), *B. megaterium* (Hassan *et al.*, 1993), *B. sporothermodurans* (Pettersson *et al.*, 1996) and *Paenibacillus lactis* (Scheldeman *et al.*, 2004).

G. stearothermophilus has a high survival potential, but is unable to grow below ~30°C and is a major problem only in hot climates. In temperate climates, *B. coagulans*, *B. subtilis* and *B. licheniformis* are the most important spoilage species, although some heat-resistant strains of *B. cereus* have also been implicated (Varnam & Sutherland, 2001). *B. sporothermodurans* is a comparatively recent discovery (Pettersson *et al.*, 1996). This organism has the unfortunate combination of properties of very high heat resistance and mesophilic growth, that is its optimum growth temperature is around room temperature. *B. sporothermodurans* does not appear to cause spoilage other than a slight discolouration of the milk and seldom reaches counts of greater than 10^5 mL^{-1}. However, it is extremely difficult to remove from contaminated equipment and has caused the closure of some UHT plants (IDF, 2000). Heating conditions required for inactivation of *B. sporothermodurans* are 148°C for 10 s or 150°C for 6 s (Hammer *et al.*, 1996). A practice which has been shown to spread this organism is reprocessing of out-of-date UHT milk. Such a practice should not be permitted.

In 2004, Scheldeman *et al.* isolated spore-forming bacteria from individual packages of UHT milk after the milk had been heated at 100°C for 30 min. They found two colony types: one was identified as *B. sporothermodurans* and the other as *Paenibacillus lactis*. This was the first time *Paenibacillus* spp. had been isolated from UHT milk, although *Paenibacillus* spores have been previously reported to survive heating at 120°C. *Paenibacillus* spores have been also isolated from silage and feed concentrates, which may be the origin of the organism in milk.

3.4.2 *Post-sterilisation contaminants*

A major consideration in the handling of milk after the high-temperature sterilisation section of a UHT plant is contamination. This may result from several sources, but two important ones are the seals in the homogeniser (if downstream) and the air supply to the aseptic packaging unit. Kessler (1994) showed that spores trapped under seals had enhanced heat stability, largely attributable to a very low water activity in their microenvironment, and could act as a reservoir of contaminating spores. The 'flat sour' defect due to contamination by *G. stearothermophilus* can arise in this way. Frequent seal changes have been found to be an effective, although expensive, way of minimising such contamination.

Bacillus spp. comprise the majority of contaminants in UHT milks (Lück *et al.*, 1978; Forschino *et al.* 1990; Skladal *et al.*, 1993, von Boekelman & von Boekelman, 1998; Coelho *et al.*, 2001). However, it is often not clear if they are post-sterilisation contaminants or heat-resistant organisms that have survived the heat treatment. The major species reported are *G. stearothermophilus*, *B. licheniformis*, *B. coagulans*, *B. circulans*, *Bacillus badius*, *B. subtilus*, *B. cereus*, *Bacillus polymyxa* and *Bacillus sphericus*. However, some of these, e.g. *B. cereus*, are unlikely to survive UHT heat treatment (Simmonds *et al.*, 2003) and must enter the milk post-sterilisation.

Non-spore-forming bacteria, both gram-positive and gram-negative, have also been isolated in several studies. For example, Lück *et al.* (1978) found that 12% of the non-sterile UHT milks in South Africa contained non-spore-forming bacteria while Coelho *et al.* (2001) found 7% of the microorganisms isolated from UHT milks in Brazil were non-spore-formers. The types of these bacteria appear to vary considerably between studies, e.g. *Micrococcus*- and *Corynebacterium*-like gram-positive non-spore-forming rods (Coelho *et al.* 2001), *Sc. lactis* and unspecified gram-negatives (von Boekelman & von Boekelman, 1998) and *S. aureus*, *Enterococcus faecalis* and *Enterobacter sakazakii* (Skladal *et al.*, 1993).

Another microbial problem which has caused problems in several companies in recent years is the filamentous fungus, *Fusarium oxysporum.* This organism can cause a flavour similar to that of blue-vein cheese in UHT milk within a few weeks and also produces gas. It is often detected when packages become swollen or 'blown'. It is a common fungus of plants and soils and can enter UHT milk packages through contaminated air in the filling machine. Negative air pressures in aseptic filling areas may facilitate contamination of the packaging equipment if there is a source of the fungus nearby. Once the fungus has contaminated a filling machine, it is difficult to eliminate (K. Scrimshaw, personal communication, 2004).

Heat-resistant enzymes

Spoilage of UHT milk results mainly from the continuing activities of heat-stable proteases and lipases produced by psychrotrophs, including *Pseudomonas* spp., *Alcaligens* spp. and *Flavobacterium* spp., in the raw milk. Residual activity after UHT sterilisation can be as high as 40% (Varnam & Sutherland, 2001). Muir (1996c) reported that, at 140°C for 5 s, proteinases of *Acinetobacter* spp. and *Aeromonas* spp. had residual activities below 10% and those of the fluorescent pseudomonads had residual activities ranging from 14 to 51%. These enzymes remained active following a heat treatment of 149°C for 10 s (Marshall, 1996).

The quality of raw milk selected for manufacture of UHT milk is critical. UHT milk prepared from raw milk containing more than 5×10^6 cfu mL^{-1} psychrotrophs is at risk of spoilage due to heat-resistant enzymes (Varnam & Sutherland, 2001). For UHT milk to have a shelf life of 1 year, the raw milk must contain less than 0.1 unit of protease mL^{-1}. According to Adams *et al.* (1975), 1 unit of enzyme by the Hull method is defined as milligrams of purified protein that would produce 1 μg of tyrosine (equivalent) mL^{-1} from a solution of 2% casein in 24 h at 45°C). Some high protease-producing bacteria can easily synthesise this amount within a day (Adams *et al.*, 1975).

It is not possible to predict the total bacterial count which will result in production of a significant amount of protease as each mix of bacteria has a different propensity to produce protease. Counts as low as 10^5 mL^{-1} can produce noticeable quantities, but also some milks with 10^7 mL^{-1} do not contain noticeable levels (Haryani *et al.*, 2003).

3.5 Microflora of ESL milk

ESL milk is produced by a heat treatment between pasteurisation and UHT. The most common conditions are in the 120–130°C range for a short time (<1 to ~4 s) (see Chapter 7). In contrast to UHT milk, it is usually not packaged aseptically. Therefore, the microflora consists of those spore-forming bacteria whose spores are resistant to the heating conditions and post-processing contaminants similar to those encountered in pasteurised milk. It does not contain the vegetative non-spore-forming thermoduric bacteria present in pasteurised milk. Since ESL milk is stored refrigerated, the major spoilage organisms will be psychrotrophic.

Mayr *et al.* (2004b) found the level of psychrotrophic spores in ESL milks to be 13–130 spores L^{-1}. However, such low numbers of psychrotrophic *Bacillus* spp. can cause spoilage after a few weeks' storage under normal refrigeration. While *B. circulans* dominated the microflora in spoiled aseptically packed milks pasteurised at 72–88°C for 15 s (Cromie *et al.*, 1989), the most frequently encountered spore-forming bacteria in ESL milk directly heat treated at 127°C for 5 s were *B. licheniformis* (73%), followed by *B. subtilis*, *B. cereus*, *Brevibacillus brevis*, and *B. pumilus*; however, of these, only *B. cereus* grows at 8°C (Mayr *et al.*, 2004b). In milks directly heated at 120–132°C for 4 s, Blake *et al.* (1995) found three spore-forming species *B. licheniformis*, *B. coagulans* and *B. cereus*. Of these, *B. cereus* is the only one capable of growing at 7°C (Ranjith, 2000). This suggests that *B. cereus* may be a significant contaminant of ESL milks; this is a concern because of the potential pathogenicity of some strains (Blake *et al.*, 1995).

In commercial ESL milk produced at 127°C for 5 s, Gram-positive non-spore-forming bacteria were found to be the most common spoilage organisms (Mayr *et al.*, 2004a). Genera included *Rhodococcus*, *Anquinibacter*, *Arthrobacter*, *Microbacterium*, *Enterococcus*, *Staphylococcus*, *Micrococcus* and coryneforms. Aerobic spore-formers and gram-negatives were found less frequently. The authors attributed the presence of the spoilers to recontamination from the air and from the packaging material. Most of the spore-formers were believed to be recontaminants and not spores which had survived the heat process. The bacteria appeared to be processor-specific as no bacterial growth occurred in ESL milk from one processor up to the expiry date, whereas ESL milk from another processor showed considerable contamination. On the basis of this paper, the post-heat-treatment contaminants in ESL milk are not largely gram-negative bacteria as in pasteurised milks.

3.6 Sources of contamination

3.6.1 *Raw milk*

It is generally accepted that milk drawn from healthy cows under hygienic milking conditions contains relatively few organisms. However, during milking, it may be subjected to many sources of microbial contamination such as the atmosphere (e.g. dust), dirty udders, unclean equipment and pipelines.

Interior of the udder

Generally, micrococci and streptococci are the main bacteria within the udder and on the teat skin (Slaghuis, 1996). In mastitis, high numbers of environmental bacteria such as *Str. uberis*, *E. coli*, coliforms and *Pseudomonas* spp. may contaminate teats especially if udders are wet and exposed to mud and manure. Counts of streptococci, staphylococci or coliforms in individual milks can be very high (up to 10^7 cfu mL^{-1}) and similar to the total plate count. The bulk milk count from these sources may be up to 10^5 cfu mL^{-1} under certain circumstances. Therefore, ineffective cleaning of teats before milking can contribute to high populations of fermentative bacteria in raw milk (Bramley & McKinnon, 1990).

Lactic acid bacteria are normal inhabitants of the skin and streak canal of the cow's teat. Consequently, all raw milk contains at least low numbers of these organisms (Bramley & McKinnon, 1990).

Exterior of the udder

When cows are housed, bedding material and feedstuffs can be contamination sources. Plant material such as grass, hay, barley and oats used for animal feed may contain from 5×10^5 to 2×10^8 cfu of psychrotrophs g^{-1} (Cousin, 1982). Lactic acid bacteria are also associated with silage and other animal feeds (Bramley & McKinnon, 1990). Contamination of bedding material can be very high due to absorption of urine and faeces. For mastitis-causing bacteria, bedding materials can be a vehicle of contamination. Teats of straw-bedded cows contain higher levels of streptococci than those of cows bedded on sawdust and shavings (Slaghuis, 1996).

The groups of microorganisms on teats which enter milk during milking are mainly aerobic spore-formers and micrococci. The aerobic thermoduric organisms on teat surfaces are almost entirely *Bacillus* spores, with spore counts ranging from 10^2 to 10^5 per teat depending on environmental conditions (Underwood *et al.*, 1974; Muir, 1996a; Slaghuis, 1996). Soil and faeces on teat surfaces are the major contamination sources, although other sources such as water and silage can play a role in increasing the spore content of raw milk (Cook & Sandeman, 2000). Bramley & McKinnon (1990) stated that wood shavings, straw and sand bedding can harbour *Bacillus* spores between 1.5×10^5 and 5.4×10^6 cfu g^{-1}. Weather-related factors, particularly those affecting the moisture of the soil, are important for contamination of milk with spores of *B. cereus* during the grazing period (Christiansson *et al.*, 1996). Although the total spore count of milk in summer is markedly lower than in winter, the psychrotrophic spore count remains the same because the proportion of psychrotrophs within the total spore population increases. The psychrotrophic spore count in summer is mainly derived from soil contaminating the teat surface. As a result, the udder becomes contaminated, resulting in the transfer of these organisms to the raw milk (Giffel *et al.*, 1996).

Water

Water used for the dairy farm has been found to contain psychrotrophic bacteria, even when it has been chlorinated (Thomas & Thomas, 1973). Treated farm water supplies may have psychrotrophic counts of 10^2 mL^{-1}. Although farm water usually contains only low populations of psychrotrophic microorganisms, its use for cleaning and rinsing milking equipment provides a direct means of entry into the milk. Psychrotrophic bacteria isolated from water are often very active producers of extracellular enzymes and grow rapidly at refrigeration temperature. *Pseudomonas* spp., *Achromobacter* spp., *Alcaligenes* spp. and *Flavobacterium* spp. dominate the psychrotrophic flora in water with *Chromobacter* spp., *Bacillus* spp. and coliforms in lower numbers (Cousin, 1982). Furthermore, heat-resistant spore-formers have been isolated from farm water supplies including hot water used for washing milking equipment (Depiazzi *et al.*, 1997).

Milk-handling equipment

Although milking equipment is generally fabricated from stainless steel, which is readily cleaned and sanitized, some parts made of rubber or other non-metallic materials are difficult to sanitise, since only moderate use results in the formation of microscopic pores or cracks. Bacteria attached to these parts are difficult to inactivate by chemical sanitisation. Milk-handling equipment, utensils and storage tanks are major sources of the gram-negative psychrotrophic spoilage bacteria (Cousin, 1982). Komorowki & Early (1993) demonstrated that high counts (10^2–10^3 cfu cm^{-2}) of thermoduric bacteria such as aerobic spore-forming Bacilli, e.g. *B. cereus*, micrococci and thermophilic strains of *Ent. faecalis* are indicative of contaminated equipment. Contributory factors include poorly designed and constructed pipeline systems, and inadequate cleaning and sanitisation between milkings (Varnam & Sutherland, 2001). Milk residues on unclean equipment provide a growth medium for psychrotrophic bacteria which can subsequently contaminate milk. Bulk tanks and silos, in particular, are notorious sources of psychrotrophs. Counts of psychrotrophic bacteria in bulk tanks may be up to 10^3 cfu cm^{-2} (Hayes, 1985).

3.6.2 *Pasteurised milk*

Possible sources for recontamination of the milk with psychrotrophs after pasteurisation are pipelines, tanks, valves and the filling machine. Other sources are the air and the packaging material (Fredsted *et al.*, 1996; Frank, 1997).

Processing equipment

Filling equipment is a common source of psychrotrophs in packaged milk. These microorganisms probably enter the filler through the vacuum system or from containers (Schroder, 1984). Even when filling equipment is effectively cleaned and sanitised, it can still become a source of psychrotrophic microorganisms which accumulate during continuous use. Additionally, tanks used for holding pasteurised milk before packaging can also be a source of psychrotrophic organisms. Tank walls may have microscopic fissures or pits which protect organisms from cleaning and sanitising procedures (Frank, 1997). Pseudomonads are particularly able to adhere strongly to the surface of milk-processing equipment. Some strains of *Pseudomonas* spp. have been shown to produce polysaccharides fibrils or fimbriae, thus enhancing their attachment to surfaces (Champagne *et al.*, 1994). Hence, they are potentially capable of colonising a processing line and remaining there unless removed by adequate cleaning and sanitising procedures (Cousin, 1982).

Some *Bacillus* spp. produce highly hydrophobic spores and can adhere firmly to various substrata such as those found during processing (stainless steel and polymers). Once this first step of adhesion has been completed, colonisation may follow when environmental conditions become favourable to spore germination. *Bacillus* spp. have been found to be involved in biofilm formation in different dairy processes. Both spores and bacteria embedded in biofilms are of concern to the food industry because of their resistance to cleaning and disinfection procedures. Moreover, adhering bacteria may detach and a further cross-contamination of products during processing may occur (Faille *et al.*, 2001). Adhered spores may also be more heat resistant than planktonic spores and hence resist sterilisation procedures (Simmonds *et al.*, 2003)

Air

The microbial population of air has been shown to be around 85% bacteria, 10% moulds and 5% yeasts (Fredsted *et al.*, 1996). The bacteria are mainly gram-positive and generally do not grow well at low temperatures. The main sources of air contamination are usually the ventilation system, air movements into the plant, floor drains and personnel. Product residues due to improper general cleaning of the plant are an important source of air contamination. Although the levels of psychrotrophic bacteria in air are generally quite low, only one viable cell per container is required to spoil the product. Product is often exposed to contaminated air during the packaging process. However, the influence of microorganisms in the air on the microbiological spoilage of milk is of minor importance if premises are well designed and maintained, and internal hygienic measures are taken (Fredsted *et al.*, 1996).

Packaging materials

Packaging materials are today manufactured under strict hygienic conditions. It has been estimated that bacterial counts on the surface of cartons are less than 5100 cfu cm^{-2} (Fredsted *et al.*, 1996). However, packaging material has been implicated in contamination of ESL milks with gram-positive bacteria (Mayr *et al.*, 2004a).

3.6.3 *UHT milk*

The major sources of spoilage bacteria in UHT milk are related to packaging problems (Varnam & Sutherland, 2001) and processing equipment (Burton, 1988). The former case usually involves the aseptic filling process or faulty seams or pinholes in the packaging itself. The nature of contamination is random and many types of microorganism may be involved. Spoilage is rapid and may occur before the pack is acquired by the consumer (Varnam & Sutherland, 2001). Heat-resistant spore-formers, including *B. cereus*, have been reported to enter from ineffectively sterilised plant downstream from the heat treatment stage of the process (Burton, 1988).

3.7 Measures to reduce bacterial contamination of raw and market milks

3.7.1 *Cleaning and sanitation*

In order to reduce/eliminate contamination by spoilage and pathogenic organisms from the farm to the dairy plant, the cow's teats and surrounding udder area, and all utensils and equipment used during milking and processing should be properly cleaned. In a study of bacteria in pasteurised milk, Craven & Macauley (1992c) concluded that the standard of hygiene was a major factor contributing to differences in the quality of milk produced by different milk processors. They also suggested that the degree of sanitation and cleaning may influence the type of pseudomonads, which contaminate milk (Craven & Macauley, 1992b).

3.7.2 *Cooling the milk during storage*

Rapid cooling of milk after collection is paramount, since contamination of the product with psychrotrophic bacteria is unavoidable. This is a major challenge for farm equipment and storage systems (Frank, 1997). Storage of raw milk at 2°C has been shown to be effective for shelf life extension compared with storage at 4 and 7°C (Muir, 1996a; Haryani *et al.*, 2003). Furthermore, the storage temperature of raw milk will influence the quality of the resulting products; the shelf life of UHT milk is much longer when processed from raw milk stored at 2°C than when processed from raw milk stored at 6°C (Griffiths *et al.*, 1988).

The initial population level has an important influence on growth rates. The higher the initial contamination level of milk, the smaller effect low temperature has on limiting bacterial growth (Champagne *et al.*, 1994).

3.7.3 *Addition of carbon dioxide*

Carbon dioxide (CO_2) at a concentration of 20–30 mM can be added as a preservative in milk. When the air in a sealed container is replaced with CO_2, the HCO_3^- ion is produced, which has antimicrobial properties against psychrotrophic, lactic acid and coliform bacteria (Champagne *et al.*, 1994). CO_2 may influence enzyme synthesis also. Sorhaug & Stepaniak (1997) reported that carbon dioxide and nitrogen reduce proteinase secretion at low temperatures.

3.7.4 *Applications of bio-preservatives*

Lactoperoxidase system

Lactoperoxidase (LPO), naturally present in milk, catalyses the oxidation of SCN^- at the expense of H_2O_2 to generate hypothiocyanite ($OSCN^-$):

$$2SCN^- + H_2O_2 \rightarrow (SCN)_2 + 2H_2O$$

$$(SCN)_2 + H_2O \rightarrow HOSCN + SCN^- + H^+$$

$$HOSCN \rightarrow OSCN^- + H^+$$

Thiocyanate may also be oxidised directly:

$$SCN^- + H_2O_2 \rightarrow OSCN^- + H_2O$$

Hypothiocyanite is antimicrobial against a wide range of bacteria, fungi and viruses. The effect is caused by oxidation of thiol groups (−SH) of cytoplasmic enzymes and damage to other cellular functions essential for the microbial metabolism (Shin *et al.*, 2001). The antimicrobial effect of the system is influenced by pH, temperature, the test medium and the ratios of the oxidation products per target cell (Touch *et al.*, 2004). LPO can be found up to 30 mg L^{-1} in milk, and a much lower quantity of the enzyme (i.e. 0.5–1 mg L^{-1}) is needed to initiate the catalytic reaction. Although both thiocyanate and H_2O_2 occur normally in milk, the quantities vary; H_2O_2 at about 0.3 mM is required in the inhibitory system (Haddadin *et al.*, 1996), but it is normally present at much lower concentrations in milk. An effective level of thiocyanate is around 0.25 mM, whereas in milk the concentration varies between 0.02 and 0.25 mM (Jay, 1996). Therefore, in order for the system to be effective in killing bacteria in raw milk as well as pasteurised milk, exogenous sources of hydrogen peroxide and thiocyanate are required. For H_2O_2, this can be achieved either by adding H_2O_2 directly or by adding xanthine oxidase or glucose oxidase with glucose to generate H_2O_2.

The system is very useful for preserving raw milk in developing countries where milking hygiene is poor and refrigeration is rarely available. Bjorck (1980) reported that only 10 ppm (0.3 mM) H_2O_2 needs to be added to prevent bacterial multiplication in warm milk during storage and transport in Africa. The shelf life of pasteurised milk at 10°C was increased by 20 days after activation of the LPO system by addition of 2.4 mM SCN and 0.6 mM H_2O_2 followed immediately by batch pasteurisation of 63°C for 30 min (O'Toole, 1995).

Milk pasteurised at 76–88°C has a shorter shelf life than milk pasteurised under the minimum HTST conditions of 72°C for 15 s (Schmidt *et al.*, 1989). This has been attributed inter alia to destruction of LPO at the higher temperatures (Barrett *et al.*, 1999).

Lactoferrin

Lactoferrin, present in bovine milk at 0.02–0.35 mg mL^{-1}, is an iron-binding protein that has antimicrobial activity (Ekstrand, 1994). The protein exerts its antimicrobial effects through depriving microorganisms of multivalent metals, namely, Fe, Mg and Ca. These three ions play essential roles in the growth and survival of most microorganisms. Iron is used for the growth, replication, DNA synthesis, and respiration of aerobic and facultative anaerobic species. Magnesium is required for the activity of many membrane-bound enzymes and also acts as a salt-bridge in the supramolecular structure in the lipopolysaccharides (LPS) of gram-negative bacteria. Calcium also has a role in stabilising LPS and regulating cellular processes involving nucleoid structure, protein phosphorylation, and alterations in transverse and lateral distributions of membrane lipids (Norris *et al.*, 1996). Therefore, once the ions are limited or deprived, susceptible organisms may not be able to grow and/or may lose viability. There is, however, evidence that the protein still retains its antimicrobial activity even after being saturated with metal ions, suggesting that direct interactions between the protein molecule, presumably due to its cationic properties, with cell membrane components (including LPS) may also be implicated in its antimicrobial effects (Shimazaki, 2000). Furthermore, lactoferricin, a cationic peptide derived from lactoferrin is also antimicrobial (Tomita *et al.*, 1994).

However, the citrate present in milk can exchange the iron chelated by lactoferrin, thus reducing and/or depleting its bacteriostatic activity (Vasavada & Cousin, 1993). On the other hand, lactoferrin inhibits only bacteria with high iron requirement, such as coliforms, but has no effect on bacteria with a low iron requirement. The bacteriostatic effect on bacteria is temporary because some gram-negative bacteria can adapt to low iron and synthesise iron chelators (Ekstrand, 1994).

Lactic acid bacteria (LAB)

Addition of lactic acid starter culture to raw milk has been shown to inhibit growth of psychrotrophic bacteria. Reportedly, 5×10^6 cfu lactococci mL^{-1} were required to significantly reduce the growth of psychrotrophs (Champagne *et al.*, 1990) and 10^8 cfu mL^{-1} of a commercial culture of LAB were needed to obtain complete inhibition of the gram-negative flora of raw milk stored at 6°C (Griffiths *et al.*, 1991). After 120 h at 7°C, no extracellular proteinase was detected in UHT milk inoculated with 10^6 cfu mL^{-1} of LAB and 10^5 cfu mL^{-1} of *P. fluorescens* NCDO 2085 (Jaspe *et al.*, 1995).

The mechanism by which LAB inhibit sensitive microflora is not well understood. It is generally thought that the LAB act against psychrotrophs through the effect of H_2O_2, produced by the organisms. However, the levels of H_2O_2 produced are probably insufficient to be effective per se (Champagne *et al.*, 1994). Muir (1996a) demonstrated that gram-negative flora may be affected by activation of the natural flora LPO complex in raw milk of many lactic acid strains producing hydrogen peroxide. Griffiths *et al.* (1991) reported that at least part of the inhibitory action of the starter is associated with a fall in pH of milk to below 6.0. While the starter does not multiply at refrigeration temperature, it is metabolically active and lactic acid production occurs. The fact that citrate-fermenting strains are effective suggests that citrate utilisation could be involved. Furthermore, LAB produce bacteriocins, such as nisin (Stiles, 1996), although nisin is not effective against gram-negative psychrotrophs (Muir, 1996a) unless chelating agents are added to sensitise them. Honer (1981) suggested that LAB could assimilate essential nutrients in milk and thus act as competitors to the psychrotrophic bacteria.

Addition of bacteriocins

Bacteriocins are proteinaceous compounds produced by bacteria which are lethal to bacteria other than the producing strains. Of the bacteriocins identified to date, only nisin, produced by *Lac. lactis* subsp. *lactis*, and related compounds such as pediocin are currently authorised for use as food preservatives. Nisin is a 34-residue antibacterial protein containing lanthionines and uncommon amino acids. Nisin acts on vegetative bacteria by a four-step process of binding, insertion, aggregation and pore formation. It binds to the target membrane by electrostatic interactions with the anionic phospholipid and penetrates the membrane via its hydrophobic patches (Bruelink *et al.*, 1997). Nisin, however, acts on the sulfhydryl membrane groups for inactivation of germinated spores. It is active against gram-positive bacteria and spores but not gram-negative bacteria and yeast or filamentous fungi. Addition of nisin (40–50 IU mL^{-1}) to milk prior to pasteurisation extended the shelf life of milk up to 41 days (Wirjantoro & Lewis, 1996).

3.7.5 *Thermal treatments*

A range of thermal treatments is used to reduce the bacterial population of milk. These include thermisation, batch and HTST pasteurisation, high temperature pasteurisation (ESL), UHT treatment and in-container sterilisation (Kelly *et al.*, 2005). The treatments applicable to market milks are discussed in Chapter 7.

3.7.6 *Non-thermal treatments*

Several non-thermal treatments can be used to destroy or remove microorganisms in foods (Datta & Deeth, 2002a,b; Deeth & Datta, 2002). These include high-pressure treatment, pulsed electric field technology, ultrasonication, centrifugation and microfiltration; however, only the last two technologies are used commercially for milk. Novel methods of milk processing are discussed in Chapter 8.

3.7.7 *Multitarget attack/integrative approaches*

Integration of several hurdles in the preservation of milk and milk products has been found to be a promising approach because different hurdles may simultaneously and/or synergistically act on different targets (e.g. cell membrane, DNA, enzyme system and other cellular functions) within the microbial cells so that survival and cell repair would become more difficult (Leistner, 2000). Such approaches may be achieved by combining the effects of antimicrobials and/or other physical treatments.

The use of two or more bio-preservatives to yield a synergistic effect on the target organisms has been extensively investigated as means to prolong the shelf life of raw and market milks (Ross *et al.*, 2003). Practically, this synergism not only allows very low doses of the antimicrobials to be used effectively, but also expands the range of organisms that may be inhibited; that is organisms normally resistant to each component of the mixture when used separately can also be usefully controlled (Leistner, 2000).

Another approach, which has recently received a growing interest from the dairy industry, is the use of natural antimicrobials with other non-thermal preservation methods, including high hydrostatic pressure, pulse electric fields, ultrasonication and irradiation as a potential pathogen intervention strategy for raw and market milks. The synergistic action of such combination preservation systems may offer several advantages: (a) improvement of the rate of inactivation, (b) development of cost-effective mild preservation and (c) reduction of the commercial problems associated with sub-lethal injury and survivor tails (Ross *et al.*, 2003).

Likewise, it is now being realised that the effect of bio-preservatives is pronounced when combined with certain physical processes (e.g. mild heating, chilling, freezing, drying or homogenisation) because microbial cells sub-lethally injured by such treatments may become more susceptible to the antimicrobials to which the healthy cells are resistant. This approach may be of economic consideration to the dairy industry, not only because of the better image of the products but also because of the reduced processing costs. An interesting instance of such treatments is the activation of the LPO systems in milk either before or after heat treatment to reduce the *D* values and to increase the keeping quality of the product (Barrett *et al.*, 1999). Similarly, pulsed electric field technology has been combined with pasteurisation to extend the shelf life of milk to up to 78 days (Sepulveda *et al.*, 2005).

3.8 Conclusion

Knowledge of the microbiology of raw milk and of milk after different heat treatments is essential for ensuring the safety and quality of milk at consumption. The different storage and treatment conditions have a major effect on the type of bacteria present and their effect on the product.

As milk leaves the cow, it is dominated by LAB but during refrigerated storage, psychrotrophic gram-negative bacteria introduced from the environment become dominant and are the major cause of spoilage of raw milk. A mild heat treatment such as thermisation destroys a large proportion of the spoilage bacteria but not all pathogenic bacteria. The commonly used pasteurisation process eliminates most, if not all, of the pathogenic bacteria and a high proportion of spoilage bacteria. Pasteurisation does not destroy bacterial endospores of bacteria, such as *Bacillus* spp. or some thermoduric non-spore-formers. Spoilage of pasteurised milk is, however, largely due to post-pasteurisation contaminants picked up from the equipment, air and packaging material. A high-pasteurisation or ESL treatment eliminates all vegetative bacteria and most of the endospores. As ESL milk is not packed aseptically, post-pasteurisation contamination as well as the residual spore-formers causes the milk to ultimately spoil. UHT sterilisation effectively destroys all bacteria, including most endospores, which could cause spoilage at room temperature. However, low numbers of heat-resistant spores and post-process non-spore-forming contaminants may be present in UHT milk.

The primary controls of the microbes in raw and processed milk are limiting the time and temperature of storage, ensuring any processing is performed effectively and paying close attention to equipment cleaning and sanitation. Secondary controls, such as the use of carbon dioxide, bacteriocins, LAB and antimicrobial proteins, are appropriate in certain circumstances. In the dairy industry, the ultimate control is by heat treatment. Some non-thermal treatments are also effective and may find commercial application in the future.

References

Adams, D.M., Barach, J.T. & Speck, M.L. (1975) Heat resistant proteases produced in raw milk by psychrotrophic bacteria of dairy origin. *Journal of Dairy Science*, **58**, 828–834.

Adams, M.R. & Moss, M.O. (1995) *Food Microbiology*, The Royal Society of Chemistry, UK.

Altekruse, S.F., Timbo, B.B., Mowbray, J.C., Bean, N.H. & Potter, M.E. (1998) Cheese-associated outbreaks of human illness in the United States, 1973–1992; sanitary manufacturing practices protect consumers. *Journal of Food Protection*, **61**, 1405–1407.

Barrett, N.E., Grandison, A.S. & Lewis, M.J. (1999) Contribution of the lactoperoxidase system to the keeping quality of pasteurized milk. *Journal of Dairy Research*, **66**, 73–80.

Bjorck, L. (1980) Enzymatic stabilisation of milk – utilisation of the milk peroxidase for the preservation of raw milk, In: *Special Addresses at IDF Sessions 1979*, Bulletin 126, pp. 5–7. International Dairy Federation, Brussels.

Blake, M.R., Weimer, B.C., McMahon, D.J. & Savello, P.A. (1995) Sensory and microbial quality of milk processed for extended shelf life by direct steam injection. *Journal of Food Protection*, **58**, 1007–1013.

Bramley, A.J. & McKinnon, C.H. (1990) The microbiology of raw milk. In: *Dairy Microbiology*, Vol. 1 (ed. R.K. Robinson), pp. 163–208, Elsevier Applied Science, New York.

Bruelink, E., van Kraaij, C., Demel, R.A., Siezen, R.J., Kuipers, O.P. & de Kruijff, B. (1997) The C-terminal region of nisin is responsible for the initial interaction of nisin with the target membrane. *Biochemistry*, **36**, 6968–6976.

Burton, H (1986) Microbiological aspects. In: *Monograph on Pasteurized Milk*, Bulletin 200, pp 9–14, International Dairy Federation, Brussels.

Burton, H. (1988) *Ultra-High Temperature Processing of Milk and Milk Products*, Elsevier Applied Science, London.

Cerf, O. & Davey, K.R. (2001) An explanation of non-sterile (leaky) milk packs in well-operated UHT plants. *Transactions of the Institute of Chemical Engineers*, **79**(Part C), 197–210.

Chambers, J.V. (2002) The microbiology of raw milk. In: *Dairy Microbiology Handbook*, 3rd edn (ed. R.K. Robinson), pp. 39–90, John Wiley & Sons, New York.

Champagne, C.P., Girard, F. & Morin, N. (1990) Inhibition of psychrotrophic bacteria in raw milk by immobilised lactic bacteria. *Journal of Food Protection*, **53**, 400.

Champagne, C.P., Laing, R.R., Roy, D. & Mafu, A.A. (1994) Pscyhrotrophs in dairy products: their effect and their control. *Critical Reviews in Food Science and Nutrition*, **34**, 1–30.

Christiansson, A. (1993) The toxicology of *Bacillus cereus*. In: Bacillus cereus *in Milk and Milk Products*, Bulletin 275, International Dairy Federation, Brussels.

Christiansson, A., Nilsson, J. & Bertilsson, J. (1996) *Bacillus cereus* spores in raw milk. Factors affecting the contamination of milk during the grazing period. In: *Symposium on Bacteriological Quality of Raw Milk*, Wolfpassing, Austria, 13–15 March, International Dairy Federation, Brussels.

Coelho, P.S., Silva, N., Brescia, M.V. & Siqueira, A.P. (2001) [Microbiological aspects of UHT whole milk commercialized in Belo Horizonte, Brazil]. *Arquivo Brasileiro de Medicina Veterinária e Zootecnia*, **53**(2), 256–262.

Cook, G.M. & Sandeman, R.M. (2000) Sources and characteristics of spore-forming bacteria in raw milk. *Australian Journal of Dairy Technology*, **55**, 119–126.

Cousin, M.A. (1980) Proteolytic activity of psychrotrophic microorganisms on milk and dairy products. In: *Psychrotrophic Microorganisms in Spoilage and Pathogenicity* (eds T.A. Roberts, G. Hobbs, J.H.B Christian & N. Skovgaard), pp. 63–72, Academic Press, London.

Cousin, M.A. (1982) Presence and activity of psychrotrophic microorganisms in milk and dairy products – a review. *Journal of Food Science*, **45**, 172–207.

Craven, H.M., Forsyth, S.R., Drew, P.G. & Macauley, B.J. (1994) A new technique for early detection of Gram-negative bacteria in milk. *Australian Journal of Dairy Technology*, **49**, 54–56

Craven, H.M. & Macauley, B.J. (1992a) Microorganisms in pasteurised milk after refrigerated storage. 1. Identification of types. *Australian Journal of Dairy Technology*, **47**, 38–45.

Craven, H.M. & Macauley, B.J. (1992b) Microorganisms in pasteurised milk after refrigerated storage. 2. Seasonal variation. *Australian Journal of Dairy Technology*, **47**, 46–49.

Craven, H.M. & Macauley, B.J. (1992c) Microorganisms in pasteurised milk after refrigerated storage. 3. Effect of milk processors. *Australian Journal of Dairy Technology*, **47**, 50–55.

Cromie, S.J., Dommett, T.W. & Schmidt, D. (1989) Changes in the microflora of milk with different pasteurisation and storage conditions and aseptic packaging. *Australian Journal of Dairy Technology*, **44**, 74–77.

Datta, N. & Deeth, H.C. (2002a) Alternatives to heat treatment: High pressure processing. In: *Encyclopedia of Dairy Sciences* (eds H. Roginski, J.W. Fuquay & P.F. Fox), pp. 1327–1333, Academic Press, London.

Datta, N. & Deeth, H.C. (2002b) Alternatives to heat treatment: other non-thermal technologies. In: *Encyclopedia of Dairy Sciences* (eds H. Roginski, J.W. Fuquay & P.F. Fox), pp. 1339–1346, Academic Press, London.

Deeth, H.C. & Datta, N. (2002) Alternatives to heat treatment: pulsed energy technologies. In: *Encyclopedia of Dairy Sciences* (eds H. Roginski, J.W. Fuquay & P.F. Fox), pp. 1333–1339, Academic Press, London.

Deeth, H.C., Khusniati, T., Datta, N. & Wallace, R.B. (2002) Spoilage patterns of skim and whole milks. *Journal of Dairy Research*, **68**, 228–242.

Depiazzi, L.J., Martin P.A.J. & Bell J.R.M. (1997) *Farm Sources of Spore Forming* Bacillus *in Bulk Tank Milk*. Paper presented at the 30th National Promaco Convention, Perth, Western Australia.

Ekstrand, B. (1994) Lactoperoxidase and lactoferrin. In: *Natural Antimicrobial Systems and Food Preservation* (eds V.M. Dillon & R.G. Board), pp. 15–63, CAB International, Wallingford, UK.

El-Gazzar, F.E. & Marth, E.H. (1992) *Salmonella*, salmonellosis, and dairy foods: a review. *Journal of Dairy Science*, **75**, 2327–2343.

Faille, C., Fontaine, F. & Benezech, T. (2001) Potential occurrence of adhering living *Bacillus* spores in milk product processing lines. *Journal of Applied Microbiology*, **90**, 892–900.

Fields, M.L. (1979) *Fundamentals of Food Microbiology*, Avi Publishing Company, Westport, CT.

Forschino, R., Gaill, A. & Ottogalli, G. (1990) Research on the microflora of UHT milk. *Annali di Microbiologia ed Enzymologia*, **40**, 47–59.

Frank, J.F. (1997) Milk and dairy products. In: *Food Microbiology: Fundamentals and Frontiers* (eds M.P. Doyle, L.R. Beuchat & T.J. Montville), ASM Press, Washington, DC.

Frank, J.F. & Hassan, A.N. (2003) Microorganisms associated with milk. In: *Encyclopedia of Dairy Sciences* (eds H. Roginski, J.W. Fuquay & P.F. Fox), Academic Press, London.

Fredsted, L.B., Rysstad, G. & Eie, T. (1996) Pure-Lac: the new milk with protected freshness and extended shelf life. In: *Heat Treatment and Alternative Methods*, Document 9602, pp. 104–125, International Dairy Federation, Brussels.

Giffel, M.C., Beumer, R.R., Slaghuis, B.A. & Rombout, F.M. (1996) Occurrence and characterisation of psychrotrophic *Bacillus cereus* on farms in the Netherlands. *Symposium on Bacteriological Quality of Raw Milk*, Wolfpassing, Austria, 13–15 March, International Dairy Federation, Brussels.

Grant, I.R. (2004) Inactivation of *Mycobacterium avium subsp. paratuberculosis* (MAP) in milk during commercial HTST pasteurisation. In: *Proceedings of an International Workshop on Heat Resistance of Pathogenic Organisms*, Bulletin 392, pp. 62–67, International Dairy Federation, Brussels.

Grant, I.R., Rowe, M.T., Dundee, L. & Hitchings, E. (2001) *Mycobacterium avium* subsp. *paratuberculosis*: its incidence, heat resistance and detection in milk and dairy products. *International Journal of Dairy Technology*, **54**, 2–13.

Griffiths, M.W., Banks, J.M., McIntyre, L. & Limond, A. (1991) Some insight into the mechanisms of inhibition of psychrotrophic bacterial growth in raw milk by lactic acid bacteria. *Journal of the Society of Dairy Technology*, **44**, 24–29.

Griffiths, M.W., Philips, J.D., West, I.G. & Muir, D.D. (1988) The effect of extended low temperature storage of raw milk on the quality of pasteurised and UHT milk. *Food Microbiology*, **5**, 75–87.

Haddadin, M.S., Ibrahim, S.A. & Robinson, R.K. (1996) Preservation of raw milk by activation of the natural lactoperoxidase systems. *Food Control*, **7**, 149–152.

Hammer, P., Lembke, F., Suhren, G. & Heesschen, W. (1996) Characterization of a heat-resistant mesophilic *Bacillus* species affecting the quality of UHT milk. In: *Heat Treatments and Alternative Methods*, Document 9602, pp. 9–16, International Dairy Federation, Brussels.

Haryani, S., Datta, N., Elliott, A.J. & Deeth, H.C. (2003) Production of proteinases by psychrotrophic bacteria in raw milk stored at low temperature. *Australian Journal of Dairy Technology*, **58**, 16–20.

Hassan, A.N., Zahnran, A.S., Metwalli, N.H. & Shalabi, S.I. (1993) Aerobic spore-forming bacteria isolated from UHT milk produced in Egypt. *Egyptian Journal of Dairy Science*, **21**, 109–121.

Hayes, P.R. (1985) *Food Microbiology and Hygiene*, Elsevier Applied Sciences, London.

Hillerton, J.E. & Berry, E.A. (2005) Treating mastitis in the cow – a tradition or an archaism. *Journal of Applied Microbiology*, **98**, 1250–1255.

Honer, C. (1981) The disappearing bacteria. *Prepared Foods*, **157**(5), 14.

IDF (2000) *Bacillus sporothermodurans – a* Bacillus *Forming Highly Heat-Resistant Spores*, Bulletin 357, pp. 1–27, International Dairy Federation, Brussels.

Jaspe, A., Oviedo, P., Ferandez, I., Palacios, P. & Sanjose, C. (1995) Cooling raw milk: change in the spoilage potential of contaminating Pseudomonas. *Journal of Food Protection*, **58**, 915–921.

Jay, J.M. (1996) *Modern Food Microbiology*, 5th edn, Chapman & Hall, New York.

Juffs H.S. & Deeth H.C. (2007) *Scientific Evaluation of Pasteurisation for Pathogen Reduction in Milk and Milk Products* Food Standards Australia New Zealand, Canberra.

Kalogridou-Vassiliadou, D. (1992) Biochemical activities of *Bacillus* species isolated from flat sour evaporated milk. *Journal of Dairy Science*, **75**, 2681–2686.
Kelly, A., Datta, N. & Deeth, H.C. (2005) Thermal processing of dairy products. In: *Thermal Food Processing: New Technologies and Quality Issues* (ed. D.-W. Sun), pp 265–298, Taylor & Francis, Boca Raton, FL.
Kessler H.G. (1994) Heat resistance of microorganisms under seals. In: *Proceedings of the 24th International Dairy Congress, Melbourne*, International Dairy Federation, Brussels.
Komorowki, E.S. & Early, R. (1993) Liquid milk and cream. In: *The Technology of Dairy Products* (ed. R. Early), pp. 1–23, VCH Publishers, New York.
Lehmann, F.L. (1996) Recontamination of industrial ultrafiltration units and pasteurisers by thermoduric – thermophilic bacteria. In: *Heat Treatments and Alternative Methods*, Document 9602, pp. 126–128, International Dairy Federation, Brussels.
Lehmann, F.L., Russell, P.S., Solomon, L.S. and Murphy, K.D. (1992) Bacterial growth during continuous pasteurization. *Australian Journal of Dairy Technology*, **47**, 28–32.
Leistner, L. (2000) Basic aspects of food preservation by hurdle technology (Review). *International Journal of Food Microbiology*, **55**, 181–186.
Lewis, M.J. (1999) Microbiological issues associated with heat-treated milks. *International Journal of Dairy Technology*, **52**, 121–125.
Lim, D. (1998) *Microbiology*, 2nd edn, WCB McGraw-Hill, Boston, MA.
Lück, H., Mostert, J.F. & Husmann, R.A. (1978) Bacteriological evaluation of UHT milk. *South African Journal of Dairy Technology*, **10**, 83–85.
Lund, B.M., Gould, G.W. & Rampling, A.M. (2002) Pasteurization of milk and the heat resistance of *Mycobacterium avium* subsp. *paratuberculosis*: a critical review of the data. *International Journal of Food Microbiology*, **77**, 135–145.
Marshall, R.T. (1996) Critical point to control contamination, growth and enzymatic activity. In: *Symposium on Bacteriological Quality of Raw Milk*, Wolfpassing, Austria, 13–15 March, International Dairy Federation, Brussels.
McKellar, R.C. (1989) Regulation and control of synthesis. In: *Enzymes of Psychrotrophs in Raw Food* (ed. R.C. McKellar), CRC Press, Boca Raton, FL.
Mayr, R., Gutser, K., Busse, M. & Seiler, H. (2004a) Gram-positive non-spore-forming bacteria are frequent spoilage organisms of German retail ESL (Extended Shelf Life) milk. *Milchwissenschaft*, **59**, 262–266.
Mayr, R., Gutser, K., Busse, M. & Seiler, H. (2004b) Indigenous aerobic spore-formers in high heat treated (127°C, 5 s) German ESL (Extended Shelf Life) milk. *Milchwissenschaft*, **59**, 143–146.
Meer, R.R., Baker, J., Bodyfelt, F.W. & Griffiths, M.W. (1991) Psychrotrophic *Bacillus* spp. in fluid milk products – a review. *Journal of Food Protection*, **54**, 969–979.
Meier, J., Rademacher, B., Walenta, W. & Kessier, H.G. (1995) Heat resistant spores under UHT treatment. *Heat Treatments and Alternative Methods*, Document 9602, pp. 17–25, International Dairy Federation, Brussels
Muir, D.D. (1990) The microbiology of heat treated fluid milk products. In: *Dairy Microbiology*, Vol. 1 (ed. R.K. Robinson), pp. 209–243, Elsevier Applied Science, New York.
Muir, D.D. (1996a) The shelf life of dairy products: I. Factors influencing raw milk and fresh products. *Journal of the Society of Dairy Technology*, **49**(1), 24–32.
Muir, D.D. (1996b) The shelf life of dairy products: II. Raw milk and fresh products. *Journal of the Society of Dairy Technology*, **49**(1), 44–48.
Muir, D.D. (1996c) The shelf life of dairy products: III. Factors influencing intermediate and long life dairy products. *Journal of the Society of Dairy Technology*, **49**(1), 67–72.
Norris, V., Grant, S., Freestone, P., Canvin, J., Sheikh, F.N., Toth, I., Trinei, M., Modha, K. & Norman, R.I. (1996) Calcium signaling in bacteria. *Journal of Bacteriology*, **178**, 3677–3682.

O'Toole, D.K. (1995) Factors inhibiting and stimulating bacterial growth in milk: an historical perspective. *Advances in Applied Microbiology*, **40**, 45–83.

Pearce, L.E., Truong, H.T., Crawford, R.A. & de Lisle, G.W. (2004) Kinetic studies on the heat inactivation of *Mycobacterium avium subsp. paratuberculosis* (MAP) during turbulent-flow pasteurisation. In: *Proceedings of an International Workshop on Heat Resistance of Pathogenic Organisms*, Bulletin 392, pp 49–52, International Dairy Federation, Brussels.

Pettersson, B., Lembke, F., Hammer, P., Stackebrant, E. & Priest, G. (1996) *Bacillus sporothermodurans*, a new species producing highly heat-resistant endospores. *International Journal of Systematic Bacteriology*, **46**, 759–764.

Rademacher, B., Walenta, W. & Kessler, H.G. (1995) Contamination during pasteurisation by biofilms of thermophilic streptococci. *Heat Treatments and Alternative Methods*, Document 9602, pp. 26–33, International Dairy Federation, Brussels.

Ranjith, H.M.P. (2000) High temperature pasteurization. In: *Proceedings of the Conference on Extended Shelf Life Milk*, Dresden, September 2000, International Dairy Federation, Brussels.

Robertson, G.L. (2003) Ultra-high temperature treatment (UHT) – aseptic packaging. In: *Encyclopedia of Dairy Sciences* (eds H. Roginski, J.W. Fuquay & P.F. Fox), pp. 2637–2642, Academic Press London.

Rohrbach, R.W., Draughon, F.A., Davidson, P.M. & Oliver, S.P. (1992) Prevalence of *Listeria monocytogenes*, *Campylobacter jejuni*, *Yersiniia enterocolitica* and *Salmonella* in bulk tank milk: risk factors and risk of human exposure. *Journal of Food Protection*, **55**, 93–97.

Rothwell, J. (1985) Microbiology of frozen dairy products. In: *Microbiology of Frozen Foods* (ed. R.K. Robinson), Elsevier Applied Science, London.

Rosen, B., Merin, U. & Rosenthal, I. (1989) Evaluation of clostridia in raw milk. *Milchwissenschaft*, **44**, 355–357.

Ross, A.I.V., Griffiths, M.W., Mittal, G.S. & Deeth, H.C. (2003) Combining nonthermal technologies to control foodborne microorganisms. *International Journal of Food Microbiology*, **89**, 125–138.

Scheldeman, P., Goossens, K., Rodriguez-Diaz, M., Pil, A., Goris, J., Herman, L., De Vos, P., Logan, N.A. & Heyndrickx, M. (2004) *Paenibacillus lactis* sp. nov. isolated from raw and heat-treated milk. *International Journal of Systematic & Evolutionary Microbiology*, **54**, 885–891.

Schroder, M.J.A. (1984) Origins and levels of post pasteurisation contamination of milk in the dairy and their keeping quality. *Journal of Dairy Research*, **51**, 59–67.

Schmidt, D., Cromie, S.J. & Dommett, T.W. (1989) Effect of pasteurisation and storage conditions on the shelf-life and sensory qualities of pasteurised milk. *Australian Journal of Dairy Technology*, **44**, 19–24.

Sepulveda, D.R., Gongora-Nieto, M.M., Guerrero, J.A. & Barbosa-Canovas, G.V. (2005) Production of extended shelf life milk by processing pasteurized milk with pulsed electric fields. *Journal of Food Engineering*, **67**, 81–86.

Shehata, A.E., Magdoub, M.N.I., Sultan, N.E. & El-Samragy, Y.A. (1983) Aerobic mesophilic and psychrotrophic spore-forming bacteria in buffalo milk. *Journal of Dairy Science*, **66**, 1228–1231.

Shelley, A.W., Deeth, H.C. & MacRae, I.C. (1987) A numerial taxonomic study of psychrotrophic bacteria associated with lipolytic spoilage of raw milk. *Journal of Applied Bacteriology*, **62**, 197–207.

Shimazaki, K.I. (2000) Lactoferrin: a marvelous protein in milk? *Animal Science Journal*, **71**, 329–347.

Shin, K., Hayasawa, H. & Lönnerdal, B. (2001) Inhibition of *Escherichia coli* respiratory enzymes by the lactoperoxidase-hydrogen peroxide-thiocynate antimicrobial system. *Journal of Applied Microbiology*, **90**, 489–493.

Simmonds, P, Mossel, B.L., Intaraphan, T. & Deeth, H.C. (2003) The heat resistance of *Bacillus* spores when adhered to stainless steel and its relationship to spore hydrophobicity. *Journal of Food Protection*, **66**, 2070–2075.

Skladal, P., Mascini, M., Salvadori, C. & Zannoni, G. (1993) Detection of bacterial contamination in sterile UHT milk using an L-lactate biosensor. *Enzyme & Microbial Technology*, **15**, 508–512.

Slaghuis, B.A. (1996) Sources and significance of contaminants on different levels of raw milk production. In: *Symposium on Bacteriological Quality of Raw Milk*, Wolfpassing, Austria, 13–15 March, pp. 19–27, International Dairy Federation, Brussels.

Smithwell, N. & Kailasapathy, K. (1995) Psychrotrophic bacteria in pasteurised milk: problems with shelf life. *Australian Journal of Dairy Technology*, **50**, 29–31.

Sorhaug, T. (1992) Temperature control. In: *Encyclopedia of Microbiology*, Vol. 4 (ed. J. Lederberg), pp. 201–211, Academic Press, London.

Sorhaug, T. & Stepaniak, L. (1997) Psychrotrophs and their enzymes in milk and dairy products: quality aspects. *Trends in Food Science and Technology*, **8**, 35–41.

Stafford, R. (2005) Outbreaks of enteric infection associated with the consumption of pasteurised and unpasteurised milk in Australia and overseas. Attachment 2 of Report *Scientific Evaluation of Milk Pasteurisation* (H.S. Juffs & H.C. Deeth, 2005, *loc cit*).

Steele, M.L., McNab, W.B., Poppe, C., Griffiths, M.W., Chen, S., Degrandis, S.A., Fruhner, L.C., Larkin, C.A., Lynch, J.A. & Odermeru, J.A. (1997) Survey of Ontario bulk tank milk for food-borne pathogens. *Journal of Food Protection*, **60**, 1341–1346.

Stepaniak, L. & Abrahamsen, R.K. (1995) Effect of sampling and storage temperature on microbial quality of pasteurised milk. *Milchwissenschaft*, **50**, 22–26.

Stiles, M.E. (1996) Biopreservation by lactic acid bacteria. In: *Lactic Acid Bacteria: Genetics, Metabolism and Applications* (eds G. Venema, J.J. Huis & J. Hugenholtz), Kluwer Academic, Dordrecht, The Netherlands

Suhren, G. (1989) Producer microorganisms. In: *Enzymes of Psychrotrophs in Raw Food* (ed. R.C. McKellar), pp. 3–34, CRC Press, Boca Raton, FL.

Sutherland, P.S. (2003) Mycobacteria. In: *Foodborne Microorganisms of Public Health Significance*, 6th edn (ed. A.D. Hocking), pp 597–604, Australian Institute of Food Science and Technology Incorporated, Waterloo, Australia.

Thomas, S.B. & Thomas, B.F. (1973) Psychrotrophic bacteria in refrigerated bulk collected raw milk, Part I. *Dairy Industries*, **38**, 11–15.

Tomita, M., Takase, M., Wakabayashi, H. & Bellamy, W. (1994) Antimicrobial peptides of lactoferrin. *Advances in Experimental Medicine & Biology*, **357**, 209–218.

Touch, V., Hayakawa, S., Yamada, S. & Kaneko, S. (2004) Effects of a lactoperoxidase-thiocyanate-hydrogen peroxide system on *Salmonella enteritidis* in animal or vegetable foods. *International Journal of Food Microbiology*, **93**, 175–183.

Underwood, H.M., McKinnon, C.H., Davies, F.L. & Cousins, C.M. (1974) Sources of *Bacillus* spores in raw milk. In: *Proceedings of the 19th International Dairy Congress*, Paris, p. 373, International Dairy Federation, Brussels.

Vasavada, P.C. & Cousin, M.A. (1993) Dairy microbiology and safety. In: *Dairy Science and Technology Handbook, Vol. 2: Product Manufacturing*, (ed. Y.H. Hui), pp. 301–426, VCH Publishers, New York.

Varnam, A.H. & Sutherland, J.P. (2001) *Milk and Milk Products – Technology, Chemistry and Microbiology*, Aspen Publishers, Gaithersburg, MD.

von Boekelman, B. & von Boekelman, I.A. (1998) *Long-life Products: Heat-treated, Aseptically Packed: A Guide to Quality*, Dr Bernard von Boekelmann, Åkarp, Sweden.

Wirjantoro, T.R. & Lewis, M.J. (1996) Effect of nisin and high temperature pasteurisation on the shelf life of whole milk. *Journal of the Society of Dairy Technology*, **49**(4), 99–102.

4 Quality Control

J. Belloque, R. Chicón and I. Recio

4.1 Introduction

Quality control has the objective of ensuring the quality and safety of the milk offered to the consumer. The nature and manipulation of the raw milk, the hygiene conditions at the farm and the industry, the process to which it is subjected and the conditions of storage change the properties of the product. Quality control of milk is done at different levels, by the farmer, by the industry and by the government. The farmer needs to have control on the raw milk in order to improve and maintain the quality of the product that is sold to the transforming industry. The dairy industry needs to control the raw milk supplied by the farmers and sets up controls on the process and/or the end product in order to ensure the safety and quality of the product going out to the market. The government agencies control the raw milk, to obtain information of hygiene and safety, and the end product, to monitor the overall manufacturing process and to prevent fraud or mislabelling.

4.2 Quality control of raw milk

The best way to manufacture milk with good quality is to start with a good raw material. Testing raw milk is thus essential to ensure safety and quality. Raw milk is analysed for the presence of macroscopic abnormalities, addition of water, microbial quality, presence of milk from mastitic cows, presence of residues, and composition. Microbial contamination is a major issue, since pathogens can compromise the safety and spoilage microorganisms can limit the shelf life. The contamination of raw milk at the farm can be due to poor udder preparation or milking conditions, insufficient cleaning, failure in milk cooling systems or the presence of milk from mastitic cows. Another safety issue is the presence of residual veterinary drugs, plaguicides or mycotoxins. A very important aspect of raw milk quality is its composition. The dry matter, fat and protein are determined by farmers and processors for payment. Other milk components can be analysed by the industry for processing performance, labelling and improving quality.

In some countries, drinking raw milk is allowed. The European Union (EU) establishes limits for total bacterial counts ($\leq$50 000 colony-forming units (cfu) mL^{-1}), *Staphylococcus aureus* and *Salmonella* spp. Other pathogens or their toxins should not be present in amounts that can become a health hazard.

4.3 Quality control of processed milk

The control of the quality of processed milk depends on the type of product. Fluid milks differ by their processing treatments: pasteurisation, high temperature short time (HTST), extended shelf life (ESL), sterilisation, ultra-high temperature (UHT) and other processes that include microfiltration. Since there are limitations on the heating process of each type of milk, tests for evaluating the heat load are done to detect insufficient or excessive heating process.

In short-shelf-life milks, the limiting shelf-life factor is post-heat treatment contamination, i.e. spoilage by bacteria that enter milk after the heat treatment, usually in the filling/packaging line, and grow at refrigeration temperatures. Routine determinations are total microbial count, coliforms, psychrotrophs and pathogens. An important source of spoilage in short-shelf-life milk is the presence of heat-resistant spore-forming bacteria, mainly *Bacillus* spp., ubiquitous in raw milk, that survive heating processes and grow at refrigeration temperatures.

Long-shelf-life fluid milks (i.e. sterile and UHT milks) should not contain living microorganisms. The major shelf-life limiting factor for UHT milk is the presence of heat-resistant enzymes, particularly lipases and proteases, produced by psychrotrophs during refrigerated storage of the raw milk, which cause age gelation and off-flavour. If the psychrotroph count is high in the raw milk, a more severe heat treatment can decrease the enzyme activity, but it causes loss of nutrients and organoleptic properties. The organoleptic characteristics are an important quality attribute of UHT milk, as the high temperature of processing leads to important changes in colour (Maillard reaction), odour and taste (e.g. cooked flavour). In addition, the heat process can cause destabilisation and precipitation of micellar caseins. The microbial content, appearance and organoleptic properties are routinely tested by an 'accelerated' shelf-life test, performed by incubation of UHT milk in a closed container for 15 days at 30°C (or for 7 days at 55°C) after being processed.

The inspection of the end products by the industry is not always sufficient to ensure the quality of the product. In addition, at this stage, corrections cannot be done on the production line. Because of this, the industry has switched from the evaluation of the end products to the control of the process through the introduction of hazard analysis of critical control points (HACCP) and good manufacturing practice programs. Current EU legislation (EU, 2004a–e) is not strict with specific testing, but leaves to the manufacturer the responsibility of ensuring quality and safety of the products by implementing an HACCP system. Testing the end product provides support on the effectiveness of the process control. Therefore, for the modern industry, the evaluation of the raw materials and/or suppliers and the continuous monitoring of the processing line are the two key points for the production of a safe product with good quality. Another key point for the control of milk in the EU is the traceability system, which has become mandatory (EU, 2002). This system, implemented through a huge database system, makes it possible to track the milk to its origins through a bar code printed on the package. This allows procedures to identify problems and to take action rapidly.

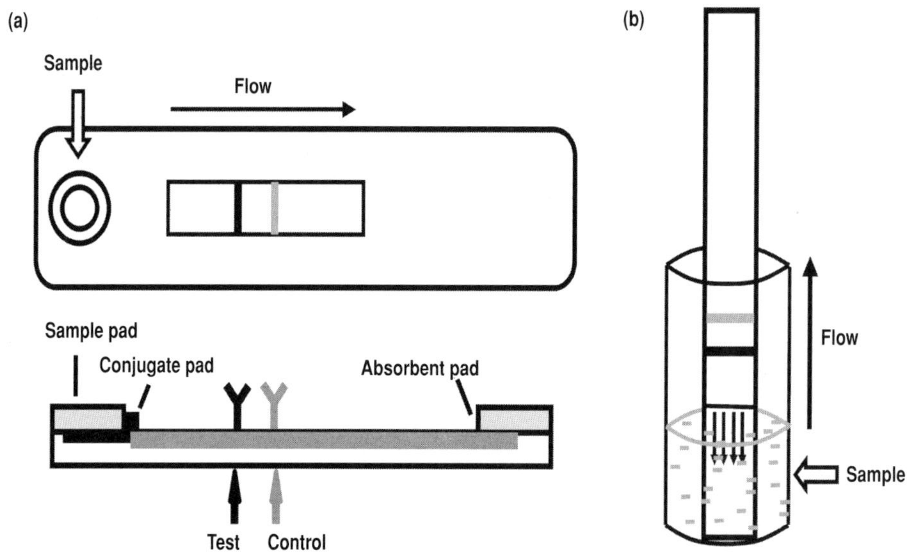

Fig. 4.1 Rapid test formats based on immunochromatography. Note: (a) lateral flow device and (b) dipstick. Reprinted Feinberg *et al.* (2006), with permission from Elsevier.

4.4 Methods of analysis

Farmers, producers and control laboratories do milk testing with different means. The analytical method used depends on the objective of the analysis, the need for a fast result, the instrumentation available, the specialised personnel available and the cost. On the farm, methods have to be rapid, easy to use and low-cost, not requiring trained personnel nor advanced instrumentation. Formats such as dipsticks or lateral flow devices that give a coloured visual signal are very handy (Figure 4.1), but sensing systems for online detection are increasingly used in automatic milking systems. Rapid online detection systems that work in real time are ideal for process control, e.g. physical sensors, spectroscopic sensors, biosensors (Figure 4.2). Fast online methods frequently gain speed at the expense of sensitivity, accuracy and/or precision. For some tests, the farmers and the industry do not have the capabilities to perform precise measurements, and they need to send their samples to control laboratories to perform these tasks. The methods used in control laboratories must be precise and accurate, at the expense of cost, specialised personnel and expensive instrumentation (Figure 4.3). Certification and accreditation of control laboratories is highly recommended to support legal claims. Reference methods have been developed and published by the major standard associations, e.g. Association of Official Analytical Chemists (AOAC), International Standards Organisation (ISO) and International Dairy Federation (IDF). Many reference methods are sometimes viewed as old-fashioned. However, they work well, they are very much in use, and what it is most important is that they are used to calibrate other routine methods.

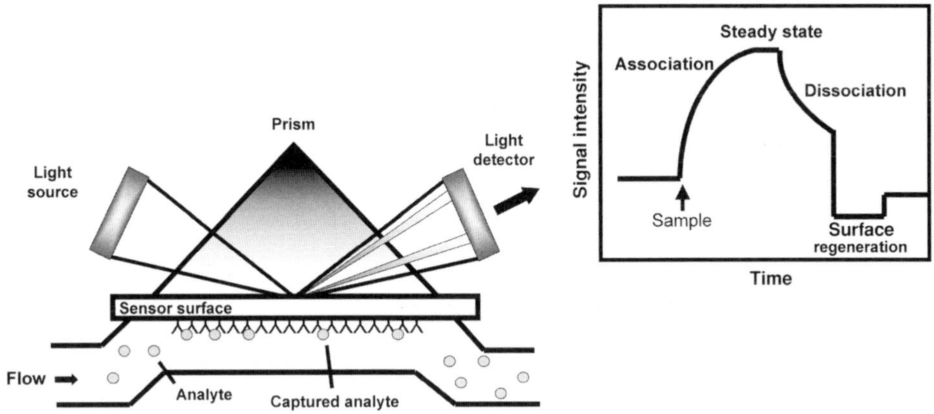

Fig. 4.2 Scheme of an optical biosensor and its signal output.

4.4.1 *Analysis of abnormal milk*

Abnormal milk, containing clots or blood, should not reach the bulk tank. To avoid it, milk can be tested visually after filtration to observe clots. For online detection, colour sensors are used for blood detection, and conductivity sensors are used to test for abnormal milk. However, the results obtained from conductivity sensors give false positives and negatives (De Mol, 2000). Because of this, a number of methods have been proposed (Reinemann & Helgren, 2004), such as an optical sensor designed to detect clots and flakes (Maasen-Franke *et al.*, 2004).

4.4.2 *Microbial analysis*

Microbial tests are applied to detect poor hygiene conditions and to evaluate the microbial quality of raw and processed milk. Total bacterial load is used to get an overview of the microbial contamination; psychrotrophs are assessed to prevent potential spoilage, coliforms to evaluate the hygiene history of milk, specific pathogens (e.g. *Salmonella* spp., *S. aureus*, *Listeria monocytogenes*, *Mycobacterium paratuberculosis*) to prevent potential health hazards and thermoduric bacteria to assess the hygiene of the production system.

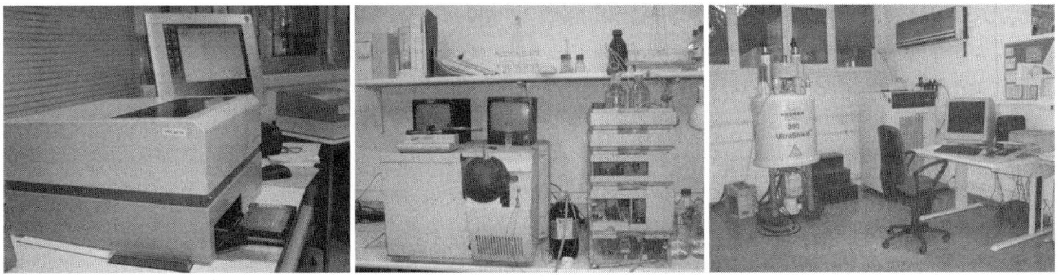

Fig. 4.3 Advanced instruments used for laboratory quality control – ELISA reader, HPLC-MS/MS and NMR spectrometer (from left to right).

Physical methods such as conductivity and titratable acidity are indirect but rapid sensing methods to estimate the microbial content. Measuring the pH is a routine practice in milk testing. These methods lack sensitivity and specificity, although they are handy and easily set up for online measurements.

Metabolite testing is also a rapid, non-specific and low-sensitive method that takes advantage of the presence of metabolites such as ATP, a cofactor in many reactions in living cells metabolism. The enzyme luciferase and its substrate luciferin use this cofactor in the reaction, resulting in a bioluminescence product that can be measured. It is commonly used for hygiene testing.

Counting methods have been the most popular methods. The standard plate count (SPC) has been working successfully for a long time and is the reference method used (IDF, 1991a) to test other techniques due to its sensitivity, accuracy and low cost. It is based on the capability of microbes to grow and form colonies that can be counted visually or with the aid of automatic colony counters. It measures only viable cells. In order to distinguish among the different types of bacteria, selective media and/or conditions are needed. For total counts, milk is incubated on non-selective medium at 30°C. Psychrotrophs are determined by preliminary incubation at 6°C and then SPC at 21°C. Selective media are used to detect coliforms. There are alternative methods for the enumeration of bacteria in milk, such as

- The plate loop (IDF, 2004a) or the Petrifilm methods; however, all of them, including SPC take a long time to produce results, from days to weeks.
- Direct microscopic count can shorten the assay time; cells are stained with a universal dye that binds DNA and then they are inspected under the microscope; however, it does not distinguish between live and dead cells.
- Flow cytometry is an automated and fast technique, able to perform bacterial counts in 20 min; many dairy industries rely currently on these instruments (e.g. BactoScan) for microbial enumeration; the cells are stained with a universal dye and counted one by one with an optical system; this method does not distinguish between viable and non-viable cells and it is not very sensitive, only being able to detect $>10^4$–10^5 cfu mL^{-1}.

Concentration/enrichment of microorganisms can be applied when the number of bacteria is too low. This can be done by a concentration or a preliminary incubation step. Enrichment by filtration allows concentration of all bacteria, but immunomagnetic separation allows concentration of a specific microorganism. Preliminary incubation is useful to increase the number of viable bacteria over dead bacteria. If selective conditions (temperature, media) are used, selected bacterial types can be assessed. Preliminary incubation is very useful to determine pathogens that are found in low numbers, and it is also advantageous to enrich viable bacteria when using methods that do not distinguish between live and dead cells.

Immunodetection methods are based on the ability of antibodies to bind specific proteins on bacteria. They are useful to detect pathogens. The selection of the antibody and the type of assay (i.e. indirect, direct, and/or competitive) is crucial. There are commercial immunoassays in the form of dipsticks or lateral flow devices, easy to use, reliable and affordable (Figure 4.1). For instance, a lateral flow system for *Escherichia coli* O157:H7 has been validated (Capps *et al.*, 2004). Immunoassay-based biosensor systems have been developed for the detection of pathogens, such as *E. coli* O157:H7, *Salmonella* spp. and

L. monocytogenes (Fratamico *et al.*, 1998; Koubova *et al.*, 2001; Brokken *et al.*, 2003; Leonard *et al.*, 2004). Biosensors are rapid methods of detection that combine a specific binding surface (based on a biological interaction, e.g. antibody or DNA) and a transducer (electrochemical, piezoelectric or optoelectronic) that converts the biological response into an electrical signal (Figure 4.3). Optical-based surface plasmon resonance biosensors (Biacore) are widely used. In general, biosensors do not have much sensitivity, although some studies with electrochemical-based sensors have shown to detect *L. monocytogenes* down to 10^3 cfu mL^{-1} (Benhar *et al.*, 2001). Currently, immunodetection methods require an enrichment or concentration step before analysis.

DNA-based methods are all based on the detection of specific DNA sequences. Techniques based on nucleic acid hybridisation and polymerase chain reaction (PCR) are useful to detect specific pathogens or groups of microorganisms. PCR can detect minute numbers of microorganisms in relatively short time. It is based on the amplification of DNA fragments flanked by two primers, whose choice is essential for the success of the method. There are primers that target universally conserved regions of DNA, such as 16S rDNA, and others that allow the detection of groups of bacteria down to specific genotypes. Multiplex PCR uses several pairs of primers to make multiple detections (Tamarapu *et al.*, 2001; Gillespie & Oliver, 2005), saving cost and time. Real-time PCR provides the means to do quantitative analysis by using a fluorogenic reporter that can be sequence specific (TaqMan) or non-specific (SYBR Green). The resulting signal becomes evident after a certain number of amplification cycles, which depend on the initial number of DNA molecules. Many PCR protocols are stepping up to the real-time version. It has been applied to milk pathogens, such as *M. paratuberculosis* (Khare *et al.*, 2004; O'Mahony & Hill, 2004), *Salmonella* spp. (Kessel *et al.*, 2003), *Campylobacter jejuni* (Yang *et al.*, 2003), *L. monocytogenes* (Hein *et al.*, 2002), *E. coli* O157:H7 (Fortin *et al.*, 2001) and thermophilic bacilli in dry milk powders (Rueckert *et al.*, 2006). Commercial kits for different pathogens are available. PCR detects viable and non-viable cells and require a pre-enrichment step that increases sensitivity and reduces false positives.

More sophisticated tools have been developed for DNA-based methods to identify microorganisms, by the combination of PCR and powerful techniques such as mass spectrometry that gives high specificity and sensitivity. By using primers that target universally conserved sequences (rDNA), the broad range of products obtained can be analysed by electrospray ionisation-mass spectrometry, and the microorganism identified by specialised software and a database of known microorganisms (Hofstadler *et al.*, 2005).

DNA-based subtyping methods, such as pulse-field gel electrophoresis (PFGE) and ribotyping are used to identify pathogenic and spoilage bacterial isolates. 'Rybotypes', which are band patterns of small fragments of DNA, specific for rRNA encoding genes, have been able to discriminate between *Pseudomonas* spp. subtypes with different spoilage potential (Dogan & Boor, 2003). There are commercially available automated standardised systems for rybotyping (e.g. RiboPrinter).

4.4.3 *Mastitic milk – somatic cell count (SCC)*

Mastitic milk contains pathogens and spoilage microorganisms, and it is characterised by an increase in Na^+ and Cl^- as well as leucocytes. Screening tests include conductivity tests, based on changes in the ion composition, and the California mastitis test (CMT), based on the increase in leucocytes. More accurate methods are based on the enumeration of somatic cells, mostly leucocytes that should not exceed 400 000 or 750 000 cells mL^{-1} (EU and US milk quality standards, respectively). In order to determine the SCC, direct microscopy (DMSCC) can be performed after DNA staining. Automatic counters are fast and reliable. They exploit the large size and high DNA content of somatic cells to separate and count them, and use electrical sensors, i.e. Coulter counter, or fluoro-optical sensors, such as the Fossomatic method (FSCC). There are standardised methods using direct microscopy and FSCC (IDF, 1995a). ATP measurements can be performed after a concentration step by filtration and further disruption of the cells.

4.4.4 *Testing for residues compounds*

Milk contaminated with residues in levels that are potentially harmful to human health should not reach the tank. The major analytical challenge for the detection of residues is the large number of compounds and the low amounts to be detected.

Antibiotics comprise a considerable number of compounds, including β-lactams (penicillins), sulphonamides, tetracyclines, aminoglycosides and macrolides that, in turn, are families comprising different compounds. β-Lactams are widely used, and they have been the targets of most developed methods. Tests for general screening, based on the ability to inhibit the growth of *Bacillus stearothermophillus* var. *calidolactis* spores, are not adequate to detect low levels of antibiotics or to detect the range of antibiotics currently used.

Pesticides comprise a great variety of groups that include different chemical species. Two groups of insecticides, organophosphates and carbamates are widely used. They can be detected as a whole by their effect, through the acetylcholinesterase (AchE) inhibition test, which has evolved over many years. Recombinant AchEs have been developed to increase the test efficiency (Schulze *et al.*, 2003). This test can be found in different formats, including biosensors (Andreescu & Marty, 2006).

Mycotoxins, particularly aflatoxin M_1, are secreted into milk of dairy cows that have consumed aflatoxin B_1. The more reliable analytical method (IDF, 1995b) is a procedure that includes an immunoaffinity clean-up step, followed by high-performance liquid chromatography (HPLC) and fluorimetric detection. A similar method, easier to set up, employs thin-layer chromatography (TLC) after the same clean-up step (IDF, 2005a).

Rapid screening tests based on competitive immunoassays have good sensitivity, and are normally used to detect the presence of residues above a certain level, which allows taking immediate action. Developing generic antibodies by mixing antibodies with different specificities and/or by a careful selection of the hapten has made it possible to detect all chemical species within the same group. Guidelines for the standardised description of immunoassays are available (IDF, 2003a, b). Commercially available formats include dipsticks, lateral flow devices and biosensors. The use of biosensor arrays, which detect many different analytes simultaneously, has great application for multiple residue detection.

Advanced methods, carried out by specialised laboratories, are used for confirmation, as they are accurate and reliable. Confirmation methods include a sample clean-up step, a separation procedure (high-pressure liquid chromatography – HPLC, gas chromatography – GC or capillary electrophoresis – CE) and a detection system. The clean-up step is critical and time consuming. Detection with mass spectrometry is increasingly used because it is highly

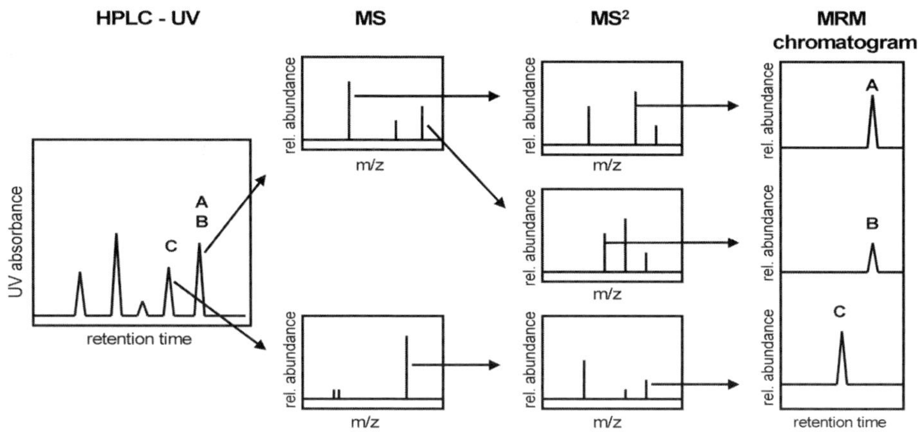

Fig. 4.4 Diagram of a multi-residue HPLC-MS/MS determination method. Note: After a clean-up step, the sample is injected in a HPLC system monitored by UV detection. The compounds that run at different retention times pass to the mass spectrometer (MS^1), where selected ions are further fragmented (MS^2). By using selective MS/MS conditions for different chromatogram retention times, each compound can be unambiguously identified. The result is a multi-residue mass (MRM) chromatogram. With the appropriate calibration, a quantitative analysis can be performed on each compound.

sensitive and specific. Multi-residue analysis with tandem mass spectrometry is a powerful technique that allows the identification and quantitation of many different compounds within complex mixtures (Figure 4.4). Even though these analysis methods are more expensive and require longer times, they can detect many residues in one run.

4.5 Major components analysis

Traditional reference methods that determine the major components in milk (dry matter, protein, fat and lactose) are based on classical chemistry. The dry matter content is measured by weighting the residue of a milk sample after being desiccated (IDF, 1987a, 1991b, 2004b). The protein content is measured by the Kjeldahl method (IDF, 2001), which determines the total nitrogen content in the milk and converts this value into protein concentration by multiplying by a factor of 6.38. Protein values may be altered by the presence of other nitrogen compounds, such as urea. The true protein concentration is calculated from the difference between total nitrogen and non-protein nitrogen. The latter is obtained from the Kjeldahl measurement of the nitrogen left in the supernatant after precipitation of protein with trichloroacetic acid (TCA). The fat content is determined by a gravimetric method (Röse-Gottlieb) (IDF, 1996, see also IDF, 1987b–d), from the matter obtained by organic solvent extraction of the milk sample. Other standard methods for fat testing are the Gerber (IDF, 1981) and the Babcock (AOAC, 2005a) methods. In addition, the measurement of fat using turbidimetry by light scattering is useful since it is automated and standardised methods are available (AOAC, 2005b, c).

Methods for determining lactose have not been paid much attention until recently, and no international reference method is available yet. However, there are gravimetric (AOAC,

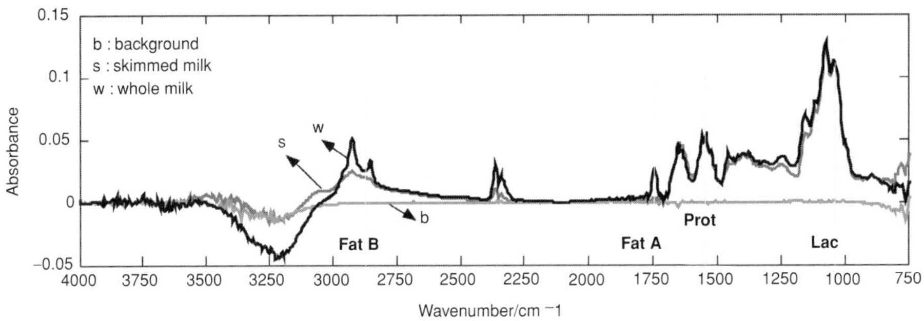

Fig. 4.5 FTIR spectra of whole and skimmed milk, highlighting the signals of major milk components (fat, protein and lactose). Reprinted from Iñón *et al.* (2004), with permission from Elsevier.

2005d), polarimetric (AOAC, 2005e) and enzymatic (IDF, 2002a) methods. The best candidates for becoming reference methods are those based on enzymatic reactions and HPLC. Enzyme assays are colorimetric methods based on the reaction of a cascade of enzymes, i.e. β-galactosidase coupled to glucose oxidase. Biosensors based on this technique are available.

Multiple components analysis can be carried out by spectroscopic methods that are rapid, non-destructive, non-invasive and can be set up online. For milk component testing, medium infrared spectroscopy (MIR) is extensively used, being able to measure simultaneously lactose, protein and fat (Figure 4.5). These values are also used for calculating the total solids, by a regression correlation to reference values for total solids, previously determined by the reference method. There are MRI standardised methods (IDF, 2000) and instruments available from specialised manufacturers (Milkoscan, Multispec, Bentley, Lactoscope, Aegys). Most current instruments rely on the Fourier transform technique (FT-MIR), and are able to get information about other components, such as casein, urea, lactic acid, citric acid, free fatty acids, carbohydrates and the freezing point. Online FT-MIR milk analysers are used for process control, i.e. standardisation of multiple components. A similar technique, near infrared spectroscopy (NIR), can be applied to other products than milk, such as cream, dairy powders and cheese. NIR can be applied to test the composition of dry milk powders, including moisture, obtaining immediate results in contrast to the 3 h needed by traditional means. It can be set up online, allowing for process control. All together, these techniques have good accuracy and precision, providing results as good as international reference methods. The instruments just need to be carefully calibrated.

Other valuable spectroscopy technique is Raman spectroscopy, which combines the specificity and optical convenience needed to become a good alternative to FTIR measurements for milk composition testing and process control. Nuclear magnetic resonance (NMR) spectrometry is a technique that has to be borne in mind, even though it is not well known nor widely used. This technique is highly versatile and it has evolved greatly in the last years, with more sensitive probes, the introduction of automated flow injection sampling systems and the coupling to chromatographic techniques. Low-resolution bench instruments, e.g. Minispec (Bruker), have been applied for many years to distinguish solid from liquid fats in dairy products. High-resolution instruments are able to detect many

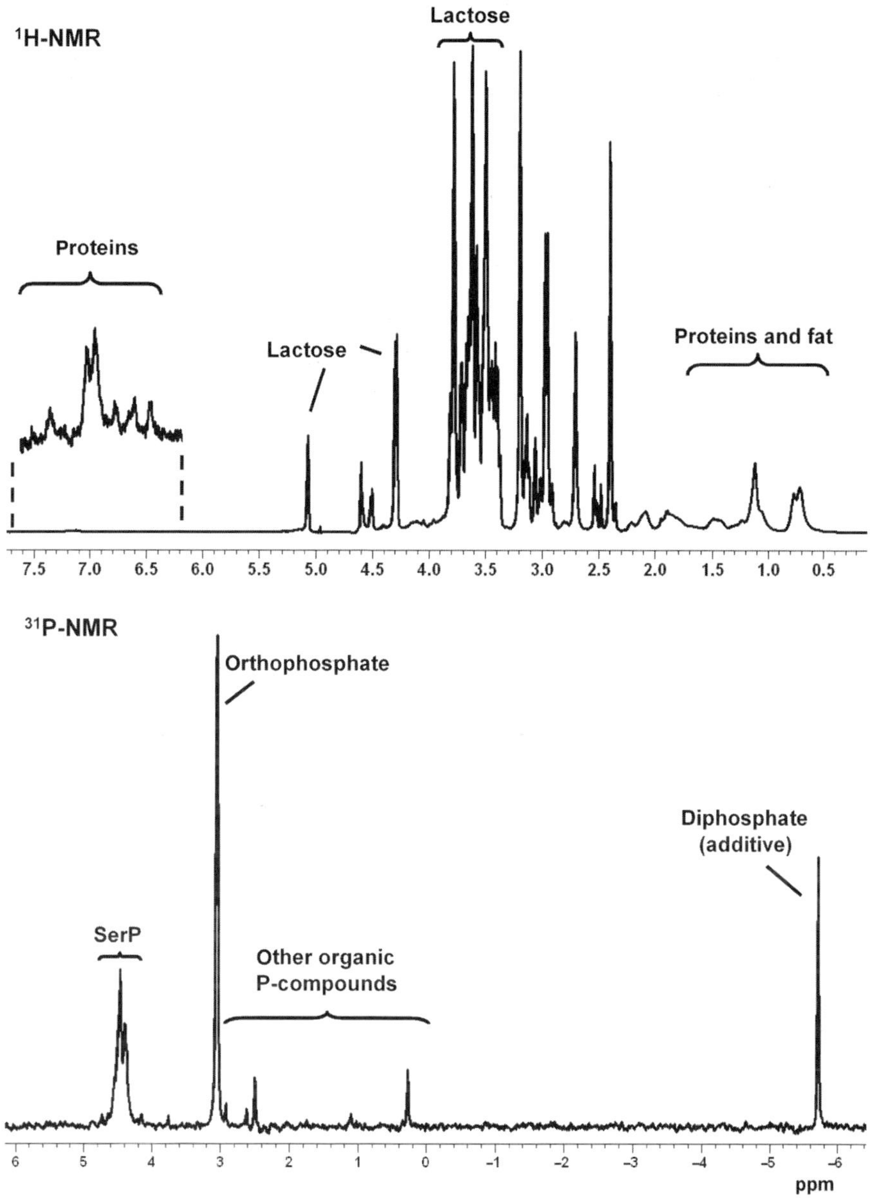

Fig. 4.6 NMR spectra of milk showing the resonances belonging to major components in milk.

components in milk (Belloque, 2006). Fat, proteins, and particularly lactose, can be simultaneously observed in a ^{1}H-NMR spectra of milk (Figure 4.6). However, high-resolution NMR spectrometers are very expensive and require much expertise. Because of this, low-resolution NMR techniques have been recently developed to analyse the major components in foods, including milk. These techniques are based on the analysis of relaxation parameters combined with chemometrics.

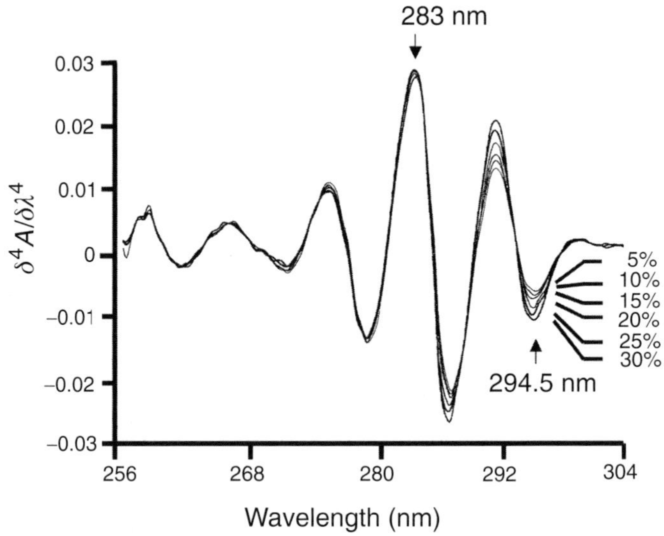

Fig. 4.7 Fourth-derivative UV spectra of a model system of caseins and whey proteins, showing differences between samples with different proportions of whey protein relative to total protein. (Reproduced by courtesy of B. Miralles.)

4.5.1 *Protein composition analysis*

Within proteins, the casein and whey fractions are a relevant quality parameter. Caseins are important from a nutritional point of view, since they are the main calcium carriers. The casein content can be analysed in raw milk by selective precipitation at pH 4.6, and further protein analysis. Processed milks are more difficult to analyse due to the heating-induced linkage between caseins and whey proteins. Determination by spectroscopic methods allows for rapid analysis and does not require fractionation. FTIR can quantify caseins (Sorensen *et al.*, 2003) and fourth-derivative UV analyses the whey protein/total protein ratio (Figure 4.7; Miralles *et al.*, 2000). These methods work well for raw and processed milks. ^{31}P-NMR has also shown to determine casein in raw and processed milks from the SerP signal (Figure 4.6; Belloque & Ramos, 2002). Other techniques have been proven useful for the analysis of the caseins/whey protein ratio, a valuable parameter to control the use of membrane technology for standardisation of milk and, therefore, to prevent frauds (see Section 4.6.2).

Chromatography and electrophoresis can give much information about the individual proteins. Whey proteins are easily quantified in raw milk by precipitation at pH 4.6 and further analysis of whey by HPLC, but it quantifies only soluble whey proteins (Resmini *et al.*, 1989). Caseins are more difficult to analyse due to their aggregate nature and require the use of denaturants and disulphide reduction reagents. Simultaneous analysis of caseins and whey proteins in raw milk has been done by HPLC and CE under denaturing conditions (Recio & Olieman, 1996; Bordin *et al.*, 2001). CE works well for raw and processed milks, and has been applied to evaluate milk proteolysis, milk protein polymorphism and whey protein/total protein ratio (Recio *et al.*, 2001). Coupling a mass spectrometry detection system provides a powerful tool to distinguish protein details such as genetic variants, glycosylation and phosphorylation (Guy & Fenaille, 2006).

4.5.2 *Fat composition analysis*

Chromatographic techniques have been commonly applied to analyse fat components (De la Fuente & Juárez, 2005; Guy & Fenaille, 2006). The fatty acid composition as well as the position occupied in the triglycerides (sn-1, sn-2, sn-3) is important from a nutritional point of view.

Fatty acids are mainly found in the triglyceride fraction, and therefore, they have to be released prior to analysis. Fatty acids in milk can be analysed by a GLC standard method (IDF, 2002b). Pre-fractionation by TLC or HPLC with further analysis by GC-MS has shown good results, although some problems have been associated with thermal degradation and volatilisation of particular fatty acids.

Triacylglycerides can be analysed in whole milk fat by GC or HPLC coupled to mass spectrometry. HPLC-MS has the advantage of requiring a minimal sample preparation, does not have problems with thermal degradation and chromatographically overlapping peaks can be resolved by MS. Mottram & Evershed (2001) used an HPLC-APCI-MS system to analyse the fatty acid composition in whole milk fat, and they were able to distinguish positional isomers. However, for this purpose it is better to use tandem mass spectrometry that allows determining masses, positional isomers and even the location of double bonds. Another powerful tool for the analysis of positional isomers is ^{13}C-NMR since this technique is structure-sensitive (Diehl & Ockels, 1995).

4.5.3 *Analysis of other milk components*

There are other milk components with nutritional importance that are normally included in the milk package labels, such as calcium, phosphorous and vitamins. These should be tested in the end product since processing can alter their contents, particularly vitamins. Vitamins are analysed mainly by HPLC and there are HPLC reference methods for vitamin A (IDF, 1990a) and vitamin D (IDF, 2002c). Other methods are based on the measurement of metabolic activity of microorganisms, but even though some fast systems are available (e.g. VitaFast), it takes 1–2 days to obtain the results. Immunoassays are also available and useful to test vitamin-enriched products.

Calcium can be detected by a titrimetric method (IDF, 1992a) or by flame atomic absorption spectrometry (IDF, 1992b). Phosphorous in milk is present in different forms, but the most common measurement is the total phosphorous content, performed by molecular absorption spectrometry (IDF, 1990b). ^{31}P-NMR is a useful means to the simultaneous determination of the major natural P-compounds (Figure 4.6), such as orthophosphate, phosphoserine, phosphorylated carbohydrates and phosphoglycerides as well as added stabilisers (diphosphates), using a simple sample preparation (Belloque & Ramos, 2002).

4.5.4 *Analysis of degradation products formed during milk storage*

Proteolysis and lipolysis are the major cause for limiting the shelf life of liquid milks, particularly UHT milk, during storage (López-Fandiño & Olano, 1999). The degradation of proteins by proteolytic enzymes cause the release of peptides into the milk whey, and therefore, this fraction shows increased peptides as proteolysis proceeds. These components can be analysed by precipitation of caseins at pH 4.6 and quantifying the nitrogen content of whey by the Kjeldahl method, or by analysing the resulting peptides by HPLC or CE (López-Fandiño *et al.*, 1993; Recio *et al.*, 1996, 2000a). Similarly, lipolysis can be evaluated by the increase of the free fatty acid fraction, particularly short-chain fatty acids, since these are the primary cause for lipolytic rancidity in milk (Bucky *et al.*, 1988). In addition, volatiles produced as a subsequent step to lipolysis can also be measured by headspace-GC (Vallejo-Córdoba & Nakai, 1994).

4.5.5 *Evaluation of heat load*

Heating milk causes different modifications in the physico-chemical state of its components, leading primarily to the denaturation of certain protein fractions (enzymes, whey proteins) and the formation of Maillard reaction products (for a review see Pellegrino *et al.*, 1995; López-Fandiño & Olano, 1999). Only some compounds are suitable as chemical markers of the heat treatment intensity: enzymes, lactose-derived compounds, Maillard reaction-derived compounds and whey proteins.

Enzyme assays are routine tests for short-shelf-life products. Pasteurised milk is obtained by heat treatment, of 72°C for 15 s, or any other temperature–time combination producing an equivalent effect. Alkaline phosphatase, a milk native enzyme, is inactivated during pasteurisation, reason why it is used as an indicator of the effectiveness of pasteurisation. Another native enzyme, lactoperoxidase, is more heat-stable than alkaline phosphatase. It remains active after pasteurisation, but loses its activity after HTST processing. Therefore, the use of the lactoperoxidase test distinguishes between the two treatments. Standard methods for alkaline phosphatase and lactoperoxidase tests include those based on spectrophotometry (EU, 1991) and fluorimetry (IDF, 2003c). Other native enzymes can also be used to evaluate heat load in milk, e.g. α-fucosidase, phosphodiesterase, α-mannosidase, but their practical use is much less extended.

Lactose-derived compounds can be used to evaluate more intensive heat treatments, i.e. direct and indirect UHT treatments and sterilisation. Lactulose is formed by isomerisation of lactose during heating of milk and has been proposed as an analytical index to distinguish UHT from sterilised milk. UHT milk should have a lactulose upper limit of 600 mg L^{-1}. Lactulose is not found in pasteurised milks, although values up to 82 mg L^{-1} have been reported in some commercial samples (Corzo *et al.*, 1986). A variety of methods have been used for lactulose determination, e.g. GC (Martínez-Castro *et al.*, 1987), HPLC (Cataldi *et al.*, 1999), CE (Soga & Serwe, 2000), enzyme methods (Kuhlmann *et al.*, 1991; Amine *et al.*, 2000a), colorimetry (Amine *et al.*, 2000b) and continuous flow amperometry (Mayer *et al.*, 1996; Moscone *et al.*, 1999). Current standard methods include a rapid enzymatic method (IDF, 2004c) and an HPLC method (IDF, 1998).

Maillard reaction products, such as lactuloselysine, results from heating and it is transformed into furosine by acid hydrolysis. Furosine is present in concentrations of 3–5 mg 100 g^{-1} protein in raw milk and 5.2–7.5 mg 100 g^{-1} protein in pasteurised milk (Clawin-Rädecker & Schlimme, 1995). In sterile milk samples, the furosine level varies within a wide range, but it is usually higher in milk processed by indirect than by direct UHT treatment (Resmini & Pellegrino, 1991; López-Fandiño *et al.*, 1993; van Renterghem & de Block, 1996). However, overlapping of values in milk submitted to different heat treatments is the main difficulty encountered when using furosine as chemical indicator to distinguish among commercial sterilised milks. Moreover, the furosine level may increase during storage of UHT milk (Corzo *et al.*, 1994). In some countries where HTST milk is widely consumed (e.g. Italy), it is established that this type of milk should give negative tests for phosphatase and peroxidase, as well as a furosine $\leq$20 mg 100 g^{-1} milk. Furosine can be determined by a number of RP-HPLC methods using different columns (Schäffer, 1988; Delgado *et al.*, 1992; Resmini *et al.*, 1990; Nicoletti *et al.*, 1997), anion-exchange chromatography with pulsed amperometric detection (Cefalu *et al.*, 1991), CE (Corradini *et al.*, 1996; Tirelli & Pellegrino, 1996) and GC (Ruttkat & Erbersdobler, 1994). Since furosine is partially degraded during gas chromatography analysis, this method cannot be recommended for routine analytical application (López-Fandiño & Olano, 1999). The international standard method is based on an ion-pair RP-HPLC system (IDF, 2004d).

Whey protein denaturation based methods are particularly suited to discriminate products with different heat treatments, even different pasteurisation conditions. The amount of undenatured β-lactoglobulin is one of the thermal indicators for distinguishing commercial milk samples submitted to different thermal treatments (Buchheim *et al.*, 1994), with levels of >2600 mg L^{-1} (pasteurised), >2000 mg L^{-1} (HTST), >50 mg L^{-1} (UHT) and <50 mg L^{-1} (sterilised).

Several chromatographic, electrophoretic and immunochemical techniques allow the quantitative analysis of individual native whey proteins that remain in solution after milk acidification at pH 4.6. Resmini *et al.* (1989) developed an RP-HPLC method on a polymeric column with UV or fluorescence detection that determined the extent of whey protein denaturation in pasteurised milk by analysing the remnant undenatured protein. International standard HPLC methods based on this method are available for the evaluation of β-lactoglobulin denaturation in liquid milk (IDF, 2005b) and for the characterisation of extra-low-heat and low-heat milk powders (IDF, 2002d). To achieve more accuracy, Recio & Olieman (1996) developed a CE method that determined directly the denatured protein and was able to discriminate among pasteurised milk samples subjected to temperatures that differed in 3°C.

The denaturation of milk proteins as a heat indicator has been incorporated into rapid systems, such as the FAST index, determined by the FAST method, which is the ratio of the maximal fluorescence emission of the total milk sample (350/440 nm), and Trp fluorescence (290/340 nm) (Birlouez-Aragon *et al.*, 2001). Another rapid method is based on the immunological detection of native and denatured α-lactalbumin, detected on a surface plasmon resonance biosensor (Biacore). In spite of the number of methods available, it has been suggested that the only possible authentication technique for heat load should be based on a multivariate analysis, based on several methods able to define an efficient discrimination approach applicable to commercial milk samples (Feinberg *et al.*, 2006). The values of these indicators for freshly processed and stored commercial milk samples are shown in Table 4.1.

4.6 Analysis of fraudulent addition of ingredients and authentication

4.6.1 *Addition of water*

Water added by accident or deliberate (i.e. to increase the volume of milk) is a fraudulent activity, and it is routinely tested. It is detected by the determination of the freezing point, as the dilution of milk causes an increase towards 0°C. The thermistor cryoscope method is the reference method (IDF, 2002e) and it is widely used. Data of the milk freezing point from the regular milk suppliers need to be collected for reference. In the industry, an FTIR method is used normally for screening. If a milk sample is found out of the reference range, then it is confirmed by the cryoscope method.

4.6.2 *Addition of whey proteins*

The addition of these proteins may become economically attractive to increase the milk protein content in liquid milk, considering the large excess and the low price of rennet whey. Determination of rennet whey is based on the detection and quantification of caseinomacropeptide (CMP), the hydrophilic fragment of κ-casein released by chymosin during milk clotting. Isoelectrofocusing (Rispoli *et al.*, 1991) and ELISA with polyclonal antibodies against κ-casein (f130–152) (Bitri *et al.*, 1993) have been used for this purpose. A method for detecting rennet whey is based on the milk treatment with TCA (8 mL 100 mL^{-1}) and analysis of CMP by RP-HPLC, using UV detection (Olieman & van Riel, 1989) or pulsed electrochemical detection (van Riel & Olieman, 1995a). This method allows resolution between CMP and other κ-casein fragments, such as pseudo-CMP (CMP lacking the N-terminal methionine). However, under certain circumstances, false positives result from the coelution of CMP with similar proteolytic products arising from the action of psychrotrophic proteinases on κ-casein. Improved resolution has been achieved by capillary electrophoresis (CE), which has become the most powerful tool for detecting CMP in powder and liquid milks (van Riel & Olieman, 1995b). This method is based on the peptide separation in a hydrophilically coated capillary with a 6 M urea and citrate buffer at low pH. A detection limit of 0.4 g 100 g^{-1} rennet whey solids in milk powder was achieved, and false positive results were prevented in some types of buttermilk powders. The investigation of the suitability of this CE method to detect whey added to UHT milk (Figure 4.8) revealed the presence of CMP and CMP-like peptides resulting from the action of psychrotroph thermostable proteinases on κ-casein (Recio *et al.*, 1996, 2000a). Therefore, although the presence of CMP in UHT milks cannot be taken as an exclusive indicator of adulteration with rennet whey, measuring the area ratio of the different CMP-like forms present makes possible to detect whether whey solids have been added to milk (Recio *et al.*, 2000b).

Other than whey rennet, such as addition of acid whey, can be detected on the base of the whey protein/total protein ratio. Because whey proteins contain significant amounts of sulphur-containing amino acids (cysteine and cystine) and aromatic amino acids

Table 4.1 Group mean and standard deviation (between parenthesis) of different heat load indicators according to the type of milk and storage conditions.

Technology	D-α-La %	D-α-La (mg L^{-1})	N-α-La (mg L^{-1})	N-β-Lg (mg L^{-1})	FAST	FTrp (AU)	Furosine (mg 100 g^{-1})[a]	Lactulose (mg L^{-1})	Lactoperoxidase (IU mL^{-1})
No storage									
Pasteurised	11.0 (3.18)	25 (7.7)	203 (12.3)	3550 (540.7)	18.1 (3.3)	5.0 (0.4)	4 (2.3)	15 (12.1)	0.52 (0.224)
High pasteurisation	32.0 (6.15)	81 (18.4)	171 (11.2)	1105 (623.2)	39.4 (10.8)	2.1 (0.7)	29 (33.6)	31 (13.6)	0.05 (0.000)
UHT (direct)	34.8 (6.33)	69 (12.8)	130 (14.3)	660 (372.4)	54.7 (13.6)	1.9 (0.5)	59 (34.8)	144 (48.6)	
UHT (indirect)	73.6 (8.25)	166 (38.0)	57 (14.7)	44 (123.8)	115.0 (19.1)	1.0 (0.1)	188 (51.8)	400 (134.4)	
Sterilised	98.6 (0.71)	244 (24.2)	3 (1.5)	0 (0.0)	257.1 (20.4)	0.5 (0.1)	371 (88.7)	834 (218.1)	
Storage 90 days at 25°C									
UHT (direct)	24.1 (3.83)	43 (6.5)	136 (15.9)	0 (0.0)	46.3 (10.7)	2.2 (0.5)	169 (30.9)	207 (83.6)	0.05 (0.000)
UHT (indirect)	60.0 (9.65)	100 (24.4)	65 (15.0)		90.0 (16.6)	1.2 (0.2)	264 (51.9)	462 (104.1)	
Sterilised	97.5 (1.31)	176 (17.2)	4 (2.2)		183.1 (25.5)	0.6 (0.0)	417 (81.3)	967 (110.5)	
Storage 90 days at 35°C									
UHT (direct)	31.9 (4.45)	56 (7.5)	121 (16.0)	0 (0.0)	47.4 (11.3)	2.5 (0.5)	464 (58.4)	415 (83.6)	0.05 (0.000)
UHT (indirect)	60.5 (5.91)	88 (12.0)	58 (14.0)		80.7 (10.8)	1.4 (0.1)	529 (83.3)	609 (101.7)	
Sterilised	97.6 (1.31)	163 (14.0)	4 (2.3)		127.3 (20.2)	0.9 (0.1)	570 (81.8)	1063 (241.1)	

[a] Content of furosine (mg) referred to 100 g of protein.

D-α-La, percentage of denatured α-lactalbumin; D-α-La, denatured α-lactalbumin; N-α-La, native α-lactalbumin; N-β-Lg, native β-lactoglobulin; FAST, FAST index; FTrp, tryptophan fluorescence.

Reproduced from Feinberg *et al.* (2006), with permission of Elsevier.

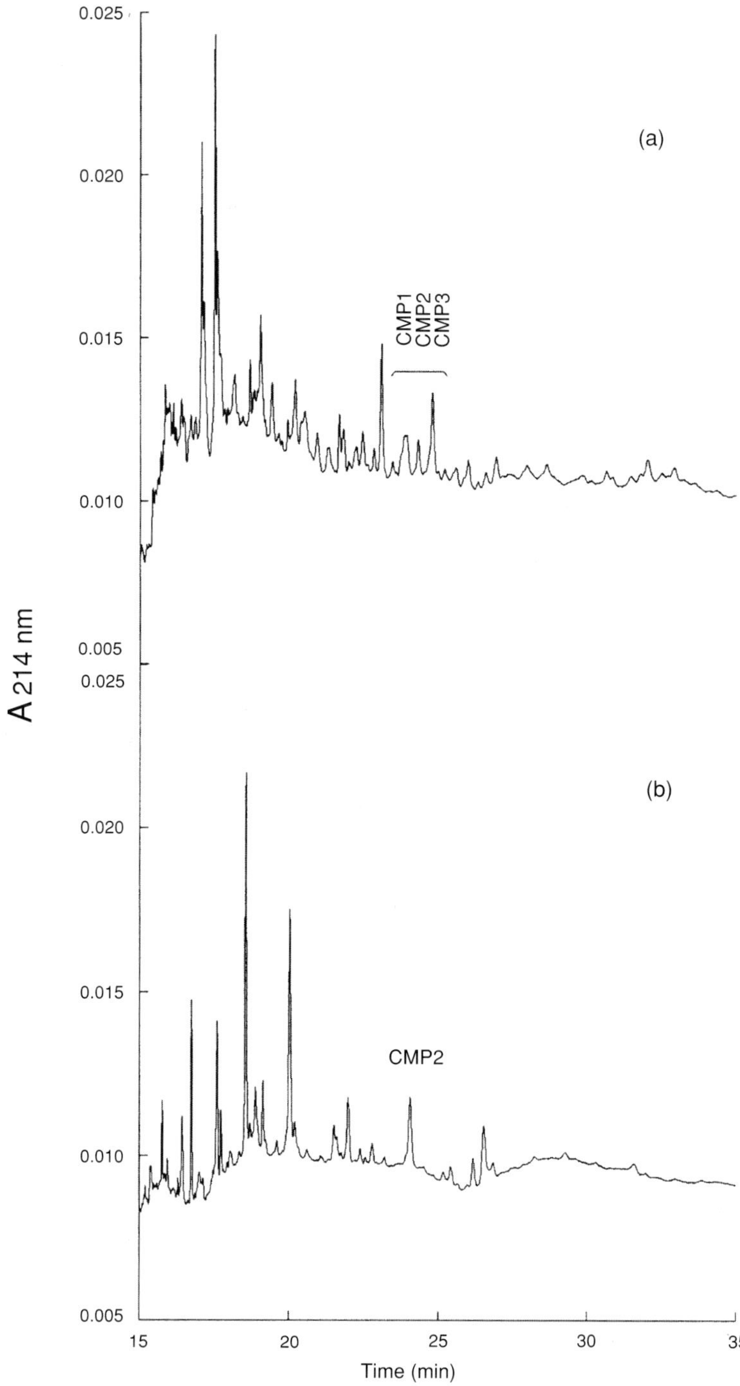

Fig. 4.8 Capillary electrophoresis (CE) analyses of whey from a genuine UHT milk sample stored for 30 days at 30°C (a) and a sample adulterated with 6.4 mL 100 mL^{-1} of whey powder reconstituted in water to 8.87 g 100 mL^{-1} immediately after UHT processing (b). Reproduced from Recio *et al.* (2000b), with permission from Elsevier.

(tryptophan, tyrosine and phenylalanine), the addition of whey proteins increase those values as compared with the genuine product. Traditional procedures to determine this addition comprise the determination of casein-bound phosphorous, measurement of casein and whey proteins by SDS-PAGE, and polarographic evaluation of cysteine/cystine. Although the polarographic method (Lechner & Klostermeyer, 1981) is very effective and reliable, it is limited by the use of hazardous chemicals (methylmercury chloride). Later, more rapid analytical methods were proposed as alternatives. Schmidt *et al.* (1999) developed a method based on pyrolysis-mass spectrometry capable of discriminating, in less than 2 min, between pure milk samples and milk samples with added whey proteins. Non-invasive analytical techniques, such as FTIR (Mendenhall & Brown, 1991), photoacoustic spectroscopy (Dóka *et al.*, 1999) and fourth-derivative UV spectroscopy (Figure 4.7), have been applied (Meisel, 1995; Luthi-Peng & Puhan, 1999; Miralles *et al.*, 2000). The latter is a method that quantifies aromatic amino acid residues of milk proteins and has shown good results, being unaffected by other absorbing non-protein material in the sample solution. Direct measurement of all relevant protein fractions by electrophoretic or chromatographic methods is more laborious but gives a more detailed insight. The whey protein/total protein ratio can be determined directly from the chromatographic trace obtained, after careful calibration using reference compounds. Miralles *et al.* (2000) compared CE, SDS-CE and UV fourth-derivative absorption to determine whey protein in total protein ratio in raw, pasteurised and UHT milks. The results confirmed that heat treatment does not influence the ratio obtained by these three methods, although under certain conditions lactosylation of proteins occurs. However, CE and SDS-CE methods need still expensive equipment and qualified operators, whereas fourth-derivative UV spectroscopy is a technique available in most laboratories, which makes it more suitable for routine analyses.

4.6.3 *Addition of non-dairy proteins*

Proteins of plant or animal origin could be added to milk, a practice that is common in countries if they are cheaper than milk proteins. A number of electrophoretic (Kanning *et al.*, 1993; Cattaneo *et al.*, 1994; López-Tapia *et al.*, 1999; Manso *et al.*, 2002a,b), chromatographic (Cattaneo *et al.*, 1994; Espeja *et al.*, 2001) and immunochemical (Turin & Bonomi, 1994; Sánchez *et al.*, 2002) methods have been used to detect the addition of non-dairy protein in milk products.

4.6.4 *Addition of reconstituted milk*

Reconstituted milk powders and other dairy products are sometimes added to liquid milk, and they are not allowed in many countries. The methods used to detect this addition include the determination of RNase activity (Ju *et al.*, 1991), the ratio of β-casein to α-lactalbumin determined by CE (Chen & Zang, 1992) and NIR spectroscopy (Pedretti *et al.*, 1993). However, the most successful determinations are based on the quantification of products formed in the milk powders as a result of the Maillard reaction or milk heat treatment. The best indicators reported in the literature are the hydroxymethylfurfural, determined by visible spectrometry (Rehman *et al.*, 2000) and furosine (Resmini *et al.*, 1992; Villamiel *et al.*, 1999, 2000). Furosine measured by HPLC was successfully conducted as an index to detect the adulteration of raw, pasteurised milks and cheese with reconstituted dried milk. However, a better indicator is the ratio of lactulose to furosine. Milk drying promotes intensive Maillard reaction but low lactulose formation and as a consequence, the lactulose/furosine ratio in UHT milk is 16 times higher than in powder milk. Ratios lower than 6.0 in UHT milk may be indicative of the addition of reconstituted milk (Montilla *et al.*, 1996), although in practice these ratios can be altered due to different processing conditions. Therefore, more research is necessary in this respect to develop a reference method to detect this fraudulent addition.

4.6.5 *Substitution of milk from a different species of mammals*

Mixing milk from different species of mammals is a fraud, and is usually committed in milk used for cheesemaking. This is done by adding cow's milk to sheep's and goat's milks, because of the lower price of the former. Hence, this adulteration does not apply to processed cow's milk, but it may occur in market-liquid milks of goat origin. Although several methods on fatty acid analysis by GC and triglycerides by NMR have been developed, most analytical methods are based on the protein fraction. The current EU reference method for the detection of cow's milk in goat's and sheep's milk is based on the separation, by isoelectric focusing, of the γ-casein peptides originated after digestion of the sample with plasmin (EU, 1996). Although the fraudulent addition of cow's milk in other species' milk is covered by this method, goat's and sheep's milk cannot be distinguished. To improve sensitivity and automate the analysis, other methods based on gel or capillary electrophoresis, chromatography, immunoassays and PCR have been proposed. These methods have been reviewed by different authors (Recio *et al.*, 2001; Moatsou & Anifantakis, 2003; Hurley *et al.*, 2004; Borkova & Snaselova, 2005; De la Fuente & Juárez, 2005; Mayer, 2005). An HPLC coupled to tandem mass spectrometry has also been applied to detect bovine in goat milk, using an easy sample preparation (Chen *et al.*, 2004).

4.6.6 *Substitution of milk with different feeding or geographical origin*

These types of milks are very difficult to analyse because the components are similar. Techniques based on isotope characterisation (i.e. stable natural isotope fractionation and isotope ratio mass spectrometry), which quantify the $^{1}H/^{2}H$ and $^{12}C/^{13}C$ ratios on selected compounds, are able to make such distinctions. Renou *et al.* (2004) have applied for the first time these techniques to the analysis of milk from different geographical origin and feeding. In addition, the fatty acid fraction is sensitive to feeding changes, and therefore, methods based on their composition can be useful. However, it is necessary to collect a large database before a method can be proposed.

4.6.7 *Mislabelling of fortified and functional milks*

Mislabelling of functional processed milks is an increasing concern because of the growing interest by consumers and the increased sales of these products in different markets. For example, vitamins, calcium and ω-3-enriched milks are commonly sold in Europe, and they should contain what it is claimed on the label. Mislabelling of these milks can be done unintentionally or on purpose. The amount of these nutrients can be assessed by the routine analytical techniques described above. However, there are other considerations; for instance, calcium-fortified milk should be manufactured by concentration of caseins through membranes, as addition of other calcium sources changes its nutritional quality. The analysis of proteins, calcium and phosphorous can be used to evaluate whether calcium salts were added or membrane technology was applied, but the addition of caseinates is difficult to assess (De la Fuente *et al.*, 2004).

The increased marketing of functional products must lead to the development of methods able to analyse biological activities and/or the compounds claimed to be the responsible for those activities. Lactose-reduced milks or ω-3-enriched milks can be analysed by methods able to determine these compounds. Bioactive peptides could be analysed by HPLC-UV or HPLC coupled to tandem mass spectrometry. However, the difficulties arise when the compounds responsible for the claimed activities are unknown or protected by patents. Another way to evaluate these products would be the analysis of the biological effect claimed, if it is possible to assay it in vitro, otherwise it would be very difficult since clinical studies should be done.

4.7 Conclusion

During the past couple decades, there has been a great improvement in the quality of market milks, which provides the consumers with better products. This progress has been a direct consequence of the way the milk is handled at dairy factories, and the quality control measures that have been taken along the milk production chain. At present, quality control measures have been advanced to provide better tools for the evaluation of different quality parameters of milk at different stages of the production. The availability of standardised methods, as well as harmonised guidelines developed by specialised international agencies, has been essential to establish method performance. Although classical methods of analysis are widely used successfully for quality control, analytical techniques have evolved greatly. Molecular techniques have contributed greatly to the detection of pathogens, and to the development of fast and easy-to-use devices. Automated methods, such as those used for the determination of microbial counts, have made it possible to obtain results in significantly shorter times, an advancement that has been crucial for the industry. The developments of online detection systems based on biosensors or physical sensors have been essential for automated milking and processing systems, moving towards real-time control. The advanced instrumentation and methods used by specialised laboratories for quality control of milk have resulted in great progress for controlling the authenticity of milks and its manufacturing process. This is particularly difficult due to the development of new products by the industry, which lead to continuous new challenges for the analyst. In addition, current computer resources provide powerful tools for data analysis, making useful enormous amounts of data that facilitate control of the raw material, processing control and control of the end products.

The price of milk and the volumes handled daily in many countries raises large amounts of money, and milk quality plays a major role in this business. Small improvements in analytical methods can lead to significant economical benefits, and for this reason, there is still great motivation to refine existing methods and to develop new ones. Analysts continue to look for methods that are fast, sensitive, precise, accurate, with better selectivity/specificity, and resulting in less false positives/negatives.

Acknowledgement

The authors thank to projects CAM S-0505/AGR/0153, CONSOLIDER-INGENIO CSD-2007-00063, AGL 2005-03384 and AGL 2007-6535 for financial support.

References

Amine, A., Moscone, D., Bernardo, R.A., Marconi, E. & Palleschi, G. (2000a) A new enzymatic spectrophotometric assay for the determination of lactulose in milk. *Analytica Chimica Acta*, **406**, 217–224.

Amine, A., Moscone, D. & Palleschi, G. (2000b) Rapid determination of lactulose in milk using Seliwanoff's reaction. *Analytical Letters*, **33**, 125–135.

Andreescu, S. & Marty, J.L. (2006) Twenty years research in cholinesterase biosensors: from basic research to practical applications. *Biomolecular Engineering*, **23**, 1–15.

AOAC (2005a) Fat in raw milk, Babcock method, Method No 989.04. In: *Official Methods of Analysis of AOAC International*, 18th edn (ed. W. Horowitz), pp. 19–21, AOAC International, Gaithersburg, MD.

AOAC (2005b) Fat in milk, Automated turbidimetric method I, Method No 969.16. In: *Official Methods of Analysis of AOAC International*, 18th edn (ed. W. Horowitz), pp. 23–25, AOAC International, Gaithersburg, MD.

AOAC (2005c) Fat in milk, Automated turbidimetric method II, Method No 973.22. In: *Official Methods of Analysis of AOAC International*, 18th edn (ed. W. Horowitz), p. 25, AOAC International, Gaithersburg, MD.

AOAC (2005d) Lactose in milk, Gravimetric method: Munson-Walker, Method No 930.28. In: *Official Methods of Analysis of AOAC International*, 18th edn (ed. W. Horowitz), p. 17, AOAC International, Gaithersburg, MD.

AOAC (2005e) Lactose in milk, Polarimetric method, Method No 896.01. In: *Official Methods of Analysis of AOAC International*, 18th edn (ed. W. Horowitz), p. 17, AOAC International, Gaithersburg, MD.

Belloque, J. (2006) High-resolution NMR of milk and milk proteins. In: *Modern Magnetic Resonance – Applications in Materials, Food and Marine Sciences*, Vol. 3 (ed. G.A. Webb), pp. 1609–1613, Springer, Amsterdam.

Belloque, J. & Ramos, M. (2002) Determination of casein content in milk by ^{31}P-NMR. *Journal of Dairy Research*, **69**, 411–418.

Benhar, I., Eshkenazi, I., Neufeld, T., Opatowsky, J., Shaky, S. & Rishpon, J. (2001) Recombinant single chain antibodies in bioelectrochemical sensors. *Talanta*, **55**, 899–907.

Birlouez-Aragon, I., Leclere, J., Quedraogo, C.L., Birlouez, E. & Grongnet, J.F. (2001) The FAST method, a rapid approach of the nutritional quality of heat-treated foods. *Nahrung*, **45**, 201–205.

Bitri, L., Rolland, M.P. & Besançon, P. (1993) Immunological detection of bovine caseinomacropeptide in ovine and caprine dairy products. *Milchwissenschaft*, **48**, 367–371.

Bordin, G., Cordeiro Raposo, F., de la Calle, B. & Rodríguez, A.R. (2001) Identification and quantification of major bovine milk proteins by liquid chromatography. *Journal of Chromatography A*, **928**, 63–76.

Borkova, M. & Snaselova, J. (2005) Possibilities of different animal milk detection in milk and dairy products – a review. *Czechoslovakian Journal of Food Science*, **23**, 41–50.

Brokken, G.C.A.M., Corbee, R.J., van Knapen, F. & Bergwerff, A.A. (2003) Immunochemical detection of Salmonella group B, D and E using an optical surface plasmon resonance biosensor. *FEMS Microbiology Letters*, **222**, 75–82.

Buchheim, W., Heeschen, W. & Schlimme, E. (1994) Definition and differentiation of heat treatments of milk. *European Dairy Magazine*, **5**(1), 42–44.

Bucky, A.R., Hayes, P.R. & Robinson, D.S. (1988) Enhanced inactivation of bacterial lipases and proteinases in whole milk by a modified ultra high temperature treatment. *Journal of Dairy Research*, **55**, 373–380.

Capps, K.L., McLaughlin, E.M., Murray, A.W.A., Aldus, C.F., Wyatt, G.M., Peck, M.W., van Amerongen, A., Ariens, R.M.C., Wichers, J.H., Baylis, C.I., Wareing, D.R.A. & Bolton, F.J. (2004) Validation of three rapid screening methods for detection of verotoxin-producing *Escherichia coli* in foods: interlaboratory study. *Journal of AOAC International*, **87**, 68–77.

Cataldi, T.R.I., Angelotti, M. & Bufo, S.A. (1999) Method development for the quantitative determination of lactulose in heat-treated milks by HPAEC with pulsed amperometric detection. *Analytical Chemistry*, **71**, 4919–4925.

Cattaneo, T.M.P., Feroldi, A., Topino, P.M. & Olieman, C. (1994) Sample preparation for selective and sensitive detection of soya proteins in dairy products with chromatographic and electrophoretic techniques. *Netherlands Milk and Dairy Journal*, **48**, 225–234.

Cefalu, W.T., Bell-Farrow, A., Wang, Z.Q. & Ralapati, S. (1991) Determination of furosine in biomedical samples employing an improved hydrolysis and high-performance liquid chromatographic technique. *Carbohydrate Research*, **215**, 117–125.

Chen, R.K, Chang, L.W., Chung, Y.Y., Lee, M.H., & Ling, Y.C. (2004) Quantification of cow milk adulteration in goat milk using high-performance liquid chromatography with electrospray ionization mass spectrometry. *Rapid Communications In Mass Spectrometry*, **18**, 1167–1171.

Chen, F.T.A. & Zang, J.H. (1992) Determination of milk proteins by capillary electrophoresis. *Journal of AOAC International*, **75**, 905–909.

Clawin-Rädecker, I. & Schlimme, E. (1995) Determination of furosine in pasteurized milk by use of ion-pair reversed-phase liquid chromatography. *Kieler Milchwirtschaftliche Forschungsberichte*, **47**, 169–175.

Corradini, D., Cannarsa, G., Corradini, C., Nicoletti, I., Pizzoferrato, L. & Vivanti, V. (1996) Analysis of e-N-2-furoylmethyl-L-lysine (furosine) in dried milk by capillary electrophoresis with controlled electroosmotic flow using N, N, N′, N′-tetramethyl-1,3-butanediamine in the running electrolyte solution. *Electrophoresis*, **17**, 120–124.

Corzo, N., Delgado, T., Troyano, E. & Olano, A. (1994) Ratio of lactulose to furosine as indicator of quality of commercial milks. *Journal of Food Protection*, **57**, 737–739.

Corzo, N., Olano, A. & Martínez-Castro, I. (1986) Diferenciación de leches sometidas a distintos tratamientos térmicos mediante el análisis de la composición de disacáridos libres por cromatografía de gases. *Revista de Agroquímica y Tecnología de Alimentos*, 26, 565–570.

De la Fuente, M.A., Belloque, J. & Juarez, M. (2004). Mineral contents and distribution between the soluble and the micellar phases in calcium-enriched UHT milks. *Journal of Science of Food and Agriculture*, **84**, 1708–1714.

De la Fuente, M.A. & Juárez, M. (2005) Authenticity assessment of dairy products. *Critical Reviews in Food Science and Nutrition*, **45**, 563–585.

De Mol, R.M. (2000) *Automated Detection of Oestrus and Mastitis in Dairy Cows*, PhD Thesis, Wageningen University, Wageningen, The Netherlands.

Delgado, T., Corzo, N., Santa-María, G., Jimeno, M.L. & Olano, A. (1992) Determination of furosine in milk samples by ion-pair reversed phase liquid chromatography. *Chromatographia*, **33**, 374–376.

Diehl, B.W.K. & Ockels, W. (1995) Fatty acid distribution by ^{13}C NMR spectroscopy. *Fett Wissenschaft Technologie*, **97**, 115–118.

Dogan, B. & Boor, K.J. (2003) Genetic diversity and spoilage potential among *Pseudomonas* spp. isolated from fluid milk products and dairy processing plants. *Applied and Environmental Microbiology*, **69**, 130–138.

Dóka, O., Bicanic, D. & Frankhuizen, R. (1999) Photoacoustic study of heated binary mixtures containing whey and skimmed-milk powders. *European Food Research and Technology*, **208**, 1–5.

EU (1991) Commission Decision of 14 February 1991 laying down certain methods of analysis and testing of raw milk and heat-treated milk – 91/180/EEC. *Official Journal of the European Communities*, **L93**, 1–48.

EU (1996) Commission Regulation (EC) No. 1081/96 of 14 June 1996 establishing a reference method for the detection of cows' milk and caseinate in cheeses made from ewes' milk, goats' milk or buffalos' milk or mixtures of ewes', goats' and buffalos' milk and repealing Regulation (EEC) No. 690/92. *Official Journal of European Communities*, **L142**, 15–25.

EU (2002) Regulation (EC) No. 178/2002 of the European Parliament and of the Council of 28 January 2002 laying down the general principles and requirements of food law, establishing the European Food Safety Authority and laying down procedures in matters of food safety. *Official Journal of the European Communities*, **L31**, 1–24.

EU (2004a) Directive 2004/41/EC of the European Parliament and of the Council of 21 April 2004 repealing certain directives concerning food hygiene and health conditions for the production and placing on the market of certain products of animal origin intended for human consumption and amending council directives 89/662/EEC and 92/118/EEC and council decision 95/408/EC. *Official Journal of the European Communities*, **L157**, 33–44.

EU (2004b) Regulation (EC) No. 852/2004 of the European Parliament and of the Council of 29 April 2004 on the hygiene of foodstuffs. *Official Journal of the European Communities*, **L139**, 1–54.

EU (2004c) Regulation (EC) No. 853/2004 of the European Parliament and of the Council of 29 April 2004 laying down specific hygiene rules for on the hygiene of foodstuffs. *Official Journal of the European Communities*, **L139**, 55–205.

EU (2004d) Regulation (EC) No. 854/2004 of the European Parliament and of the Council of 29 April 2004 laying down specific rules for the organisation of official controls on products of animal origin intended for human consumption. *Official Journal of the European Communities*, **L139**, 206–319.

EU (2004e) Regulation (EC) No. 882/2004 of the European Parliament and of the Council of 29 April 2004 on official controls performed to ensure the verification of compliance with feed and food law, animal health and animal welfare rules. *Official Journal of the European Communities*, **L165**, 1–141.

Espeja, E., García, M.C. & Marina, M.L. (2001) Fast detection of added soybean proteins in cow's, goat's and ewe's milk by perfusion reversed-phase high-performance liquid chromatography. *Journal of Sheep Science*, **24**, 856–864.

Feinberg, M., Dupont, D., Efstathiou, T., Louapre, V. & Guyonnet, J.P. (2006) Evaluation of tracers for the authentication of thermal treatments of milks. *Food Chemistry*, **98**, 188–194.

Fortin, N.Y., Mulchandani, A. & Chen, W. (2001) Use of real time polymerase reaction and molecular beacons for the detection of *Escherichia coli* O157:H7. *Analytical Biochemistry*, **289**, 281–288.

Fratamico, P.M., Strobaugh, T.P., Medina, M.B. & Gehring, A.G. (1998) Detection of *Escherichia coli* O157:H7 using a surface plasmon resonance biosensor. *Biotechnology Techniques*, **12**, 571–576.

Gillespie, B.E. & Oliver, S.P. (2005) Simultaneous detection of mastitis pathogens, *Staphylococcus aureus*, *Streptococcus uberis*, and *Streptococcus agalactiae* by multiplex real-time polymerase chain reaction. *Journal of Dairy Science*, **88**, 3510–3518.

Guy, P.A. & Fenaille, F. (2006) Contribution of mass spectrometry to assess quality of milk-based products. *Mass Spectrometry Reviews*, **25**, 290–326.

Hein, I., Krismer, H., Rieck, P. & Wagner, M. (2002) The use of real time PCR for rapid detection of *L. monocytogenes* in milk and dairy products. *Wiener Tierarztliche Monatsschrift,* **89**, 312–318.

Hofstadler, S.A., Sampath, R., Blyn, L.B., Eshoo, M.W., Hall, T.A., Jiang, Y., Drader, J.J., Hannis, J.C., Sannes-Lowery, K.A., Cummins, L.L., Libby, B., Walcott, D.J., Schink, A., Massire, C., Ranken, R., Gutierrez, J., Manalili, S., Ivy, C., Melton, R., Levene, H., Barret-Wilt, G., Li, F., Zapp, V., White, N., Samant, V., McNeil, J.A., Knize, D., Robbins, D., Rudnick, K., Desai, A., Moradi, E. & Ecker, D.J. (2005). TIGER: the universal biosensor. *International Journal of Mass Spectrometry*, **242**, 23–41.

Hurley, I.P., Ireland, H.E., Coleman, R.C. & Williams, J.H.H. (2004) Application of immunological methods for the detection of species adulteration in dairy products. *International Journal of Food Science and Technology*, **39**, 873–878.

IDF (1981) *Milk – Determination of Fat Content*, Standard No. 105, International Dairy Federation, Brussels.

IDF (1987a) *Milk, Cream & Evaporated Milk – Determination of Total Solids Content*, Standard No. 021B, International Dairy Federation, Brussels.

IDF (1987b) *Dried milk, Dried Whey, Dried Buttermilk & Dried Butter Serum – Determination of Fat Content*, Standard No. 009C, International Dairy Federation, Brussels.

IDF (1987c) *Evaporated Milk & Sweet Condensed Milk – Determination of Fat Content*, Standard No. 013C, International Dairy Federation, Brussels.

IDF (1987d) *Skimmed Milk, Whey & Buttermilk – Determination of Fat Content*, Standard No. 022B, International Dairy Federation, Brussels.

IDF (1990a) *Dried Skimmed Milk – determination of Vitamin A Content*, Standard No. 142, International Dairy Federation, Brussels.

IDF (1990b) *Milk – Determination of Total Phosphorus Content*, Standard No. 042B, International Dairy Federation, Brussels.

IDF (1991a) *Milk & Milk Products – Enumeration of Microorganisms*, Standard No. 100B, International Dairy Federation, Brussels.

IDF (1991b) *Sweet Condensed Milk – Determination of Total Solids Content*, Standard No. 0015B, International Dairy Federation, Brussels.

IDF (1992a) *Milk – Determination of Calcium Content*, Standard No. 036A, International Dairy Federation, Brussels.

IDF (1992b) *Milk and Dried Milk – Determination of Calcium Content*, Standard No. 154, International Dairy Federation, Brussels.

IDF (1995a) *Milk – Enumeration of Somatic Cells*, Standard No. 148A, International Dairy Federation, Brussels.

IDF (1995b) *Milk and Milk Powder – Determination of Aflatoxin M_1 Content*, Standard No. 171, International Dairy Federation, Brussels.

IDF (1996) *Milk – Determination of Fat Content*, Standard No. 001D, International Dairy Federation, Brussels.

IDF (1998) *Heat Treated Milk – Determination of Lactulose Content*, Standard No. 147B, International Dairy Federation, Brussels.

IDF (2000) *Whole Milk – Determination of Milkfat, protein & Lactose Content – Guide for the Operation of Mid-Infra-Red Instruments*, Standard No. 141C, International Dairy Federation, Brussels.

IDF (2001) *Milk – Determination of Nitrogen Content*, Standard No. 002–5, International Dairy Federation, Brussels.

IDF (2002a) *Caseins & Caseinates – Determination of Water Content*, Standard No. 078C, International Dairy Federation, Brussels.

IDF (2002b) *Milkfat – Determination of the Fatty Acid Composition by Gas-Liquid Chromatography*, Standard No. 184, International Dairy Federation, Brussels.

IDF (2002c) *Dried Skimmed Milk – Determination of Vitamin D Content Using High-Performance Liquid Chromatography*, Standard No. 177, International Dairy Federation, Brussels.

IDF (2002d) *Dried Milk – Assessment of Heat Treatment Intensity*, Standard No. 162, International Dairy Federation, Brussels.

IDF (2002e) *Milk – Determination of Freezing Content*, Standard No. 108, International Dairy Federation, Brussels.

IDF (2003a) *Milk and Milk Products – Guidelines for the Standardized Description of Immunoassays or Receptor Assays for the Detection of Antimicrobial Residues*, Standard No. 188, International Dairy Federation, Brussels.

IDF (2003b) *Milk and Milk Products – Determination of Nitrate and Nitrite*, Standard No. 189, International Dairy Federation, Brussels.

IDF (2003c) *Milk and Milk Products – Determination of Alkaline Phosphatase Activity*, Standard No. 155, International Dairy Federation, Brussels.

IDF (2004a) *Milk – Enumeration of Microorganisms*, Standard No. 131, International Dairy Federation, Brussels.

IDF (2004b) *Dried Milk – Determination of Moisture Content*, Standard No. 026, International Dairy Federation, Brussels.

IDF (2004c) *Milk – Determination of Lactulose Content*, Standard No. 175, International Dairy Federation, Brussels.

IDF (2004d) *Milk and Milk Products – Determination of Furosine Content*, Standard No. 193, International Dairy Federation, Brussels.

IDF (2005a) *Milk and Milk Products – Determination of Aflatoxin* M_1 *Content – Clean-Up by Immunoaffinity Chromatography and Determination by Thin-Layer Chromatography*, Standard No. 190, International Dairy Federation, Brussels.

IDF (2005b) *Liquid Milk – Determination of Acid-Soluble* β*-lactoglobulin Content – Reverse Phase HPLC Method*, Standard No. 178, International Dairy Federation, Brussels.

Iñón, F.A., Garrigues, S. & Guardia, M. (2004) Nutritional parameters of commercially available milk samples by FTIR and chemometric techniques. *Analytica Chimica Acta*, **513**, 401–412.

Ju, C.C., Chou, W.L. & Lin, C.W. (1991) Studies on detection of reconstituted milk in raw and pasteurized milk through determination of milk RNase activity. *Journal of Chinese Society of Animal Science*, **20**, 103–114.

Kanning, M., Casella, M. & Olieman, C. (1993) Milk and soy proteins analysis using capillary zone electrophoresis. *Liquid Chromatography-Gas Chromatography International*, **6**, 701–705.

Kessel, J.S., van Karns, J.S. & Perdue, M.L. (2003) Using portable real time polymerase chain reaction (RT-PCR) assays to detect Salmonella spp. in raw milk. *Journal of Food Protection*, **66**, 1762–1767.

Khare, S., Ficht, T.A., Santos, R.L., Romano, J., Ficht, A.R., Zhang, S., Grant, R., Libal, M., Hunter, D. & Adams, L.G. (2004) Rapid and sensitive detection of *Mycobacterium avium* subsp. *paratuberculosis* in bovine milk and faeces by a combination of immunomagnetic bead separation conventional PCR and real-time PCR. *Journal of Clinical Microbiology*, **42**, 1075–1081.

Koubova, V., Brynda, E., Karasova, L., Skvor, J., Homola, J., Dostalek, J., Tobiska, P. & Rosicky, J. (2001) Detection of food borne pathogens using surface plasmon resonance biosensors. *Sensors and Actuators B: Chemical*, **74**, 100–105.

Kuhlmann, B., Klostermeyer, H. & Fries, A. (1991) Analysis of lactulose 1. Automated enzymatic micromethod for the determination of lactulose. *Milchwissenschaft*, **46**, 555–558.

Lechner, V.E. & Klostermeyer, H. (1981) Nachweis einer Verfälschung von Magermichpulver mit Molkenpulver (polarographische Methode). *Milchwissenschaft*, **36**, 267–270.

Leonard, P., Heary, S., Quinn, J. & O'Kennedy, R. (2004). A generic approach for the detection of whole *Listeria monocytogenes* cells in contaminated samples using surface plasmon resonance. *Biosensors Bioelectronics*, **19**, 1331–1335.

López-Fandiño, R., Corzo, N., Villamiel, M., Delgado, T., Olano, A. & Ramos, M. (1993) Assessment of quality of commercial UHT milks by chromatographic and electrophoretic methods. *Journal of Food Protection*, **56**, 263–264.

López-Fandiño, R. & Olano, A. (1999) Selected indicators of the quality of thermal processed milk. *Food Science and Technology International*, **5**, 121–137.
López-Tapia, J., García-Risco, M.R., Manso, M.A. & López-Fandiño, R. (1999) Detection of the presence of soya protein in milk powder by sodium dodecyl sulfate capillary electrophoresis. *Journal of Chromatography A*, **836**, 153–160.
Luthi-Peng, Q.Q. & Puhan, Z. (1999) The 4th derivative spectroscopic method for the rapid determination of protein and casein in milk. *Milchwissenschaft*, **54**, 74–77.
Maasen-Franke, B., Wiethoff, M., Suhr, O., Clemens, C. & Knoll, A. (2004) A method to detect flakes and clots in milk in automatic milking systems. Automatic milking – a better understanding. In: *Proceedings of the International Symposium of Milking – Automatic Milking – A Better Understanding* (eds A. Meijering, H. Hogeveen & C.J.A.M. de Koning), pp. 251–252, Wageningen Academic Publishers, Wageningen, The Netherlands.
Manso, M.A., Cattaneo, T.M., Barzaghi, S., Olieman, C. & López-Fandiño, R. (2002a) Determination of vegetal proteins in milk powder by sodium dodecyl sulfate-capillary gel electrophoresis. Interlaboratory study. *Journal of AOAC International*, **85**, 1090–1095.
Manso, M.A., Cattaneo, T.M., Barzaghi, S., Pérez, D., Sánchez, L., Calvo, M., Olieman, C., Brett, G. & López-Fandiño, R. (2002b) Detection of vegetable proteins in milk products by electrophoretic and immunochemical methods: in-house prevalidation tests and collaborative trials. *Use of Enzyme Preparations in Cheesemaking (Other than Rennet)/Detection of Vegetable Proteins in Milk Products*, Document No. 371, pp. 24–50, International Dairy Federation, Brussels.
Martínez-Castro, I., Calvo, M.M. & Olano, A. (1987) Chromatographic determination of lactulose. *Chromatographia*, **23**, 132–136.
Mayer, H.K. (2005) Milk species identification in cheese varieties using electrophoretic, chromatographic and PCR techniques. *International Dairy Journal*, **15**, 595–604.
Mayer, M., Genrich, M., Kuennecke, W. & Bilitewski, U. (1996) Automated determination of lactulose in milk using an enzyme reactor and flow analysis with integrated dialysis. *Analytica Chimica Acta*, **324**, 37–45.
Meisel, H. (1995) Application of fourth derivative spectroscopy to quantitation of whey protein and casein in total milk protein. *Milchwissenschaft*, **50**, 247–251.
Mendenhall, I.V. & Brown, J.R. (1991) Fourier transform infrared determination of whey powder in nonfat dry milk. *Journal of Dairy Science*, **74**, 2896–2890.
Miralles, B., Bartolomé, B., Ramos, M. & Amigo, L. (2000) Comparison of three methods to determine the whey protein to total ratio in milk. *Journal of Dairy Science*, **83**, 2759–2765.
Moatsou, G. & Anifantakis, E. (2003) Recent developments in antibody-based analytical methods for the differentiation of milk from different species. *International Journal of Dairy Technology*, **56**, 133–138.
Montilla, A., Calvo, M.M., Santa-María, G., Corzo, N. & Olano, A. (1996) Correlation between lactulose and furosine in UHT-heated milk. *Journal of Food Protection*, **59**, 1061–1064.
Moscone, D., Bernardo, R.A., Marconi, E., Amine, E. & Palleschi, G. (1999) Rapid determination of lactulose in milk by microdialysis and biosensors. *Analyst*, **124**, 325–329.
Mottram, H.R. & Evershed, R.P. (2001) Elucidation of the composition of bovine milk fat triacylglycerols using high performance liquid chromatography-atmospheric pressure chemical ionisation mass spectrometry. *Journal of Chromatography A*, **926**, 239–253.
Nicoletti, I., Cogliandro, E., Corradini, C. & Corradini, D. (1997) Analysis of epsilon-N-2-furoylmethyl-L-lysine (furosine) in concentrated milk by reversed phase chromatography with a microbore column. *Journal of Liquid Chromatography and Related Technologies*, **20**, 719–729.
O'Mahony, J. & Hill, C. (2004) Rapid real-time PCR assay for detection and quantitation of *Mycobacterium avium* subsp. *paratuberculosis* DNA in artificially contaminated milk. *Applied and Environmental Microbiology*, **70**, 4561–4568.

Olieman, C. & van Riel, J.A.M. (1989) Detection of rennet whey solids in skim milk powder and buttermilk powder with reversed-phase HPLC. *Netherlands Milk and Dairy Journal*, **43**, 27–36.

Pedretti, N., Bertrand, D., Semenou, M., Robert, P. & Giangiacomo, R. (1993) Application of an experimental design to the detection of foreign substances in milk. *Journal of Near Infrared Spectroscopy*, **1**, 174–184.

Pellegrino, L., Resmini, P. & Luf, W. (1995) Assessment (indices) of heat-treatment. In: *Heat Induced Changes in Milk* (ed. P.F. Fox), Document No. 238, pp. 409–453. International Dairy Federation, Brussels.

Recio, I., García-Risco, M.R., López-Fandiño, R., Olano, A. & Ramos, M. (2000b) Detection of rennet whey solids in UHT milk by capillary electrophoresis. *International Dairy Journal*, **10**, 333–338.

Recio, I., García-Risco, M.R., Ramos, M. & López-Fandiño, R. (2000a) Characterisation of peptides produced by psychrotrophic proteinases on κ-casein. *Journal of Dairy Research*, **67**, 625–630.

Recio, I., López-Fandiño, R., Olano, A., Olieman, C. & Ramos, M. (1996) Study of the formation of caseinomacropeptides in stored ultrahigh temperature-treated milk by capillary electrophoresis. *Journal of Agriculture and Food Chemistry*, **44**, 3845–3848.

Recio, I. & Olieman, C. (1996) Determination of denatured serum proteins in the casein fraction of heat-treated milk by capillary zone electrophoresis. *Electrophoresis*, **17**, 1228–1233.

Recio, I., Ramos, M. & Lopez-Fandino, R. (2001) Capillary electrophoresis for the analysis of food proteins of animal origin. *Electrophoresis*, **22**, 1489–1502.

Rehman, Z.U., Saeed, A. & Zafar, S.I. (2000) Hydroxymethylfurfural as an indicator for the detection of dried powder in liquid milk. *Milchwissenschaft*, **55**, 256–257.

Reinemann, D.J. & Helgren, J.M. (2004). Online milk sensing issues for automatic milking. In: *ASAE/CSAE Annual International Meeting*, Paper No 04–4191, Ottawa.

Renou, J.P., Deponge, C., Gachon, P., Bonnefoy, J.C., Coulon, J.B., Garel, J.P., Vérité, R. & Ritz, P. (2004) Characterization of animal products according to geographical origin and feeding diet using nuclear magnetic resonance and isotope ratio mass spectrometry: cow milk. *Food Chemistry*, **85**, 63–66.

Resmini, P. & Pellegrino, L. (1991) Analysis of food heat damage by direct HPLC of furosine. *International Chromatography Laboratory*, **6**, 7–11.

Resmini, P., Pellegrino, L. & Batelli, G. (1990) Accurate quantification of furosine in milk and dairy products by a direct HPLC method. *Italian Journal of Food Science*, **3**, 173–183.

Resmini, P., Pellegrino, L., Hogenboom, J.A. & Andreinin, R. (1989) Thermal denaturation of whey protein in pasteurized milk – fast evaluation by HPLC. *Italian Journal of Dairy Science*, **71**, 1472–1483.

Resmini, P., Pellegrino, L., Masotti, F., Tirelli, A. & Prati, F. (1992) Determination of reconstituted dried milk in raw and pasteurized milk by HPLC of furosine. *Scienza e Tecnica Lattiero- Casearia*, **43**, 169–186.

Rispoli, S., Rivemale, M. & Sauges, R. (1991) Identification and quantitative determination of cow's milk in ewes' cheeses using isoelectric focusing of wheys. Application to cheeses at high ripening stage: Roquefort cheeses. *Lait*, **71**, 501–510.

Rueckert, A., Ronimus, R.S. & Morgan, H.W. (2006) Development of a real-time PCR assay targeting the sporulation gene, *spo0A*, for the enumeration of thermophilic bacilli in milk powder. *Food Microbiology*, **23**, 220–230.

Ruttkat, A. & Erbersdobler, H.F. (1994) Degradation of furosine during heptafluorobutyric anhydride derivatization for gas chromatographic determination. *Journal of Chromatography A*, **678**, 103–107.

Sánchez, L., Pérez, D., Puyol, P., Calvo, M. & Brett, G. (2002) Determination of vegetal proteins in milk powder by enzyme-linked immunosorbent assay: interlaboratory study. *Journal of AOAC International*, **85**, 1390–1397.

Schäffer, L. (1988) Glycated lysine increases in serum samples stored at -20°C. *Clinical Chemistry*, **34**, 1906–1908.

Schmidt, M.A.E., Radovic, B.S., Lipp, M., Harzer, G. & Anklam, E. (1999) Characterisation of milk samples with various whey protein contents by pyrolysis-mass spectrometry (Py-MS). *Food Chemistry*, **65**, 123–128.

Schulze, H., Vorlova, S., Villate, F., Bachmann, T.T. & Schmid, R.D. (2003) Design of acetylcholinesterases for biosensor applications. *Biosensors Bioelectronics*, **18**, 201–209.

Soga, T. & Serwe, M. (2000) Determination of carbohydrates in food samples by capillary electrophoresis with indirect UV detection. *Food Chemistry*, **69**, 339–344.

Sorensen, L.K., Lund, M. & Juul, B. (2003) Accuracy of Fourier transform infrared spectrometry in determination of casein in dairy cows' milk. *Journal of Dairy Research*, **70**, 445–452.

Tamarapu, S., McKillip, J.L. & Drake, M. (2001) Development of a multiplex polymerase chain reaction assay for detection and differentiation of *Staphylococcus aureus* in dairy products. *Journal of Food Protection*, **64**, 664–668.

Tirelli, A. & Pellegrino, L. (1996) Determination of furosine in dairy products by capillary zone electrophoresis: a comparison with HPLC method. *Italian Journal of Food Science*, **7**, 379–385.

Turin, L. & Bonomi, F. (1994) Immunochemical detection of ovalbumin in dairy-based foods. *Journal of Science of Food Agriculture*, **64**, 39–45.

Vallejo-Córdoba, B. & Nakai, S. (1994) Keeping quality assessment of pasteurised milk by multivariate analysis of dynamic headspace gas chromatographic data. Flavor classification by linear discriminant analysis. *Journal of Agriculture and Food Chemistry*, **42**, 994–999.

van Renterghem, R. & de Block, J. (1996) Furosine in consumption milk and milk powders. *International Dairy Journal*, **6**, 371–382.

van Riel, J.A.M. & Olieman, C. (1995a) Selective detection in RP-HPLC of Tyr- Trp- and sulfur-containing peptides by pulsed amperometry at platinum. *Analytical Chemistry*, **67**, 3911–3915.

van Riel, J.A.M. & Olieman, C. (1995b) Determination of caseinomacropeptide with capillary zone electrophoresis and its application to the detection and estimation of rennet whey solids in milk and buttermilk powder. *Electrophoresis*, **16**, 529–533.

Villamiel, M., Arias, M., Corzo, N. & Olano, A. (1999) Use of different thermal indices to assess the quality of pasteurized milks. *European Food Research and Technology*, **208**, 169–171.

Villamiel, M., Arias, M., Corzo, N. & Olano, A. (2000) Survey of the furosine content in cheeses marketed in Spain. *Journal of Food Protection*, **63**, 974–975.

Yang, C., Jiang, Y., Huang, K., Zhu, C. & Yin, Y. (2003) Application of real time PCR for quantitative detection of *Campylobacter jejuni* in poultry, milk and environmental water. *FEMS Immunology and Medical Microbiology*, **38**, 265–271.

Further Reading

van Amerongen, A., Barug, D. & Lauwaars, M., (eds) (2005) *Rapid Methods for Biological and Chemical Contaminants in Food and Feed*, Wageningen Academic Publishers, Wageningen, The Netherlands.

Goff, H.D. & Griffiths, M.W. (2006) Major advances in fresh milk and milk products: fluid milk products and frozen desserts. *Journal of Dairy Science*, **89**, 1163–1173.

Barbano, D.M. & Lynch, J.M. (2006) Major advances in testing of dairy products: milk component and dairy product attribute testing. *Journal of Dairy Science*, **89**, 1189–1194.

AOAC (2006) *Official Methods of Analysis*, Association of Official Analytical Chemists International, 18th edition, AOAC, Gaithersburg, MD (available at: http://eoma.aoac.org or http://www.aoac.org).

Fernández-Alba, A.R. (ed.) (2004) *Chromatographic – Mass Spectrometric Food Analysis for Trace Determination of Pesticide Residues*, Comprehensive Analytical Chemistry, Vol. 43 (series eds R.C. Hill, J.R. Startin & J. Fussel), Elsevier Publishers, Amsterdam.

Smit, G. (ed.) (2003) *Dairy Processing – Improving Quality*, Woodhead Publishing Cambridge.
Zadoks, R.N. (ed.) (2004) *Molecular Methods in Milk Quality*, In: Proceedings of a Symposium to Celebrate the Opening of the New Ithaca Facilities of Quality Milk Production Services (available at: qmps.vet.cornell.edu).

Web Pages of Interest

International Standards Organisation (ISO): http://www.iso.org
International Dairy Federation (IDF): http://www.fil-idf.org
European Union legislation (Eur-Lex): http://europe.eu.int/eur-lex

5 Current Legislation of Market Milks

M. Hickey

5.1 Introduction

At the beginning of the twenty-first century, market milks are widely recognised as safe and wholesome products. It may be difficult nowadays to comprehend that up to the late nineteenth century, just 130 years ago, milk was a significant vehicle of transmission for a number of foodborne diseases of humans and animals, including anthrax, diphtheria, typhoid fever, foot and mouth disease, scarlet fever and tuberculosis (Westhoff, 1978). This situation began to change following the application of pasteurisation to milk. Although named after Louis Pasteur, it is unlikely that Pasteur worked on milk, rather it was the process he developed, using a relatively mild heat treatment for wine and beer, which others such as Soxhlet later applied to milk (Westhoff, 1978). The rapid uptake of pasteurisation in its early years probably owed more to its extending shelf life of milk rather than food safety. Nonetheless, at the third International Dairy Congress held in The Hague in 1907, the following statement was adopted:

> *The Congress, while considering that milk from those dairies which have selected healthy animals and used all necessary hygienic precautions, may be sold in the raw state, recommends nevertheless that, from a hygienic standpoint in relation to the spreading of tuberculosis through milk, milk should be consumed after sterilisation, or adequate boiling or pasteurization by processes the reliability of which, as far as action on the tuberculosis bacillus is concerned, has been scientifically demonstrated, in view of the divergent resistance exhibited by the micro-organism* (Staal, 1986).

While in-bottle sterilisation was used to a limited extent from those early days of heat treatment, the development of continuous flow ultra-high-temperature (UHT) processes for milk and milk products combined with aseptic filling into cartons in the early 1960s lead to commercialisation of UHT milks. Within 10 years, at a symposium of the Society of Dairy Technology in February 1977 on the topic, it was reported that the process was now well established in Europe, if not in the UK (Burton, 1977). By the mid-1990s, the European milk market was made up of 54% UHT, 42% pasteurised and 4% sterilised milks, but with widely differing patterns in the European Member States. For example, the French, Spanish and Portuguese markets are predominantly UHT milk; the UK, the Netherlands, Irish, Danish, Swedish, Norwegian and Finnish markets are mainly fresh pasteurised milk; while the Belgian, German, Italian, Polish, Hungarian, Czech Republic and Swiss markets are divided between UHT and pasteurised (Rysstad & Kolstad, 2006; te Giffel *et al.*, 2006).

During the 1990s, another category of drinking milk emerged – extended shelf life (ESL) milk. At the International Dairy Federation's World Dairy Summit, in Dresden in 2000, a conference was held on this topic of ESL milk. While the technology to produce such milk using high-temperature treatments (called ultra-pasteurisation) was available in the USA since the 1960s (Henyon, 1999), during the 1990s, processes were developed for the production of ESL milks, which had a significantly longer shelf life at refrigerated temperatures, but which retained the favour and 'freshness' of pasteurised milk. Different production processes were commercialised, usually combining normal pasteurisation with one or more additional processing steps aimed at reducing the microbial load of the finished product. Such milks did not meet definitions of UHT or pasteurisation, but comprised a new category between pasteurised milk and UHT milk.

Market milks have also been modified as regards composition, with the reduction of fat levels, sodium content, lactose content and protein enrichment being the most common. Milks have also been supplemented by the addition of nutrients, such as vitamins and minerals.

This chapter reviews primarily the legislation related to compositional requirements and hygiene/food safety standards in the European Union (EU), UK, Ireland and USA, together with those developed with an international focus by the Codex Alimentarius Commission (CAC). Other related legislation is referred to, but discussed in less detail. The chapter also considers the legislation governing the sale of raw milk, which though small in volume, continues to be a matter of some debate and controversy. In some countries, there is an outright ban on its sale, while in others it is permitted, usually under strict control and subject to clear labelling, sometimes with warnings, to ensure that consumers are clear as to its nature.

5.2 EU legislation

5.2.1 *Access to EU legislation*

The legislation of the EU is published in the L-series of the *Official Journal of the European Union*. It may also be accessed using the EUR-LEX website (http://eur-lex.europa.eu/RECH_naturel.do). Use of this website is facilitated greatly by knowing the type (directive, regulation, decision or COM-final), the year and the number of the relevant legislation. These will be given when referring to legislation in this section. Recent legislation is usually available electronically in portable document format (PDF), but earlier legislation is given in HTML format only. Amendments to this legislation published in the official journal show the text that is being changed only, but consolidated texts of some legislation can be accessed electronically. Such consolidated legislation that incorporates amendments into the original text can be very useful as they simplify use of the documents.

5.2.2 *Compositional requirements*

In its common agricultural policy as regards dairy products, the European Commission envisaged the market organisation would apply to liquid milk and other fresh milk products. Notwithstanding this objective, it was not until 1971 that the first legislation, i.e. Regulation

Table 5.1 Milks permitted for sale as drinking milk by EU legislation.

Category	Requirement
Raw milk	Milk which has not been heated above 40°C or subjected to treatment having equivalent effect
Non-standardised whole milk	Milk with its natural fat content
Standardised whole milk	Milk with minimum fat content 3.5 g 100 g^{-1}; Member States may also agree an additional category of whole milk with fat content of 4 g 100 g^{-1} or more
Semi-skimmed milk	Milk with a fat content minimum 1.5 g 100 g^{-1} and maximum 1.8 g 100 g^{-1}
Skimmed milk	Milk with a fat content of not more than 0.5 g 100 g^{-1}

Data compiled from EU (1997).

1411/71 (EU, 1971), was adopted to address milk, and it was 1977 before the regulation was operative throughout the Community that by then consisted of nine Member States. This regulation, as amended, was in place until replaced by Regulation 2597/97 (EU, 1997). The aim of both these regulations was to establish marketing standards to guarantee quality, especially compositional quality, to consumers.

Under Regulation 2597/97 (EU, 1997), and indeed the earlier Regulation 1411/71 (EU, 1971), milk is defined as the produce of the milking of one or more cows, and drinking milk as the products falling within CN Code 04.01, intended for delivery to the consumer, without further processing. The products that are permitted for sale in the EU as drinking milk are outlined in Table 5.1. In order to meet the fat content requirements, modification of the natural fat content is permitted by the removal or addition of cream or the addition of whole milk, semi-skimmed milk or skimmed milk.

Up to an amendment in 1993, Member States had to choose between standardised milk and non-standardised milk. The UK and Ireland chose non-standardised milk, and up to that time, imported milk from other Member States also had to conform to this requirement. However, this was considered an obstacle to trade, and from 1993, all types of milk could be sold in all Member States provided the product was adequately labelled. Ireland had a derogation, under Regulation 539/94 (EU, 1994a), regarding milk whose natural content did not meet the minimum fat requirement for non-standardised milk of 3.5 g 100 g^{-1}, specified in Regulation 1411/71 (EU, 1971), for some years from 1994, but this was not continued after the adoption of the Regulation 2597/97 (EU, 1997). Sweden and Finland had derogation to the end of 2003 allowing the sale of milks with different fat contents to the above.

Additional compositional standards were established for protein content, fat-free dry matter, density and freezing point; these standards are outlined in Table 5.2.

In addition to modification of the natural fat content, the regulation permitted milk to be enriched by the addition of milk proteins, mineral salts and vitamins. Where proteins are added, the protein content must be 3.8 g 100 g^{-1} or greater. The reduction of the natural protein content of milk is not allowed. The lactose content may be reduced by conversion to glucose and galactose. However, Member States may prohibit these modifications, and

Table 5.2 Compositional standards for milk specified in EU legislation.

Parameter	Non-standardised whole milk	Standardised whole milk	Semi-skimmed milk	Skimmed milk
Fat (g 100 g^{-1})	3.5 minimum	3.5 minimum	1.5–1.8	<0.5
Protein (g 100 g^{-1})	2.9 minimum (based on 3.5-g fat content)	2.9% minimum (based on 3.5% fat content)	Equivalent to that in whole milk based on its fat content	Equivalent to that in whole milk based on its fat content
Fat-free dry matter (g 100 g^{-1})	8.5 minimum based on 3.5-g fat content	8.5 minimum based on 3.5-g fat content	Equivalent content to that in whole milk based on its fat content	Equivalent content to that in whole milk based on its fat content
Density in g L^{-1} at 20°C	1.028 based on 3.5-g fat 100 g^{-1})	1.028 based on 3.5-g fat 100 g^{-1}	Equivalent weight to that in whole milk based on fat content	Equivalent weight that in whole milk based on fat content
Freezing point	Close to the average freezing point for raw milk recorded in the area of origin of the drinking milk collected	Close to the average freezing point for raw milk recorded in the area of origin of the drinking milk collected	Close to the average freezing point for raw milk recorded in the area of origin of the drinking milk collected	Close to the average freezing point for raw milk recorded in the area of origin of the drinking milk collected

Data compiled from EU (1997).

where such modifications have been made, they must be clearly indicated on the labels; this requirement does not remove any requirements for nutritional labelling under Directive 90/496 (EU, 1990a).

5.2.3 *Amendment to existing compositional requirements*

In February 2007, the Commission published a proposal to amend Regulation 2597/97 (EU, 2007c). This proposal noted that a large number of derogations from the three categories of drinking milk (skimmed milk, semi-skimmed milk and whole milk) have been accepted in the past, in connection with the various accession treaties. The Commission proposal also noted that all those derogations were scheduled to expire in 2009, with the exception of that for Estonia which was scheduled to expire in April 2007, though Estonia had already requested that its derogation be prolonged to 2009. The proposal was adopted in September 2007 by Regulation 1153/2007 (EU, 2007a), which amends Regulation 2597/97, by adding the following wording to Article 3(1):

> *Heat treated milk not complying with the fat content requirements laid down in points (b), (c) and (d) of the first sub-paragraph* [the provisions defining whole, semi-skimmed and skimmed milks] *shall be considered drinking milk provided that the fat content is clearly indicated with one decimal and easily readable on the packaging in the form of "... % fat". Such milk shall not be described as whole milk, semi-skimmed milk or skimmed milk.*

The provisions of this amendment to the regulation apply from 1 January 2008, and it is anticipated that Member States will amend the national provisions that implement Regulation 2597/97 (EU, 1997) where necessary.

This amendment changes the requirements for fat content that have been in place since Regulation 1411/71 (EU, 1971) was first adopted. Member States have different consumption patterns for milk by fat content (see Figure 5.1; EU, 1996b), but for many years, there has been a clear and steady tendency in the Community, reflected in consumption data from the UK, towards consumption of dairy products with less milk fat and in particular as regards milk for direct consumption (see Figure 5.2). The liberalisation of the compositional standards by allowing the production of drinking milks with fat contents outside the three existing categories was made in the light of these changing nutritional habits, and to encourage the production of such lower fat products, while, at the same time, providing for clear and readable information on the fat content in the requirements for product labelling.

5.2.4 *Hygiene and food safety requirements*

Market regulation took up much of the early legislative work in the EU, and it was not until 1985 that the first Community hygiene measure for milk was adopted, i.e. Directive 85/397 (EU, 1985). This started the process of harmonising hygiene standards within the Community in order to facilitate intra-Community trade without compromising existing Member State hygiene rules. It covered all aspects of the production, transport and processing of milk from farm to the final consumer.

This was followed in 1992 by a new Milk Hygiene Directive 92/46 (EU, 1992), which became effective from 1 January 1994. This Directive contained animal health requirements

Country		<0.5 g fat 100 g^{-1}	0.5–2.5 g fat 100 g^{-1}	>2.5 g fat 100 g^{-1}
Belgium	BE	7.9	0	92.1
Denmark	DK	13.6	58.5	28
Finland	FI	21	63.8	15.2
France	FR	4.3	73.7	22
Germany	DE	5.4	30	64.6
Greece	GR	4.6	5	90.4
Ireland	IE	0	16.2	83.8
The Netherlands	NL	2	79.4	18.7
Spain	ES	19	14	67
United Kingdom	UK	2.7	51.2	46.2

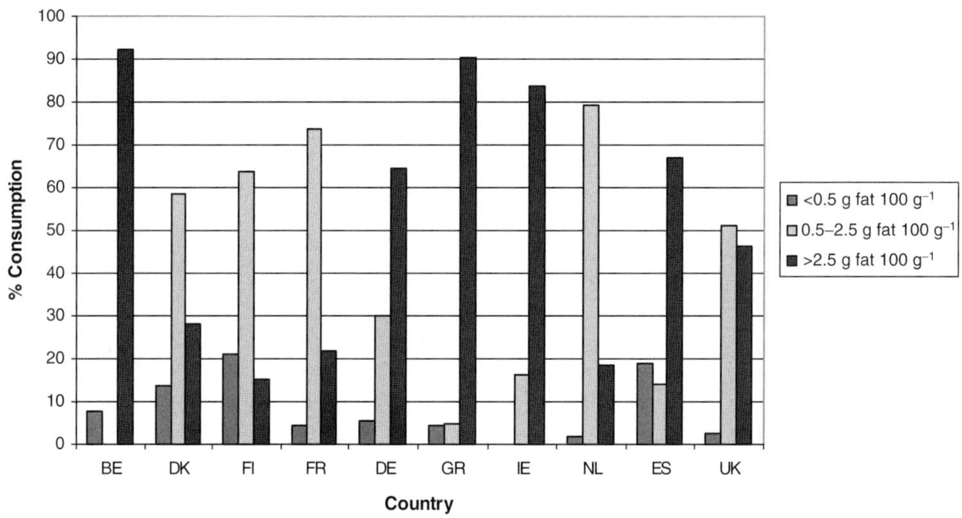

Fig. 5.1 Consumption of drinking milk by fat content in various EU countries in 1996. Adapted from EU (1996).

for raw milk, hygiene requirements for registered holdings, hygiene requirements in milking, collection and transport of milk to collection centres, standardisation centres, treatment establishments and processing establishments. For the first time, uniform EU-wide hygiene standards were created as the earlier Directive 85/397 (EU, 1985) applied to intra-Community trade only. Furthermore, Directive 92/46 (EU, 1992) laid down minimum compositional standards for milk, and also standards for the maximum plate count and somatic cell count for raw milk at collection from dairy farms intended for the production of heat-treated drinking milk as well as fermented milk, flavoured milk, cream and other specified related products.

A major review was carried out on the EU Hygiene Directives, following a recommendation in the EU White Paper on Food Safety. Prior to this review, there were a total of 16 commodity-specific EU Directives and 1 Directive on general food law, which had been gradually developed in the period from 1964 and had given a high level of protection to the consumer. However, they comprised a mixture of different disciplines (hygiene, animal health, official controls), and were detailed and complex. It was decided to overhaul the

Year	Whole milk	Semi-skimmed milk	Skimmed milk
1994	43.4	42.9	12.7
1995	39.6	46.1	13.3
1996	37.7	48.3	13.1
1997	36.3	50.1	12.7
1998	35.4	50.6	12.9
1999	33.9	52.2	13.0
2000	32.3	52.9	13.8
2001	30.3	53.7	14.9
2002	28.8	54.6	15.4
2003	27.3	57.2	15.5
2004	27.2	57.3	15.4
2005	25.6	59.0	15.4
2006	24.3	59.7	16.0

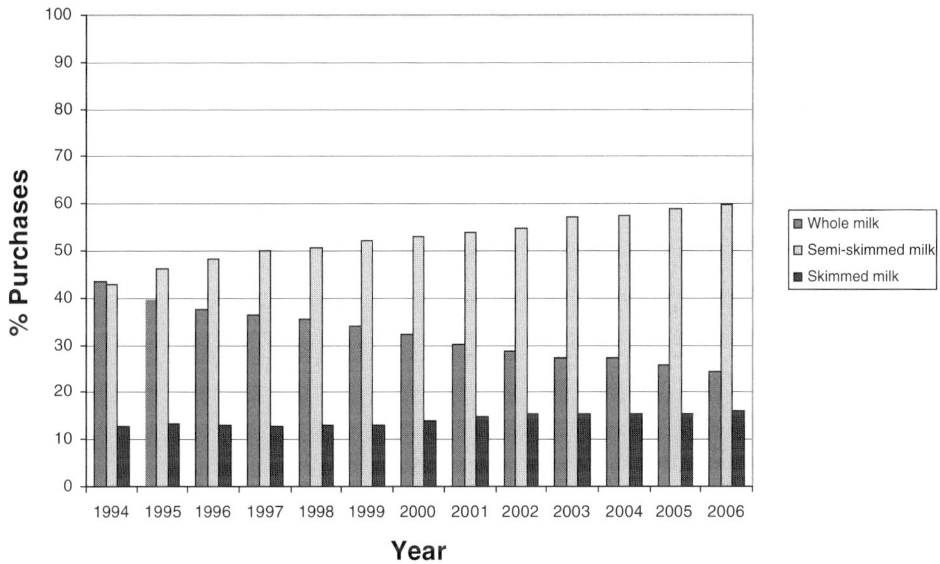

Fig. 5.2 Household purchases of drinking milk by fat content in the UK. Adapted from MDC (2004, 2007).

legislation to improve, simplify and modernise it, and separate aspects of food hygiene from animal health and food control issues. The review aimed for a more consistent and clear approach throughout the food production chain from 'farm to fork'.

A package of new hygiene rules was adopted in April 2004 by the European Parliament and the Council. They became applicable from 1 January 2006, and in the case of milk and milk products, replace Directive 92/46 (EU, 1992). The new rules are regulations and not directives, making them binding in Member States without the necessity of national legislation to be enacted to implement their provisions. However, instead of all the hygiene requirements being embodied in a single piece of legislation, hygiene requirements for the dairy sector are now contained across at least six different regulations. The three main regulations are (a) Regulation 852/2004 on the hygiene of foodstuffs (EU, 2004d), (b) Regulation 853/2004 laying down specific hygiene rules for food of animal origin; Annex III, Section XI thereof contains specific requirements for raw milk and dairy products (EU, 2004e) and (c) Regulation 854/2004 laying down specific rules for the organisation of official controls on products of animal origin intended for human consumption (EU, 2004f).

Then, in early December 2005, two important additional regulations were published, i.e. Regulation 2074/2005 (EU, 2005b) and Regulation 2076/2005 (EU, 2005c). In addition to laying down implementing measures and transitional measures, they also contain important amendments to the original regulations. Microbiological criteria for foodstuffs are laid down in Regulation 2073/2005 (EU, 2005a). A number of milk product requirements are contained therein.

Although largely self-contained, within these regulations, there is also significant cross-referencing to each other and to other legislation that affects the health, food safety and hygiene requirements of milk and milk product production. These include:

- Directive 2000/13, as amended, relating to the labelling, presentation and advertising of foodstuffs (EU, 2000).
- Regulation 2377/90 establishing maximum residue limits of veterinary medicinal products in foodstuffs of animal origin (EU, 1990b).
- The framework Food Safety Regulation 178/2002 (EU, 2002).
- Directive 97/78 relating to veterinary checks on imports from third countries (EU, 1998a).
- Directive 2002/99 relating to the animal health rules governing the production (EU, 2003b).
- Directive 96/23 on measures to monitor certain substances and residues in live animals and animal products (EU, 1996a).
- Directives 64/432 (EU, 1964) and 91/68 (EU, 1991) on animal health problems affecting intra-Community trade in bovine animals and swine.
- Directive 98/83, as amended, on the quality of potable water (EU, 1998b).

The new hygiene regulations adopt an approach based more on outcomes, or food safety objectives, hazard analysis and critical control points (HACCP) principles, than the detailed production, milking and process requirements of the earlier Directive.

This approach may be illustrated by reference to heat treatments, such as pasteurisation. The 92/46 Directive (EU, 1992) specified a minimum time/temperature requirement – '(at least 71.7°C for 15 seconds or any equivalent combination) or a pasteurising process using different time/temperature combinations to obtain an equivalent effect'.

In Regulation 853/2004 (EU, 2004e), heat treatment of dairy products was cross-referenced to Regulation 852/2004 (EU, 2004d) Chapter XI and therein states, inter alia, in addressing heat treated products in hermetically sealed containers:

> *... to ensure that the process employed achieves the desired objectives, food business operators are to check regularly the main relevant parameters (particularly temperature, pressure, sealing and microbiology), including by the use of automatic devices; and that the process used should conform to an internationally recognised standard (for example, pasteurisation, ultra high temperature or sterilisation ....*

Table 5.3 Heat treatment requirements under EU hygiene legislation.

Milk type	Requirements
Pasteurised	HTST – minimum 72°C for minimum 15 s or LHT – minimum 63°C for minimum 30 min or any other combination of time–temperature conditions to obtain an equivalent effect and in all cases, a negative result for alkaline phosphatase test immediately after treatment
Ultra-high temperature (UHT)	Continuous flow at a high temperature for a short time with not less than 135°C for a suitable holding time such that there are no viable microorganisms or spores capable of growing in the treated product when kept in an aseptic closed container at ambient temperature and sufficient to ensure that the products remain microbiologically stable after incubating for 15 days at 30°C in closed containers or for 7 days at 55°C in closed containers or after any other method demonstrating that the appropriate heat treatment has been applied

Data compiled from EU (2005b).

Regulation 853/2004 (EU, 2004e) added a further requirement that, when considering whether to subject raw milk to heat treatment, food business operators must have regard to HACCP principles, and comply with requirements imposed by the competent authority when approving the establishment or in carrying out checks under Regulation 854/2004 (EU, 2004f).

However, Regulation 2074/2005 (EU, 2005b) reintroduced prescriptive time/temperature requirements for pasteurisation and UHT treatment, by way of amendment (using additional wording) of the relevant section of Regulation 853/2004 (Section IX, II (II) 1) (EU, 2004e). These requirements are outlined in Table 5.3. The requirement in Directive 92/46 (EU, 1992) for pasteurised milk to be peroxidase positive has been removed, and the requirement to label pasteurised milk that is peroxidase negative as high-temperature pasteurised milk has also been dropped. The reasons for the reintroduction of the specific time/temperature requirements are not outlined, but may have been due to pressures for legal certainty regarding enforcement. Furthermore, it may be noted that the high-temperature short-time (HTST) process where the temperature has been increased from 71.7°C in Directive 92/46 (EU, 1992) to 72°C in Regulation 2074/2005 (EU, 2005b). Why was this done? The reason is not stated in the legislation, but the new definition is in line with the Codex Alimentarius definition of pasteurisation in the Code of Hygiene Practice for Milk and Milk Products (CAC/RCP 57–2004) (FAO/WHO, 2003c), and also ensures the European definition meets the Office International des Epizooties (OIE), the World Organisation for Animal Health, recommendations for the control of foot and mouth disease contained in Article 3.6.2.5 of the Terrestrial Animal Health Code (OIE, 2007), which may be accessed online at http://www.oie.int/eng/normes/mcode/en_sommaire.htm.

Some other specific requirements are also retained. An example in Regulation 853/2004 (EU, 2004e) is the requirement to ensure that 'during transport the cold chain must be maintained and, on arrival at the establishment of destination, the temperature of the milk must not be more than 10°C'. In fact, this requirement is stricter in the new regulation, as the derogation in the earlier Directive regarding milk collected within 2 h of milking is removed.

Chapter III of Regulation 852/2004 (EU, 2004d) encourages the development, dissemination and use of national or sectoral and community guides to good practice. The following are some key requirements from the main regulations.

Regulation 852/2004 (EU, 2004d), which sets down general hygiene requirements, states the following in Article 1:

- The main responsibility for food safety rests with the food business operator (FBO).
- Food safety should be ensured throughout the food chain starting with primary production.
- The chill chain must be maintained for food which cannot be stored at ambient temperatures.
- The FBO responsibility should be reinforced by the implementation of procedures on the basis of HACCP and on the application of good hygiene practice.
- Guides to good practice are valuable to aid FBOs compliance with food hygiene and application of HACCP principles.
- It is necessary to establish microbiological criteria and temperature control requirements on the basis of scientific risk assessment.
- It is necessary to ensure food imports are at least of the same or equivalent hygiene standard to that produced in the EU.

Guidance documents on Regulation 852/2004 (EU, 2004d) are EU (2005e, f, 2006) and Regulation 853/2004 (EU, 2005d) have also been developed by the commission. Specific hygiene and animal health requirements for raw milk and dairy products are laid down in Annex III, Section IX of Regulation 853/2004 (EU, 2004e). They cover the following:

- Animal health requirements
- Hygiene on milk production holdings – including premises, equipment, milking, collection transport and staff
- Criteria for raw milk
- Requirements for dairy products, including temperature, heat treatment requirements, raw milk prior to processing and labelling
- Identification marking – this replaces the health mark requirement contained in the 92/46 Directive (EU, 1992) (see Figure 5.3).

Table 5.4 outlines the main criteria for raw milk for the manufacture of all milk products including drinking milk. However, the microbiological criteria for pasteurised milk are laid down in Regulation 2073/2005 (EU, 2005a), and are outlined in Table 5.5. It will be noted that the standard for Enterobacteriaceae replaces that for coliforms contained in Directive 92/46 (EU, 1992), while the standard for the pre-incubation (PIC) test has been deleted.

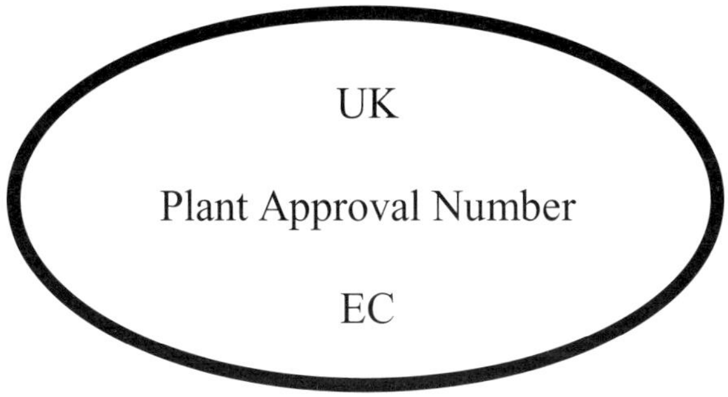

Fig. 5.3 Format of identification mark as required for milk and milk products in the European Union. Data compiled from EU (2004e).

Raw milk for direct consumption

Article 10 of Regulation 853/2004 (EU, 2004e) permits Member States to adopt national measures adapting the specific requirements which do not compromise the objectives of the regulation being achieved. Furthermore, a Member State may on its own initiative maintain or establish national rules prohibiting or restricting the placing on the market of raw milk for direct consumption. The options available in this article result in different provisions across the Member States. Some countries permit the sale, usually with strict controls, while other Member States prohibit the sale. In situations where raw milk is intended for direct consumption, it must be labelled with the words *raw milk* (Annex III, Section XI, Chapter IV, 1 (a).

Additives in market milks

The use of additives in foods in the EU is subject to horizontal legislation primarily under the additive framework Directive 89/107/EEC (EU, 1989) and under specific additive directives addressing colours Directive 94/36 (EU, 1994c), sweeteners Directive 94/35 (EU, 1994b, as amended), additives other than colours and sweeteners Directive 95/2 (EU, 1995, as amended) and flavourings Directive 88/388 (EU, 1988). In brief, colours are not permitted in market milks (other than flavoured milks, which are outside the scope of this chapter); sweeteners and flavourings are not required in plain (unflavoured) milks. Article 2.3 of Directive 95/2 (EU, 1995) includes pasteurised and sterilised (including UHT) milk (including plain, skimmed and semi-skimmed) in the list of products that are not allowed to use the additives listed in Annex I of that Directive except where specifically permitted by Annex II and plain milks are not included therein. Hence, the only additives permitted in market milks are those specifically permitted by Annex IV of Directive 95/2 (EU, 1995) in sterilised and UHT milk, namely – phosphoric acid (E338) and the phosphates (sodium phosphates (E339), potassium phosphates (E340), calcium phosphates (E341), magnesium phosphates (E342), diphosphates (E450), triphosphates (E452) and polyphosphates (E453). No additives are permitted in pasteurised whole milk, semi-skimmed or skimmed milk.

Table 5.4 Criteria for raw milk supply in EU legislation.

Parameter	Requirement	Qualification
Plate count (colony-forming units (cfu) mL^{-1}) at 30°C	≤100 000	Based on rolling geometric average over a 2-month period with at least two samples per month
Somatic cell count	≤400 000	Based on rolling geometric average over a 3-month period with at least one sample per month, unless the competent authority specifies another methodology to take account of seasonal variations in production levels
Antibiotic residues	Below levels specified in EU (1990b) or combined total of residues does not exceed any maximum permitted value	
Temperature during storage on farm	Immediately cooled to ≤8°C in case of daily collection or ≤6°C if collection is not daily	Food business operator (FBO) need not comply with the requirement if, either, milk is to be processed within 2 h of milking, or a higher temperature is necessary for technological reasons related to the manufacture of certain dairy products and the competent authority so authorises
Temperature during transport	Chill chain must be maintained and on arrival at destination ≤10°C	FBO need not comply with the requirement if, either, milk is to be processed within 2 h of milking, or a higher temperature is necessary for technological reasons related to the manufacture of certain dairy products and the competent authority so authorises
Temperature during storage prior to processing	Upon acceptance at a processing establishment, quickly cooled to ≤6°C and kept at that temperature until processing	FBO need not comply with the requirement if processing begins immediately after milking or within 4 h of acceptance at a processing establishment, or the competent authority authorises a higher temperature for technological reasons concerning the manufacture of certain dairy products
Plate count (cfu mL^{-1}) at 30°C immediately before processing	≤300 000	This applies only where the milk is to be heat treated and has not been so treated within the period of acceptance specified in the HACCP system put in place by the FBO

Data compiled from EU (2004e, 2005c).

Labelling

The labelling requirements for foodstuffs contained in the EU Labelling Directive 2000/13 (EU, 2000, as amended) apply to market milks. These requirements include the following provisions:

- Name of the food – in the case of milk the provisions of Regulation 2597/97 (EU, 1997) and amending Regulation 1153/2007 (EU, 2007a) should be respected.

Table 5.5 Microbiological criteria for pasteurised milk in EU legislation.

	Sampling	Plan	Limits	
Microorganisms	*n*	*c*	*m*	*M*
Enterobactericeae	5	2	<1 cfu mL^{-1}	5 cfu mL^{-1}
Listeria spp.	5	0	Absent in 25 mL	

n, number of units comprising the sample; *c*, number of sample units between *m* and *M*; *m*, the acceptable microbiological level in a sample unit; *M*, maximum level for any sample unit which, when exceeded in one or more samples, would cause the lot to be rejected.
Data compiled from EU (2005a).

- List of ingredients – this would not be required in pasteurised milks where no ingredients or additives are added. However, if vitamins or minerals are added, these should be indicated. If additives are added to UHT or sterilised milks, then these should be listed in an ingredient list.
- The net quantity should be given.
- The date of minimum durability – this information should be indicated. In the case of pasteurised milk, this is normally in the form '*Use by* . . . ', as this category of milk has been regarded as a foodstuff which, from the microbiological point of view, is highly perishable. Given the present day quality and shelf life of pasteurised milk, this point could be debated. ESL, UHT and sterilised milks would use the '*Best before* . . . ' form.
- Special conditions of storage and use – in the case of pasteurised milk and extended shelf life milk that require refrigeration, the instructions 'Keep Refrigerated' should be on the label in conjunction with the date of minimum durability.
- Name and address of manufacturer or seller – this information should be given in addition to the identification mark required by the hygiene regulations outlined above.

In addition, the specific labelling requirements of other applicable legislation for milk and milk products, as outlined elsewhere in this chapter, should be respected.

Packaging

Most milk on the market for direct human consumption is now packed in laminated cartons and/or plastic containers. These packages should comply with the general requirements of Regulation 1935/2004 (EU, 2004c) and the particular requirements in Directive 2002/72 (EU, 2003a), as amended by Directive 2004/19 (EU, 2004b) and Directive 2004/1 (EU, 2004a). It is normal for milk processors to specify to their packaging suppliers that their products comply with the requirements of these directives.

Range of sizes

In the 1960s, different national rules on nominal quantities of pre-packed goods, including foods, were a major barrier to trade in such products between the Member States. There was a need to harmonise these sizes, while at the same time, there was a concern not to impose additional Community rules on companies involved only on their own national market and which did not export to other Member States. The harmonising legislation, therefore, was of an 'optional nature'. Member States adopted the Community rules, but were also allowed to maintain existing national rules for the national market. Where this was the situation, only products conforming to the Community rules would benefit from free circulation.

Two directives were adopted to do this, Directive 75/106 (EU, 1975) addressed liquid products and Directive 80/232 (EU, 1980) addressed other products. The scope of these directives applied to both food and certain non-food products.

The sizes specified for milk in Directive 75/106 (EU, 1975) were 100 mL, 200 mL, 250 mL, 500 mL, 750 mL, 1 L, 2 L, 3 L and 4 L, with temporary allowance for 220 mL, 330 mL and 600 mL. For most products, including milk, national nominal quantities were allowed to exist alongside Community nominal quantities. However, for certain products (such as wine, spirits), total harmonisation was introduced. In the case of these latter products, Community sizes were made mandatory for everyone and all existing national sizes were abolished.

In the period from the adoption of these directives to the present time, changes in consumer preferences, packaging innovations and changing retail patterns at the Community and national levels led to a need to reassess whether existing legislation was still appropriate. This reassessment led to the conclusion that nominal quantities should generally not be subject to regulation at Community or national level, and it should be possible to place pre-packed goods on the market in any nominal quantity. This review led to the adoption of Directive 2007/45 (EU, 2007b) in September 2007, which replaces the existing legislation, i.e. Directives 75/106 and 80/232 (EU, 1975, 1980), and abolishes Community nominal quantities of package sizes for most sectors while maintaining obligatory nominal quantities for a very limited number of goods, mainly wine and spirits.

The provisions of the main articles of the Directive apply from 11 April 2009, but by way of derogation, Member States that currently prescribe mandatory nominal quantities for milk and other products, such as butter, dried pasta and coffee, are allowed to continue to do so, if they so wish, until 11 October 2012, and those Member States, which currently prescribe mandatory nominal quantities for white sugar, may continue to do so until 11 October 2013. These derogations do not impose any obligation on other Member States to regulate nominal quantities.

Extended shelf life milk

At this time, there is no specific legislation in the EU governing ESL milk. Indeed some ESL processes were developed to allow milk to meet the definition of standard pasteurisation in the earlier Hygiene Directive 92/46 (EU, 1992), in that the milk is phosphatase negative, but peroxidase positive, while the milk from other processes, which use higher heating temperatures, would be peroxidase negative and designated as high-temperature pasteurised. However, the distinction between high-temperature pasteurisation and normal pasteurisation has not been maintained in the new EU hygiene package.

In a paper given at the International Dairy Federation World Dairy Summit 2000 in Dresden, Dr Glazier of the European Commission stated that legislation should not impair technological progress, but should focus on consumer safety and adequate consumer information (Mann, 2001). It may be held that the processes used at this time do not compromise consumer safety, rather they enhance it. It could be argued that the requirement of Article 5.1(a) of the Food Labelling Directive, 2000/13 (EU, 2000), would require some additional information to be given on the label to explain the longer than normal shelf life of ESL milk. In practice, this information is often supplied as ESL milk is sold with a price premium over 'normal' pasteurised milk with the reasons for this premium (e.g. the processes used) being given on the label.

5.3 UK legislation

5.3.1 *Background*

A paper presented to the Scottish Section of the Society of Dairy Technology in 1947 outlines the early history of pasteurisation in the UK with particular focus on developments in Scotland (Smillie, 1948). It is stated that even before the end of the Great War (1914–1918) official recognition had been granted to Grade A (Certified) Milk and Grade A Milk and this continued in the Ministry Food (Continuance) Act of 1920 (HMSO, 1920). However, it was also mentioned that there was sufficient hidden opposition to the recognition of pasteurised milk to prevent its official introduction.

Up to recent times, England and Wales had common legislation, signed by the appropriate minister of the UK government and the Secretary of State for Wales. Separate but similar legislation is enacted for Scotland and Northern Ireland; however, some differences could and sometimes did occur. Then from 2000, with the establishment of the Welsh Assembly, separate but similar legislation for Wales was enacted.

By the mid 1950s, the production and sale of pasteurised milk was firmly established and controlled by the general provisions of the Food and Drugs Act 1955 (HMSO, 1955). In England and Wales, the production of milk, at farm level, was controlled by the Milk and Dairies (General) Regulations 1959 (HMSO, 1959), which were amended a number of times. In the 1970s, separate regulations, the Milk (Special Designations) Regulations 1977 (HMSO, 1977), controlled the pasteurisation of whole milk, while the earlier Milk and Dairies (Skimmed and Semi-skimmed Milk) (Heat Treatment and Labelling) Regulations 1973 (HMSO, 1973) applied to semi-skimmed and skimmed milks (Wilton, 1983).

When the UK joined the then European Economic Community on 1 January 1973, European legislation began to have a major role in shaping the evolving national legislation. However, as outlined in the discussion of European legislation earlier, harmonisation of legislation on hygienic aspects of milk production did not begin until the mid-1980s. Details of the current legislation in the UK as well as the separate legislation applicable to Scotland, Wales and Northern Ireland may be found via the relevant link on the (UK) Foods Standards Agency website (http://www.foodstandards.gov.uk) or that of the Office of Public Sector Information (OPSI) website (http://www.opsi.gov.uk/legislation/uk.htm). In searching the OPSI website, it is necessary to know the year and number of the Act or Statutory Instrument of interest.

5.3.2 *Present legislation on composition*

UK Statutory Instruments (SI) related to milk composition, hygiene and food safety are now based on EU regulations. Since the Community regulations are binding on Member States, in such cases the SI reference the requirements therein and outline particular elements, such as interpretations/definitions, specify the competent authority, address administration, detail offences, defences and penalties, and specify certain schedules. Where the European regulations specify general provisions, the UK SI may lay down more specific requirements, and may address national provisions where discretion or optional provisions are delegated to Member States.

The compositional requirements of Regulation 2597/97 (EU, 1997) are implemented in England, Wales and Scotland under The Drinking Milk Regulations 1998 (HMSO, 1998), which replaced earlier 1976 regulations. The amendment to Regulation 2597/97 (EU, 1997) that is recently introduced by EU Regulation 1153/2007 (EU, 2007a), as outlined above, shall apply from 1 January 2008, and an amendment to the UK regulations is anticipated.

5.3.3 *Present legislation on hygiene*

The new EU hygiene and food safety regulations, which are outlined above, are implemented in England by The Food Hygiene (England) Regulations 2006 (HMSO, 2006). Wales and Northern Ireland have similar but separate Regulations, for example the Food Hygiene (Wales) Regulations 2006 (The Stationery Office, 2006b) and The Food Hygiene Regulations (Northern Ireland) 2006 (The Stationery Office, 2006c). In Scotland, implementation is by The Food Hygiene (Scotland) Regulations 2006 (The Stationery Office, 2006a). The latter, while similar to the other three regulations, differs in one particular aspect – while the English, Welsh and Northern Irish regulations permit the sale of raw milk under strict requirements, such sale is not permitted in Scotland.

Regulation 32 and Schedule 6 of the English, Welsh and Northern Irish regulations lay down the following microbiological requirements for raw cows' milk for direct consumption:

Plate count at 30°C	$<$20 000 colony-forming units (cfu) mL^{-1}
Coliforms	$<$100 cfu mL^{-1}

The following conditions are also specified:

- Only the registered producer can sell such milk
- At or on the farm where the cows are maintained
- To a temporary guest or visitor to the farm or
- To a distributor (elsewhere defined as a milk roundsman)

In addition, the distributor may sell only raw cows' milk intended for direct human consumption: (a) which he or she has purchased as outlined above, (b) in the containers which he or she has received the milk, with the fastenings of the containers unbroken, (c) from a vehicle which is lawfully used as a shop premises and (d) direct to the final consumer. Any other person who sells raw cows' milk intended for direct human consumption shall be guilty of an offence. Hence, sale in retail stores is not permitted. The amount of raw cows' milk that is sold in England and Wales is quite small. The UK authorities have estimated that such sales account for only about 0.01% of total cow's milk consumption. Clear health warnings about the risks associated with its consumption must be given in the labelling of the container in which it is sold, or a notice must be displayed at a farm restaurant, which state that 'This milk has not been heat treated and may therefore contain organisms harmful to health'. Also, there is thought to be some limited consumption of raw milk from other species (e.g. goat, sheep and buffalo) (Food Standards Agency, 2005).

On three occasions since 1984, the UK Government has assessed the health risks of raw milk and proposed a ban on its sale, most recently in 2007. However, on each occasion, it was decided not to give effect to the proposal in the face of opposition from stakeholders, including some producers and consumer groups. Nonetheless, after each review, additional requirements were introduced to protect consumers. These included restricting sales of raw cow's drinking milk at the farm gate, introducing health warning labels and increasing the frequency of inspection and microbiological sample testing of raw cow's drinking milk at registered production holdings. In 2000, the Wales Food Advisory Committee (WFAC) also considered a ban on raw drinking milk, but following consultations, concluded that a ban would not be appropriate and the Welsh Assembly accepted this advice. However, they agreed to the introduction of additional labelling requirements applying to all raw drinking milk which, in addition to the existing health warnings, should provide advice that vulnerable groups should avoid consuming raw milk because of the risks to health. The Food Standard Agency (FSA) website now makes clear the risks associated with consuming raw milk, particularly by vulnerable groups. It also advises that, despite being popular with some people, unpasteurised milk (and cream) could be harmful. Although sale of raw milk for direct consumption is permitted in Northern Ireland under the same conditions as in England and Wales, there are no known sales there.

In contrast, the situation in Scotland is different. In the Scottish regulations, Schedule 6 states that no person shall place on the market raw milk, or raw cream, intended for direct human consumption (The Stationery Office, 2006a). It is a defence if the person accused of breaching this regulation can prove that the raw milk or raw cream was intended for export to England, Wales, Northern Ireland or another EU Member State; or a third country in accordance with Article 12 of Regulation 178/2002 (EU, 2002) as read with Article 11 of Regulation 852/2004 (EU, 2004d).

In fact, the sale of raw cows' milk in Scotland has been prohibited since 1983. The ban was introduced at that time following a number of milk-related illnesses and 12 deaths, which could be associated with to these illnesses. The introduction was successful in that it resulted in a marked decline in milk-related illness, and the ban has been maintained in subsequent years. When the dairy hygiene Directive 92/46 (EU, 1992) was being implemented, the policy was reviewed, but following consultation and scientific and medical advice, the ban was retained. In 1998, the ban was extended to raw drinking milk from other species (i.e. sheep, goat and buffalo), but for a number of reasons, including uncertainty about the legal basis underpinning such action, legislation giving effect to this decision was not made.

In 2004, Scottish ministers reconfirmed their wish to maintain and extend the ban so that all raw drinking milk and cream sales in Scotland would be prohibited. It is anticipated that provisions to address this matter will be introduced by regulations implementing the new EU hygiene package in Scotland scheduled for late 2008.

5.3.4 *The Dairy UK Code of Practice for HTST pasteurisation*

Dairy UK is an organisation that represents the UK dairy industry, and in 2006, it published an updated code of practice to apply best practice, and to seek to ensure that adequate controls and monitoring procedures are in place for HTST milk pasteurisation (Anonymous, 2006). This updates the previous UK Dairy Industry Federation Code of Practice (DIF, 1995) by addressing the requirements in the new EU hygiene regulations, and in particular, the requirement in Regulation 852/2004 (EU, 2004d) that FBOs should apply the principles of HACCP principles to ensure product safety. Furthermore, it applies the findings of research undertaken into the effectiveness of HTST processing conditions on the destruction of *Mycobacterium avium* subsp. *paratuberculosis* (MAP) in the intervening period. In this regard, it recommends that the minimum holding time of 15 s at 72°C used for drinking milk be increased to a minimum of 25 s on the basis of experimental evidence that this increases the effectiveness of the destruction of MAP under certain conditions (Grant *et al.*, 2005). While this recommendation has been adopted widely by the UK industry, and supported by many retailers, it is a recommendation that is voluntary and is not a legal requirement for HTST heat treatment, which remains at 72°C for 15 s.

Pasteurisation has long been recognised as the most important critical control point (CCP) in ensuring the safety of milk and dairy products. The Dairy UK code aims to help ensure that products have been adequately and correctly pasteurised and post-pasteurisation contamination has been avoided. One example is a discussion on the phosphatase test, used to verify the adequacy of milk pasteurisation. This code applies to 'normal' HTST pasteurisation, but not to high-temperature pasteurisation or UHT processing, where different considerations would apply.

5.3.5 *Permitted range of packaging sizes for milk in the UK*

As mentioned elsewhere when discussing EU legislation, Member States were allowed to maintain existing national rules for the national market. In the UK, milk can be pre-packed only if it is made up in one of the quantities specified in The Weights and Measures (Miscellaneous Foods) Order 1988 (HMSO, 1988), as amended by SI 1994 No. 2868 (HMSO, 1994). The consolidated quantities in force are outlined in Table 5.6. Milk in returnable containers (glass bottles) is allowed to continue to be sold in pints.

Table 5.6 Prescribed quantities for pre-packaged milk in the UK.

Food	Prescribed quantities (pre-packed foods)	Exemptions from prescribed quantities
Milk other than milk in a returnable container (e.g. in cartons, plastic bottles, etc.)	189 mL, 200 mL, 250 mL, 284 mL, 500 mL, 750 mL or a multiple of 284 or 500 mL	50 mL or less
Milk in a returnable container (e.g. glass bottles)	$\frac{1}{3}$ pint, $\frac{1}{2}$ pint or a multiple of $\frac{1}{2}$ pint; or 200 mL, 250 mL, 500 mL, 750 mL, 1 L, 2 L or thereafter a multiple of 500 mL	50 mL or less

Data compiled from HMSO (1988, 1994).

5.4 Irish legislation

5.4.1 *Introduction*

From the founding of the state until 1990, the hygiene, production and marketing of milk and milk products in Ireland were regulated by the Dairy Produce Acts 1924–1984, the Creameries Acts 1928–1934, the Milk and Dairies Acts 1935–1956 and the Regulations and Orders issued under these acts. The Compendium of Food Law in Ireland (Food Safety Authority of Ireland, 1998) contains a list of these acts and their implementing regulations and orders. These provisions had laid down a definition of Grade 'A' pasteurised milk, including prescribed heat treatment options, certain hygienic requirements for milk production and animal health, inspections of animals and dairies and other such details.

The European Communities (Standards for Heat-Treated Milk in Intra-Community Trade) Regulations 1990 (The Stationery Office Dublin, 1990) were adopted primarily to implement and update legislation following the adoption of the Directive 85/397 (EU, 1985), but also cross-referenced the compositional requirements of the European Regulation 1411/71 (EU, 1971) and its amendments. When the EU Hygiene Directive 92/46 (EU, 1992) and its amendments were adopted in the period 1992–1994, the European Communities (Hygienic Production and Placing on The Market of Raw Milk, Heat-Treated Milk and Milk-Based Products) Regulations 1996 (The Stationery Office Dublin, 1996) were enacted to implement the EU provisions. The text of the Irish Regulations was a consolidated text containing the amendments that had been made to the original 92/46 Directive and was thus a useful reference.

5.4.2 *Present legislation on hygiene*

The 1996 Regulations (The Stationery Office Dublin, 1996) remained in force until 2006 when they were replaced by the European Communities (Food and Feed Hygiene) Regulations 2005 (The Stationery Office Dublin, 2005), and its amendments SI 387 of 2006 (The Stationery Office Dublin, 2006a) and SI 56 of 2007 (The Stationery Office Dublin, 2007), which implement the EU Regulations 852/2004, 853/2004, 854/2004, 882/2004, 2073/2005 and 2076/2005 (EU, 2004d–g, 2005a, c, respectively) insofar as they relate to the Irish Department of Agriculture and Food, who are responsible for foods of animal origin; these regulations include the provisions governing milk and milk products. It should be noted that the European Communities (Hygiene of Foodstuffs) Regulations 2006 (The Stationery Office Dublin, 2006b) implement EU Regulation 852/2004 (EU, 2004d) for those foods under the control of the Irish Department of Health and Children.

As the European legislation concerned was in the form of regulations that were directly applicable in all Member States, the national implementing regulations address aspects, such as commencement, enforcement, administration, offences, penalties, while referring to the specific regulations concerned for the detailed provisions.

A publication from the then Department of Agriculture, Food and Rural Development, Ireland in 2001 (DAFRDI, 2001) stated that the sale of raw milk for human consumption had been prohibited in Ireland since 1997. Furthermore, the relevant Irish implementing regulations do not contain any provisions or standards for raw milk. At this time, there is no raw milk for sale for direct human consumption on the Irish market. It is understood that the authorities are considering whether it should avail of the provision in Chapter IV Article 10.8 of Regulation 853/2004 (EU, 2004e) that, on its own initiative, it should establish national rules explicitly prohibiting the placing on the market of raw milk for direct consumption.

5.4.3 *Present legislation on composition*

The Community regulation on requirements for milk composition, i.e. Regulation 2597/97 (EU, 1997), as discussed previously, apply directly to drinking milk for human consumption. The recent amendment to Regulation 2597/97 (EU, 1997) introduced by Regulation 1153/2007 (EU, 2007a), outlining greater flexibility as regards fat content outlined above, shall apply from 1 January 2008, and an amendment to the Irish regulations is anticipated. It is expected that Irish dairies will avail of this flexibility to produce milks of 1 and 2 g 100 g^{-1} fat content.

5.4.4 *Range of packaging sizes for fresh drinking milk in Ireland*

As discussed earlier, under EU legislations, Member States were allowed to maintain existing national rules for the national market. In Ireland, the provisions of the Merchandise Marks (Pre-packed Goods) (Marking and Quantities) (Amendment) Order 1981 (The Stationery Office Dublin, 1981) permitted milk to be pre-packed only if it is made up in one of the following quantities: 100 mL, 190 mL, 200 mL, 250 mL, 284 mL, 500 mL, 568 mL, 750 mL, 1 L, 1.136 L and 2 L. This order deleted 16 other sizes permitted under the earlier Merchandise Marks (Prepacked Goods) (Marking and Quantities) Order 1973 (The Stationery Office Dublin, 1973). It will be noted that sizes above 2 L are not currently allowed for retail sale under the 1981 Order (The Stationery Office Dublin, 1981). As a result of the deregulation of range of sizes by EU Directive 45/2007 (EU, 2007b), and due to differences between the sizes permitted in the UK, including Northern Ireland, Irish dairies may wish to have the deregulation of the range of sizes expedited.

5.5 USA legislation

5.5.1 *Background*

At the outset, it should be pointed out that this chapter shall address US federal legislation. The United States Public Health Service (USPHS), through its agency the Food and Drug Administration (FDA), does not have direct legal jurisdiction in the enforcement of milk hygiene standards throughout the country, except where milk moves across state boundaries. Therefore, the USPHS serves only in an advisory capacity as regards intrastate control, and its functions are intended primarily to assist the state and local regulatory agencies in their local situations. The aims of the USPHS are as follows:

- To promote the establishment of effective and well-balanced milk hygiene programmes in each state.
- To stimulate the adoption of adequate and uniform state and local milk control legislation.
- To encourage the application of uniform enforcement procedures through appropriate legal and educational measures.

The interest in the USPHS milk hygiene programme derives from two important public health considerations. First, the USPHS has promoted increased milk consumption, due to its importance as a major nutritional source for the maintenance of good health, especially that for young children and the elderly. Second, the recognition that milk had the potential, historically, to serve as a significant vehicle for the transmission of disease. Indeed, in the past, milk has been associated with major disease outbreaks.

5.5.2 *Historical details*

The USPHS and FDA activities in the area of milk hygiene began at the start of the twentieth century with studies on the role of milk in the spread of disease. These studies led to the recognition that effective control required the application of hygiene procedures and practices throughout the full product chain from production, through handling, processing and product distribution. Subsequent research identified and evaluated the hygiene requirements that should be adopted and enforced to control disease. These included studies that led to improvement of the pasteurisation process. One such example was in 1950 when it was proposed that raw milk containing the causative organism of Q fever (*Coxiella burnettii*) might be the cause of a significant number of cases of this disease in California (Bell *et al.*, 1950). The effect of pasteurisation on this organism was investigated (Lennette *et al.*, 1952), and based on a joint research project by USPHS and University of California (Enright *et al.*, 1957), which showed that there could be some survivors if high numbers were present in raw milk, led to the recommendation that the low-heat-treatment (LHT) pasteurisation temperature be raised from 61.7°C for 30 min to 62.8°C for 30 min to ensure adequate destruction. No change in HTST Pasteurisation was necessary as a result of this research.

Over the years, the incidence of milkborne illness in the USA has been reduced significantly. Such illnesses constituted 25% of all disease outbreaks due to infected foods and contaminated water in 1938, but the most recent information shows that milk and fluid milk products are now associated with less than 1% of such reported outbreaks (US Public Health Service, 2005). The FDA is recognised as having made a major contribution to the improvement of the national milk supply through its technical assistance, training, research, standards development, evaluation and certification activities.

Despite the progress that has been made, occasional milkborne outbreaks still occur, and this requires continued vigilance throughout the full product chain from farm to table. The situation has been complicated by the introduction of new products and processes, the use of new packaging materials and new marketing patterns. This has led to considerable efforts being expended in the development and use of the HACCP-based systems throughout the industry.

5.5.3 *The Pasteurised Milk Ordinance (PMO) and the Code of Federal Regulations*

Federal USA legislation on milk for human consumption is laid down in the Grade 'A' Pasteurised Milk Ordinance (US Public Health Service, 2005) and in the Code of Federal Regulations (US National Archives and Records Administration, 2007a, b, c, d). In 1924, the USPHS developed the model regulation that became known as the Standard Milk Ordinance for voluntary adoption by state and local milk control agencies. To provide for the uniform interpretation of this ordinance, an accompanying code was published in 1927, which provided administrative and technical details on compliance. Ongoing revisions of the PMO incorporate new knowledge and technology into effective and practicable legislation. The ordinance has been revised and updated many times in the intervening period, and the current PMO is the 25th revision, the title changing to the present one in 1965. This model milk regulation, now titled the Grade 'A' Pasteurized Milk Ordinance (Grade 'A' PMO) (US Public Health Service, 2005), incorporates the provisions governing the processing, packaging and sale of Grade 'A' milk and milk products, including buttermilk and buttermilk products, whey and whey products, and condensed and dry milk products.

The USPHS and the FDA do not produce the PMO on their own. It is developed with input and assistance from Milk Regulatory and Rating Agencies at every level of Federal, State and Local Government, including both Health and Agriculture Departments, all segments of the dairy industry, including producers, milk processors, equipment manufacturers and representative dairy associations, inputs from educational and research institutions and with comments from many individual hygienic experts and others.

The Grade 'A' PMO is the basic standard used in the voluntary State-USPHS/FDA Programme for the Certification of Interstate Milk Shippers. This is a programme in which 50 states participate, together with the District of Columbia and US Trust Territories. The National Conference on Interstate Milk Shipments (NCIMS) recommends changes and modifications to the FDA at its biennial conferences.

The PMO is used as the hygiene regulation for milk, and milk products, for interstate carriers; it is recognised by the public health agencies, the milk industry and many others as the national standard for milk hygiene. It represents a consensus of current knowledge and experiences and, as such, is said to represent a practical and equitable milk hygiene standard for the USA. It has been adopted by 46 of the 50 states for their own standards – California, Pennsylvania, New York and Maryland being the exceptions, and these have adopted similar standards. However, as stated earlier, where it is adopted locally, its enforcement becomes a function of the local or state authorities.

5.5.4 *Definitions and standards for milk in the USA*

Milk is defined in §131.110 of the Code of Federal Regulations (US National Archives and Records Administration, 2007a) as:

> *the lacteal secretion, practically free of colostrum, obtained by the complete milking of one or more healthy cows. Milk in the final package for beverage use shall be pasteurised and shall contain not less than 8.25% milk solids not fat and not less*

than 3.25% milk fat. Milk may have been adjusted by separating part of the milk fat therefrom or by adding thereto cream, concentrated milk, dry whole milk, skim milk, concentrated skim milk or non fat dry milk. Milk may be homogenised. Optional addition of Vitamin A and Vitamin D is allowed.

Milk must be labelled pasteurised, ultra-pasteurised, homogenised or with added vitamin A and vitamin D where these processes and modifications apply. The following is a list of milk products given in the PMO: half-and-half, sour half-and-half, acidified sour half-and-half, cultured sour half-and-half, reconstituted or recombined milk and milk products, non-fat (skimmed) milk, reduced-fat milk, low-fat milk, cultured non-fat (skimmed) milk, acidified milk, acidified reduced fat or low-fat milk, acidified non-fat (skimmed) milk, low-sodium milk, low-sodium reduced fat or low-fat milk, low-sodium non-fat (skimmed) milk, lactose-reduced milk, lactose-reduced fat or low-fat milk, lactose-reduced non-fat (skimmed) milk, aseptically processed and packaged milk and milk products, reduced-fat milk, low-fat milk or non-fat (skimmed) milk with added safe and suitable microbial organisms and any other milk product made by the addition or subtraction of milk fat or addition of safe and suitable optional ingredients for protein, vitamin or mineral fortification.

Although there are many different products, the most common types of milk sold in the USA are skimmed milk, which has <0.5 g fat 100 g^{-1} milk, and can also be designated as fat-free milk; 1 g fat 100 g^{-1} milk, which can be designated as low-fat milk; 2 g fat 100 g^{-1} milk, which can be designated as reduced-fat milk; and 3.5 g fat 100 g^{-1} milk. Nutrient claims related to fat, such as fat-free, low-fat and/or any other categories, are governed by the labelling requirements contained in Title 21 §101.62 of the Code of Federal Regulations (US National Archives and Records Administration, 2007b). Other nutrient claims are governed by either general or specific sections of Part 101 of the same regulation. Half-and-half milk, which is a blend of milk and cream, contains between 10.5 and 18 g 100 g^{-1} milk fat.

The USA standards for Grade 'A' raw milk for processing are outlined in Table 5.7. Pasteurisation is defined as the process of heating every particle of milk in properly designed and operated equipment, and held at or above the temperatures for at least the times specified

Table 5.7 Standards for grade 'A' raw milk and milk products for pasteurisation, ultra-pasteurisation and aseptic processing supply in USA legislation of Pasteurised Milk Ordinance (PMO).

Criterion	Requirement
Temperature	Cooled to ≤10°C within 4 h or less of commencement of first milking and to ≤7°C within 2 h of completion of milking, provided that the blend temperatures after the first milking and subsequent milkings do not exceed 10°C
Bacterial limits – individual producer	≤100 000 cfu mL^{-1} prior to mixing with other producers milk
Bacterial limits – mixed milk prior to pasteurisation	≤300 000 cfu mL^{-1}
Drugs	No positive results on drug residue detection methods β-Lactam detection methods acceptable
Somatic cell count – individual producer	≤750 000 mL

cfu, colony-forming units.
Data compiled from US Public Health Service (2005).

Table 5.8 Heat treatment requirements specified in USA legislation under the Pasteurised Milk Ordinance (PMO) and the Code of Federal Regulations.

Type	Temperature (°C) minimum	Time (minimum)
Pasteurisation	63[a]	30 min
	72[a]	15 s
	89	1.0 s
	90	0.5 s
	94	0.1 s
	96	0.05 s
	100	0.01 s
Ultra-pasteurisation	130	2 s

[a] If the fat content is $\geq$10 g 100 g^{-1}, then these minimum temperatures should be increased by 3°C.
Data compiled from US Public Health Service (2005) and US National Archives and Records Administration (2007c).

in Table 5.8. The USA standards for Grade 'A' Pasteurised Milk and Milk Products are shown in Table 5.9. Ultra-pasteurisation means the milk or milk product shall be heat treated as outlined in Table 5.8 either before or after packaging so as to produce a product which has extended shelf life under refrigerated conditions. The definition of aseptic processing is a milk product that has been subjected to sufficient heat processing and packaged in a hermetically sealed container so that the product meets the definition of commercial sterility as outlined in §113.3(e)1 of the Code of Federal Regulations (US National Archives and Records Administration, 2007c). This requires the milk or milk product to be free of microorganisms capable of reproducing therein under normal non-refrigerated storage and distribution and also free viable microorganisms (including spores) of public health significance.

5.5.5 *Sale of raw milk for direct consumption in the USA*

One significant area where the PMO differs from the legislation of certain individual states is as regards the sale of raw milk. The PMO (US Public Health Service, 2005) and Title 21 of the

Table 5.9 Standards for grade 'A' pasteurised milk and milk products in the USA.

Criterion	Requirement
Temperature	Cooled to $\leq$7°C and maintained thereat
Bacterial limits – individual producer	$\leq$20 000 colony forming units (cfu) mL^{-1}
Coliforms count	$\leq$10 cfu mL^{-1}
Phosphatase	<350 milliunits L^{-1} for fluid products by approved electronic phosphatase procedures; this is not applicable to ultra-pasteurised products that have been thermally processed $\geq$138°C
Drugs	No positive results on drug residue detection methods. β-Lactam detection methods deemed acceptable for heat-treated milk and milk products

US Public Health Service (2005).

Code of Federal Regulations §1240.61 (US National Archives and Records Administration, 2007d) require that milk in interstate commerce, in its final package for direct human consumption, must be pasteurised. A survey of all 50 states in May 1995 revealed that 28 of the 50 states permit the sale of raw milk with varying levels of restriction. In the states where sale is allowed, it is estimated that the volume sold was less than 1% of total milk sales. (Headrick *et al.*, 1998). In most cases, where allowed, such sales are directly from the farm and only 7 states (Arizona, California, Connecticut, Maine, New Mexico, Pennsylvania and South Carolina) allow sale of raw milk from stores (for details, see http://www.realmilk.com/milk-laws-1.html). In some states, it is left to city or town boards to determine if this is allowed.

As regards the consumption of raw milk, it should be stated that the USA Food and Drugs Association (FDA), the Centres for Disease Control and Prevention (CDC), the National Conference on Interstate Milk Shipments (NCIMS), the National Association of State Departments of Agriculture (NASDA) and the Association of Food & Drug Officials (AFDO) have all developed national policies or position statements that caution strongly against drinking raw milk, endorse regulatory efforts to prohibit or restrict its sale and/or support mandatory pasteurisation milk for direct consumption (FDA, 2007).

5.6 The international perspective – Codex Alimentarius

5.6.1 *What is Codex Alimentarius?*

In the period of 1961–1962, the Joint FAO/WHO Food Standards Conference established the CAC, and asked it to implement the joint FAO/WHO foods standards programme, and to create the Codex Alimentarius, the name which derives from Latin and translates as the food code. The Codex Alimentarius has now become the global reference point of national food control agencies, food producers and processors and international food trade. Its influence extends to all continents, and has contributed to the protection of human health and safety and fair trade practices worldwide.

The Codex Alimentarius system provides the opportunity for all member countries to join the international community in formulating and harmonising food standards and ensuring their implementation. It also allows them a role in the development of guidelines and codes of good practice related to hygienic processing and recommendations relating to compliance with those standards. In addition, the Codex Alimentarius has an increasing relevance to the international food trade. The Agreement on the Application of Sanitary and Phytosanitary Measures (SPS Agreement) and the Agreement on Technical Barriers to Trade (TBT Agreement) both encourage the international harmonisation of food standards. These Agreements cite international standards, guidelines and recommendations as the preferred measures for facilitating international trade in food. As such, Codex Alimentarius standards have become the benchmarks against which national food measures and regulations are evaluated within the legal parameters of the World Trade Organization (WTO) Agreements.

5.6.2 *Codex Alimentarius Commission (CAC) membership and structures*

The CAC is an international intergovernmental body. Its membership is open to member nations and associate members of the FAO and/or the WHO. As of July 2007, it had 175 member countries and 1 member organisation (the European Community). In general, the CAC meets every year, and the venue alternates between the FAO headquarters in Rome and WHO headquarters in Geneva. Nominated senior officials represent member governments at Codex Alimentarius meetings. National delegations may also include representatives of the industry, consumers and academia. A significant number of other international governmental organisations, for example the Office International des Epizooties (OIE), the World Trade Organisation (WTO) and recognised international non-governmental organisations (NGOs), such as the International Dairy Federation (IDF) and the Confederation of the Food and Drink Industries of the EU (i.e. Confédération des Industries Agro-Alimentaires – CIAA), may also attend in an observer capacity. Observers are allowed to contribute to meeting at all stages except in final decisions. This is the exclusive prerogative of member governments.

The CAC has established two types of subsidiary committees: (a) Codex committees, and (b) coordinating committees. The former type committee is subdivided into general subject committees (currently 9 in number), so-called because of the horizontal nature of their work, and commodity committees (currently 21 in number of which 16 are active), which develop the standards for specific foods or classes of foods. There are five regional coordinating committees whose role is to ensure that the CAC is responsive to regional interests, and the needs of developing countries. The CAC also establishes ad hoc intergovernmental task forces given stated tasks on specific topics. Currently, there are three such task forces, on foods derived from biotechnology, antimicrobial resistance, and quick frozen foods. The Codex Alimentarius structure is shown in Figure 5.4.

The main aims of the Codex Alimentarius are as follows:

- The protection of consumer health.
- Ensuring fair trading practices.
- Facilitating international trade.

5.6.3 *Codex Alimentarius standards*

The Codex Alimentarius consists of 13 volumes, which contain general principles, general standards, commodity standards, definitions, codes, methods and recommendations and, as of July 2006, its content is shown in Table 5.10. As may be seen, in addition to individual food commodity standards, it encompasses food labelling, food additives, food hygiene, contaminants, nutrition and foods for special dietary uses, and methods of analysis and sampling.

The CAC has established a number of principles on the scientific basis for its decision-making (Randell & Race, 1996). These principles ensure that the quality and food safety provisions shall be based on sound science, and that in establishing food standards, other legitimate factors may be considered that are relevant to consumer's health and the promotion of fair trade. The Codex Alimentarius standards and related texts are subject to revision, as and when deemed necessary by the CAC and its subsidiary bodies, to ensure that they are consistent with, and reflect current scientific knowledge. Any member of CAC may identify and present new scientific or other information to the relevant body that may warrant a revision.

The Uruguay round of multilateral trade negotiations held under the General Agreement on Tariffs and Trade (GATT), which took place between 1986 and 1994, led to the formation

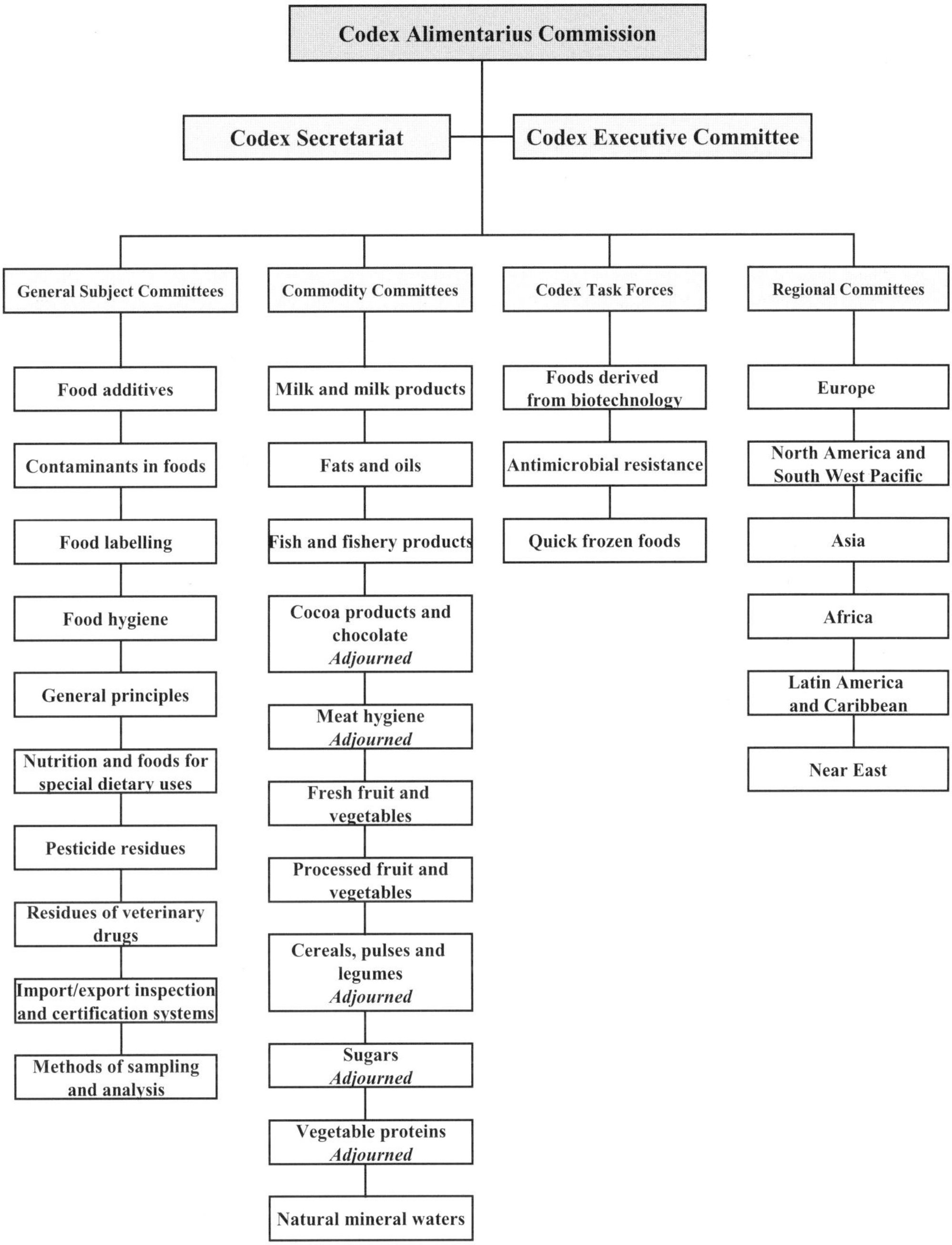

Fig. 5.4 Structure of the Codex Alimentarius Commission. Note: not all the commodity committees are included.

Table 5.10 Content of the Codex Alimentarius (as of July 2006).

Category	Number of Codex standards, guidelines and codes of practice
Food commodity standards	186
Food commodity related texts	46
Food labelling	9
Food hygiene	5
Food safety risk assessment	3
Sampling and analysis	15
Inspection and certification procedures	8
Animal food production	6
Contaminants in foods (maximum levels, detection and prevention)	12
Food additives provisions (covering 292 food additives)	1112
Food additives related texts	7
Maximum limits for pesticide residues (covering 218 pesticides)	2930
Maximum limits for veterinary drugs in foods (covering 49 veterinary drugs)	441
Regional guidelines	3

Data compiled from FAO/WHO (2006).

of the World Trade Organization (WTO) on January 1995. For the first time, GATT agreements included agriculture and food in its scope; however, the Marrakesh agreement of 1994 also included the agreements on sanitary and phytosanitary measures (commonly referred to as the SPS Agreement) and on technical barriers to trade (commonly referred to as the TBT Agreement). These agreements acknowledge the need for the harmonisation of international standards to minimise the risk of sanitary, phytosanitary and other technical standards becoming barriers to international trade. Thus, the SPS and TBT agreements gave formal recognition to international standards, guidelines and recommendations of international organisations, including the CAC, as reference points for facilitating international trade and resolving disputes. Hence, the role of Codex Alimentarius in this regard is now well recognised.

5.6.4 *Codex Alimentarius standards and milk for human consumption*

The CAC has not developed any specific standards for milk for direct human consumption. It has, however, developed a number of general standards and codes of practice that are of relevance to such products. In this regard, four such standards are as follows:

- The Codex General Standard for the Use of Dairy Terms CODEX STAN 206 – 1999 (FAO/WHO, 1999) – often referred to by the acronym GSUDT.
- The Recommended International Code of Practice General Principles of Food Hygiene CAC/RCP 1 – 1969, Revision 4 (2003) (FAO/WHO, 2003c).
- The Codex Code of Hygiene Practice for Milk and Milk Products CAC/RCP 57–2004 (FAO/WHO, 2004).
- The Codex General Standard for Food Additives (GSFA) CODEX STAN 192–1995 (FAO/WHO, 1995).

Compositional standards and compositional modification of milk

From the time of its adoption in 1958 until 1999, the Codex Code of Principles for Milk and Milk Products formed the basis for the identity of milk products, and sought to prevent confusion arising between milk products and imitation milk products. Following the establishment of the Codex Committee for Milk and Milk Products (CCMMP) in 1994, a review was started to update the code of principles to ensure conformance to the Codex requirements for food standards. The revision was finalised at the third session of the CCMMP in Montevideo – Uruguay in 1998, and the standard was adopted by the CAC in Rome in June 1999 as the GSUDT (International Dairy Federation, 2005). It contained definitions of milk, milk products, composite milk products, reconstituted and combined milk products and also defines dairy terms.

The GSUDT defines milk in Section 2.1 as follows: 'Milk is the normal mammary secretion of milking animals obtained from one or more milking without either addition to it or extraction from it, intended for consumption as liquid milk or for further processing'. It is clarified in Section 4.2 on use of the name milk, that the product meeting this definition and sold as such should be designated as 'raw milk' or some other such appropriate name that would not confuse consumers as to its nature.

Section 4.2.2 states that milk which is modified in composition by the addition and/or withdrawal of milk constituents may be identified with a name using the term 'milk', provided that a clear description of the modification to which the milk has been subjected is given in close proximity to the name. Section 4.2.3 which states that milk which is adjusted for fat and/or protein contents and which is intended for direct consumption may also be named 'milk' provided that:

- It is sold only where such adjustment is permitted in the country of retail sale.
- The minimum and maximum limits of fat and/or protein contents (as the case may be) of the adjusted milk are specified in the legislation of the country of retail sale. Where the protein content is adjusted this shall be within the limits of natural variation within that country.
- The adjustment has been performed according to methods permitted by the legislation of the country of retail sale, and only by the addition and/or withdrawal of milk constituents, without altering the whey protein to casein ratio may be allowed.
- The nature of the adjustment is declared in close proximity to the product name.

It should be noted that the above just sets the conditions to be observed in the use of the term milk for products if fat and or protein modification is to be allowed. It neither prohibits nor promotes such modifications.

Codex hygiene codes of practice

As outlined earlier, Codex Alimentarius has developed a general hygiene code to cover all foodstuffs and a specific code for milk and milk products. These hygiene codes consist of ten sections, common to both horizontal and specific codes, as outlined in Table 5.11. The aim of the general code (the Codex Recommended International Code of Practice General

Table 5.11 Section headings of Codex Alimentarius codes of hygiene.

	Introduction
Section I	Objectives
Section II	Scope, Use and Definitions
Section III	Primary Production
Section IV	Establishment: Design and Facilities
Section V	Control of Operation
Section VI	Establishment: Maintenance and Sanitation
Section VII	Establishment: Personal Hygiene
Section VIII	Transportation
Section IX	Product Information and Consumer Awareness
Section X	Training
Annexes	Where necessary in specific codes

Data compiled from FAO/WHO (2003c, 2004).

Principles of Food Hygiene; FAO/WHO, 2003c) is to provide a sound basis for ensuring food hygiene, and should be used in conjunction with any specific hygiene code for the sector concerned, together with the Codex Guidelines for the Establishment and Application of Microbiological Criteria for Foods (FAO/WHO, 2003a) and the Codex Principles and Guidelines for the Conduct of Microbiological Risk Assessment (FAO/WHO, 2003b). An appendix to the general code addresses the HACCP system, and contains guidelines for its application.

The Code of Hygiene Practice for Milk and Milk Products should not be looked at in isolation, but in conjunction with the general code and the other hygiene texts referred to above. References are also made therein to a Codex Guidelines for the Validation of Food Hygiene Control Measures that is still under development. The objective of this code is to provide specific guidance on achieving the general hygiene requirements of Codex commodity standards for milk products. The code uses a food safety objective approach, outlining hygiene principles (in bold font), explanatory narratives (in italic font) and guidelines for the application of the principles (in normal font). Forty pages long, it expands in particular on the requirements of Section 3 (Primary Production) and Section 5 (Control of Operations). The guidelines outline what is required, but does not go into detail on how to achieve the requirement. Annex I contains guidelines for the primary production of milk, additional provisions are given for the production of milk to be used in raw milk products. Annex II gives guidelines for the management of control measures during and after processing. In many cases, the guidelines are of a general nature; for example, it states that for perishable products, the storage temperature should be sufficient to maintain product safety and suitability throughout the shelf life, a specific storage temperature is not indicated.

Appendix A outlines typical Microbiostatic Control Measures that are used, such as refrigeration, water activity control, pH reduction, use of preservatives, modified atmosphere packaging; whilst Appendix B outlines typical Microbiocidal Control Measures that are used, such as pasteurisation, microfiltration, high-pressure treatment and commercial sterilisation. Performance criteria are established for pasteurisation as the heat treatment designed to achieve at least a 5 $\log_{10}$ reduction of *Cloxiella burnetii* in whole milk of 4 g 100 g^{-1} fat. The process criteria state that, according to validations carried out on whole milk, the minimum pasteurisation conditions are those having bactericidal effects equivalent to heating every particle of milk to 72°C for 15 s (continuous flow) or 63°C for 30 min (batch pasteurisation). It goes on to say that similar conditions can be obtained by joining the line connecting these points on a log time versus temperature graph. It cautions that extrapolation of this graph to temperatures outside the temperature range of 63–72°C, and in particular above 72°C, must be treated with care, as the ability to have them scientifically validated is beyond current experimental techniques. It also states that where there are changes in composition, processing or use of the end product, the necessary changes to the scheduled heat treatment should be established and the efficiency evaluated. In this regard, the New Zealand Food Safety Authority have produced a table of equivalent time and temperature combinations for heat treatments for the pasteurisation of milk of up to 10 g fat 100 g^{-1} (see Table 5.12) (New Zealand Food Standards Agency, 1993).

Table 5.12 Equivalent time and temperature heat treatments for pasteurisation of milk of up to 10 g fat 100 g^{-1}.

Minimum holding time (s)	Minimum temperature (°C)
0.01	100
0.05	96
0.1	94
0.5	90
1.0	89
8	73.4
9	73.1
10	72.8
11	72.7
12	72.5
13	72.3
14	72.1
15	72.0
16	71.9
17	71.8
18	71.7
19	71.6
20	71.5
22	71.3
24	71.1
26	70.9
28	70.8
30	70.7
35	70.4
40	70.1
45	69.9
50	69.7
55	69.5
60	69.3

Data compiled from New Zealand Food Standards Agency (1993).

Food additives and Codex Alimentarius

Codex Alimentarius is also developing a General Standard for Food Additives (GSFA) and an associated Food Category System (FCS), which is contained in Annex B of the GSFA. The latter being a means for assigning food additive uses in the standard (FAO/WHO, 1995). All foods are included in the system, but the food category descriptors used are not intended to be legal product designations nor are they intended for labelling purposes. The FCS is hierarchical, and an additive permitted in the general category is taken as permitted in all sub-categories, unless otherwise stated. All dairy products and dairy analogues, except those that are fat emulsions (e.g. butter and dairy spreads), form the general category 01.0 and milk and dairy-based drinks in sub-category 01.1. The following scheme outlines the FCS for this category for milk and dairy-based drinks:

01.1 Milk and dairy-based drinks – this is defined to include all plain and flavoured fluid milk products based on skimmed, part-skimmed, low-fat and whole milk.

- **01.1.1** Milk and buttermilk (plain) – this is defined to include plain fluid products only; this category of products also includes reconstituted plain milk that contains only dairy ingredients.
  - **01.1.1.1** Milk (plain) – this is defined to include fluid milk obtained from all milking animals; milk is usually heat-treated by pasteurisation, UHT treatment or sterilisation; this category of products also includes skimmed, part-skimmed, low-fat and whole milk.
  - **01.1.1.2** Buttermilk (plain)
- **01.1.2** Dairy-based drinks, flavoured and/or fermented (e.g. chocolate milk, cocoa, eggnog, drinking yoghurt, whey-based drinks).

The scope of this section of the FCS refers to food category 01.1.1.1, and discussion on permitted additives shall be confined to this category. The GSFA is made up of three tables: (a) Table 1 lists additives alphabetically that are permitted for use under specified conditions in certain food categories or individual food items, (b) Table 2 lists food categories or individual food items in which food additives are permitted and (c) Table 3 lists additives permitted for use in food in general, unless otherwise specified, in accordance with Good Manufacturing Practice (GMP). It should be pointed out that Annex 3 to Table 3 specifies the food category 01.1.1 (i.e. milk and buttermilk) as one in which the use of additives is governed by Tables 1 and 2 only; therefore, Table 3 does not apply. At this time, there are no additives adopted in Table 2 for the food category 01.1.1. However, this is likely to change. Development of the GSFA is work in progress, and a total of 24 additives are at various stages of the Codex eight-step procedure for the development of standards; of these, 19 additives are at Step 7 and may be adopted in the near future. If and when these are adopted, they will be incorporated into the GSFA and published on the Codex website at http://www.codexalimentarius.net/gsfaonline/index.html?lang=e.

5.7 Conclusions and possible future developments

Over the last 130 years, since the development of the pasteurisation process and its application to the dairy sector, milk for human consumption has become one of the safest and most wholesome of foods for all ages, from the young to the elderly. Combined with improvements in hygiene in the production and processing of milk, this has also resulted in an increase in the shelf life of pasteurised milk from the 2 days at the start of the twentieth century (Westhoff, 1978) to 10–12 days, when stored at maximum 6°C, while the shelf life of ESL milk can range from 14 to more than 45 days at the same storage temperatures (Rysstad & Kolstad, 2006). The development of UHT milks from the 1960s has resulted in milk which can be stored at ambient temperatures for 6 months or more. The emergence of new food safety issues, such as enteropathogenic, enterohemolytic and verotogenic species of *Escherichia coli*, such as *E. coli* O157:H7 in the early 1980s and *Enterobacter sakasakii* in infant formulae in the 1990s, which both have very low infective doses for susceptible populations, created new challenges. Legislators responded to these and other challenges by incorporating the requirement for a HACCP approach in food safety management systems. This replaced the earlier approach of laying down specific and detailed requirements in legislation.

It is always more difficult to anticipate what may occur in the future than to review the past. Legislative developments usually lag behind technology. Nowadays, they tend to be initiated to address issues of food safety, trade problems or misleading consumers, with a preference for horizontal rather than vertical legislation. Amendments to existing legislation are likely to continue if and as necessary. The development of sectoral codes of practice, not just for hygiene but also for labelling and compositional requirements, may replace or substitute for specific and detailed legislation.

The Codex Committee on Milk and Milk Products is scheduled to complete its current programme of work in 2010, and is unlikely to take on the development of new international standards in the near future. However, the completion of the Codex General Standard for Food Additives is likely to take quite a few years to complete.

One area where further legislation may emerge is in the field of the environment. Discussions on climate change, food miles and carbon footprints at national, regional and international level may lead to further legislative initiatives. Also, nutritional concerns regarding health, obesity, disease risks and the like may well result in increased levels of clearer and simpler nutritional labelling. Such demands are likely to be balanced by a corresponding concern to avoid increasing the legislative burden on the food industry, where operating margins are narrow. Furthermore, new and unexpected problems will continue to arise, preventing any sense of complacency setting in.

References

Anonymous (2006) *Code of Practice on HTST Pasteurisation*, Dairy UK, London.

Bell, J.A., Beck, M.D. & Huebener, R.J. (1950) Epidemiological studies of Q-fever in southern California. *Journal of the American Medical Association*, **142**, 868–872.

Burton, H. (1977) An introduction to ultra-high-temperature processing and plant. *Journal of the Society of Dairy Technology*, **30**, 135–142.

DAFRDI (2001) *The Safe Food Chain – Every Link is Vital*, Department of Agriculture, Food and Rural Development, Dublin.

DIF (1995) *DIF Guidelines for Good Hygienic Practice in the Manufacture of Dairy-Based Products*, Ministry of Agriculture, Fisheries & Food (MAFF) Publications, London.

Enright, J.B., Sadler, W.W. & Thomas, R.C. (1957) Thermal inactivation of *Coxiella burnetii* and its relation to pasteurization of milk. *American Journal of Public Health*, **47**, 695–700.

EU (1964) Council Directive 64/432/EEC of 26 June 1964 on animal health problems affecting intra-Community trade in bovine animals and swine. *Official Journal of the European Commission*, **121**, 1977–2012.

EU (1971) Regulation (EEC) No. 1411/71 of the Council of 29 June 1971 laying down additional rules on the common organisation of the market in milk and milk products for products falling within tariff heading No 04.01. *Official Journal of the European Commission*, **L148**, 4–7.

EU (1975) Council Directive 75/106/EEC of 19 December 1974 on the approximation of the laws of the Member States relating to the making-up by volume of certain prepackaged liquids. *Official Journal of the European Commission*, **L42**, 1–13.

EU (1980) Council Directive 80/232/EEC of 15 January 1980 on the approximation of the laws of the Member States relating to the ranges of nominal quantities and nominal capacities permitted for certain prepackaged products. *Official Journal of the European Commission*, **L51**, 1–7.

EU (1985) Council Directive 85/397/EEC of 5 August 1985 on health and animal-health problems affecting intra-Community trade in heat-treated milk. *Official Journal of the European Commission*, **L226**, 13–29.

EU (1988) Council Directive 88/388/EEC of 22 June 1988 on the approximation of the laws of the Member States relating to flavourings for use in foodstuffs and to source materials for their production. *Official Journal of the European Commission*, **L184**, 61–66.

EU (1989) Council Directive 89/107/EEC of 21 December 1988 on the approximation of the laws of the Member States concerning food additives authorized for use in foodstuffs intended for human consumption. *Official Journal of the European Commission*, **L40**, 27–33.

EU (1990a) Council Directive 90/496/EEC of 24 September 1990 on nutrition labelling for foodstuffs. *Official Journal of the European Commission*, **L276**, 40–44.

EU (1990b) Council Regulation (EEC) No. 2377/90 of 26 June 1990 laying down a Community procedure for the establishment of maximum residue limits of veterinary medicinal products in foodstuffs of animal origin. *Official Journal of the European Commission*, **L224**, 1–8.

EU (1991) Council Directive 91/68/EEC of 28 January 1991 on animal health conditions governing intra-Community trade in ovine and caprine animals. *Official Journal of the European Commission*, **L46**, 19–36.

EU (1992) Council Directive 92/46/EEC of 16 June 1992 laying down health rules for the production and placing on the market of raw milk, heat-treated milk and milk-based products. *Official Journal of the European Commission*, **L268**, 1–32.

EU (1994a) Commission Regulation (EC) No 539/94 of 10 March 1994 authorizing Ireland to derogate from the minimum fat content of drinking milk. *Official Journal of the European Commission*, **L68**, 21.

EU (1994b) European Parliament and Council Directive 94/35/EC of 30 June 1994 on sweeteners for use in foodstuffs. *Official Journal of the European Commission*, **L237**, 3–12.

EU (1994c) European Parliament and Council Directive 94/36/EC of 30 June 1994 on colours for use in foodstuffs. *Official Journal of the European Commission*, **L237**, 13–29.

EU (1995) European Parliament and Council Directive No 95/2/EC of 20 February 1995 on food additives other than colours and sweeteners. *Official Journal of the European Commission*, **L61**, 1–40.

EU (1996a) Council Directive 96/23/EC of 29 April 1996 on measures to monitor certain substances and residues thereof in live animals and animal products and repealing Directives 85/358/EEC and 86/469/EEC and Decisions 89/187/EEC and 91/664/EEC. *Official Journal of the European Commission*, **L125**, 10–32.

EU (1996b) *Evaluation of School Milk Scheme*, http://ec.europa.eu/agriculture/eval/reports/schoolmilk/8.pdf.

EU (1997) Council Regulation (EC) No. 2597/97 of 18 December 1997 laying down additional rules on the common organization of the market in milk and milk products for drinking milk. *Official Journal of the European Commission*, **L351**, 13–15.
EU (1998a) Council Directive 97/78/EC of 18 December 1997 laying down the principles governing the organisation of veterinary checks on products entering the Community from third countries. *Official Journal of the European Commission*, **L24**, 9–30.
EU (1998b) Council Directive 98/83/EC of 3 November 1998 on the quality of water intended for human consumption. *Official Journal of the European Commission*, **L330**, 32–54.
EU (2000) Directive 2000/13/EC of the European Parliament and of the Council of 20 March 2000 on the approximation of the laws of the Member States relating to the labelling, presentation and advertising of foodstuffs. *Official Journal of the European Commission*, **L109**, 29–42.
EU (2002) Regulation (EC) No 178/2002 of the European Parliament and of the Council of 28 January 2002 laying down the general principles and requirements of food law, establishing the European Food Safety Authority and laying down procedures in matters of food safety. *Official Journal of the European Commission*, **L31**, 1–24.
EU (2003a) Corrigendum to Commission Directive 2002/72/EC of 6 August 2002 relating to plastic materials and articles intended to come into contact with foodstuffs. *Official Journal of the European Commission*, **L39**, 1–42.
EU (2003b) Council Directive 2002/99/EC of 16 December 2002 laying down the animal health rules governing the production, processing, distribution and introduction of products of animal origin for human consumption. *Official Journal of the European Commission*, **L18**, 11–20.
EU (2004a) Commission Directive 2004/1/EC of 6 January 2004 amending Directive 2002/72/EC as regards the suspension of the use of azodicarbonamide as blowing agent. *Official Journal of the European Commission*, **L7**, 45–46.
EU (2004b) Commission Directive 2004/19/EC of 1 March 2004 amending Directive 2002/72/EC relating to plastic materials and articles intended to come into contact with foodstuffs. *Official Journal of the European Commission*, **L71**, 8–21.
EU (2004c) Regulation (EC) No. 1935/2004 of the European Parliament and of the Council of 27 October 2004 on materials and articles intended to come into contact with food and repealing Directives 80/590/EEC and 89/109/EEC. *Official Journal of the European Commission*, **L338**, 4–17.
EU (2004d) Regulation (EC) No. 852/2004 of the European Parliament and of the Council of 29 April 2004 on the hygiene of foodstuffs. *Official Journal of the European Commission*, **L226**, 3–21.
EU (2004e) Regulation (EC) No. 853/2004 of the European Parliament and of the Council of 29 April 2004 laying down specific rules for food of animal origin. *Official Journal of the European Commission*, **L226**, 22–82.
EU (2004f) Regulation (EC) No. 854/2004 of the European Parliament and of the Council of 29 April 2004 laying down specific rules for the organisation of official controls on products of animal origin intended for human consumption. *Official Journal of the European Commission*, **L139**, 83–127.
EU (2004g) Regulation (EC) No. 882/2004 of the European Parliament and of the Council of 29 April 2004 on official controls performed to ensure the verification of compliance with feed and food law, animal health and animal welfare rules. *Official Journal of the European Commission*, **L165**, 1–141.
EU (2005a) Commission Regulation (EC) No 2073/2005 of 15 November 2005 on microbiological criteria for foodstuffs. *Official Journal of the European Commission*, **L338**, 1–26.
EU (2005b) Commission Regulation (EC) No. 2074/2005 of 5 December 2005 laying down implementing measures for certain products under Regulation (EC) No. 853/2004 of the European Parliament and of the Council and for the organisation of official controls under Regulation (EC) No. 854/2004 of the European Parliament and of the Council and Regulation (EC) No. 882/2004 of the European Parliament and of the Council, derogating from Regulation (EC) No. 852/2004

of the European Parliament and of the Council and amending Regulations (EC) No. 853/2004 and (EC) No. 854/2004. *Official Journal of the European Commission*, **L338**, 27–58.

EU (2005c) Commission Regulation (EC) No. 2076/2005 of 5 December 2005 laying down transitional arrangements for the implementation of Regulations (EC) No. 853/2004, (EC) No. 854/2004 and (EC) No. 882/2004 of the European Parliament and of the Council and amending Regulations (EC) No. 853/2004 and (EC) No. 854/2004. *Official Journal of the European Commission*, **L338**, 83–88.

EU (2005d) Guidance document on the implementation of certain provisions of Regulation (EC) No. 853/2004 on the hygiene of food of animal origin. EU Health & Consumer Protection Directorate-General, Brussels.

EU (2005e) Guidance document on the implementation of certain provisions of Regulation (EC) No 852/2004 on the hygiene of foodstuffs. EU Health & Consumer Protection Directorate-General, Brussels.

EU (2005f) Guidance document on the implementation of procedures based on the HACCP principles, and on the facilitation of the implementation of the HACCP principles in certain food businesses. EU Health & Consumer Protection Directorate-General, Brussels.

EU (2006) Guidelines for the development of Community guides to good practice for hygiene or for the application of the HACCP principles, in accordance with Article 9 of Regulation (EC) No. 852/2004 on the hygiene of foodstuffs and Article 22 of Regulation (EC) No. 183/2005laying down requirements for feed hygiene. EU Health & Consumer Protection Directorate-General, Brussels.

EU (2007a) Council Regulation (EC) No. 1153/2007 of 26 September 2007 amending Regulation (EC) No. 2597/97 laying down additional rules on the common organisation of the market in milk and milk products for drinking milk. *Official Journal of the European Commission*, **L258**, 6.

EU (2007b) Directive 2007/45/EC of the European Parliament and of the Council of 5 September 2007 laying down rules on nominal quantities for prepacked products, repealing Council Directives 75/106/EEC and 80/232/EEC, and amending Council Directive 76/211/EEC. *Official Journal of the European Commission*, **L247**, 17–20.

EU (2007c) Proposal for a Council Regulation amending Regulation (EC) No 2597/97 laying down additional rules on the common organisation of the market in milk and milk products for drinking milk. Available at http://eur-lex.europa.eu/LexUriServ/LexUriServ.do?uri=CELEX:52007PC0058(03):EN:HTML.

FAO/WHO (1995) *Codex General Standard for Food Additives*, CODEX STAN 192 – 1995. Available at http://www.codexalimentarius.net/gsfaonline/CXS_192e.pdf.

FAO/WHO (1999) *The Codex General Standard for the Use of Dairy Terms*, CODEX STAN 206–1999. Available at http://www.codexalimentarius.net/web/more_info.jsp?id_sta=332.

FAO/WHO (2003a) Codex Guidelines for the Establishment and Application of Microbiological Criteria for Foods, CAC/GL-30 (1999). In: *Food Hygiene Basic Texts*, pp. 45–52, Secretariat of the Codex Alimentarius Commission Joint FAO/WHO Food Standards Programme, Rome.

FAO/WHO (2003b) Codex Principles and Guidelines for the Conduct of Microbiological Risk Assessment CAC/GL-30 (1999). In: *Food Hygiene Basic Texts*, pp. 53–62, Secretariat of the Codex Alimentarius Commission Joint FAO/WHO Food Standards Programme, Rome.

FAO/WHO (2003c) The Recommended International Code of Practice General Principles of Food Hygiene CAC/RCP 1 –1969, Revision 4 (2003). In: *Food Hygiene Basic Texts*, pp. 1–30, Secretariat of the Codex Alimentarius Commission Joint FAO/WHO Food Standards Programme, Rome.

FAO/WHO (2004) *Code of Hygienic Practice for Milk and Milk Products*, CAC/RCP 57 – 2004. Available at http://www.codexalimentarius.net/web/more_info.jsp?id_sta=10087.

FAO/WHO (2006) *Understanding The Codex Alimentarius*, 3rd edn, pp. 11, Secretariat of the Codex Alimentarius Commission Joint FAO/WHO Food Standards Programme, Rome. Available at ftp://ftp.fao.org/codex/Publications/understanding/Understanding_EN.pdf.

FDA (2007) *Raw Milk-Associated Public Health Risks*. Available at http://www.cfsan.fda.gov/~dms/rawmilk2.html.

Food Safety Authority of Ireland (1998) *Compendium of Food Law in Ireland*, 1998 edition, Food Safety Authority of Ireland, Dublin.

Food Standards Agency (2005) *Historical background to the different control requirements for raw drinking milk and raw cream in different countries of the UK*. Available at http://www.food.gov.uk/aboutus/ourboard/boardmeetoccasionalpapers/

Grant, I.R., Rowe, M.T., Williams, A.G. & Muir, D.D. (2005) Efficacy of various pasteurisation time/temperature conditions in combination with homogenisation on the inactivation of Mycobacterium avium subsp. paratuberculosis in milk. *Applied and Environmental Microbiology*, **71**, 2853–2861.

Headrick, M.L., Korangy, S., Bean, N.H., Angulo, F.J., Altekruse, S.F., Potter, M.E. & Klontz, K.C. (1998) The epidemiology of raw milk-associated foodborne disease outbreaks reported in the United States – 1973 through 1992. *American Journal of Public Health*, **88**, 1219–1221.

Henyon, D. (1999) Extended shelf-life milks in North America: a perspective. *International Journal of Dairy Technology*, **52**, 95–101.

HMSO (1920) *The Ministry Food (Continuance) Act* (10 & 11 Geo.V c.47). Her Majesty's Stationery Office, London.

HMSO (1955) *The Food and Drugs Act 1955* (4 & 5 Eliz. II c16). Her Majesty's Stationery Office, London.

HMSO (1959) *The Milk and Dairies (General) Regulations 1959*, SI 1959 No. 277. Her Majesty's Stationery Office, London.

HMSO (1973) *The Milk and Dairies (Skimmed and Semi-skimmed Milk) (Heat Treatment and Labelling) Regulations 1973*, SI 1973 No. 1064. Her Majesty's Stationery Office, London.

HMSO (1977) *The Milk (Special Designations) Regulations 1977*, SI 1977 No.1033. Her Majesty's Stationery Office, London.

HMSO (1988) *The Weights and Measures (Miscellaneous Foods) Order 1988*, SI 1988 No. 2040. Her Majesty's Stationery Office, London.

HMSO (1994) *The Weights and Measures (Metrication) (Miscellaneous Goods) (Amendment) Order 1994*, SI 1994 No. 2868. Her Majesty's Stationery Office, London.

HMSO (1998) *The Drinking Milk Regulations 1998*, SI 1998 No. 2424. Her Majesty's Stationery Office, London.

HMSO (2006) *The Food Hygiene (England) Regulations 2006*, SI 2006 No. 14. Her Majesty's Stationery Office, London.

IDF (2005) *The Codex General Standard on the Use of Dairy Terms – Its nature, Intent and Implications*, Document No. 397. International Dairy Federation, Brussels.

Lennette, E.H., Clark, W.H., Ahinanti, M.M., Baumetti, O. & Covert, J.M. (1952) Q Fever Studies. XII. The Effect of pasteurization on *Coxiella burnetii* in naturally infected milk. *American Journal of Hygiene*, **55**, 246–253.

Mann, E. (2001) Extended shelf life milk. *Dairy Industries International*, **66**(5), 17–18.

MDC (2004) *Dairy Facts and Figures 2003*, pp. 132. Milk Development Council, Cirencester, Gloucestershire, UK. Available at http://www.dairyuk.org/pdf/MDC_DFactsFig_1612.pdf.

MDC (2007) *Retail & Consumer Data – GB Milk Purchases by Fat Content*, Milk Development Council, Cirencester, Gloucestershire, UK. Available at http://www.mdcdatum.org.uk/ backdata/DFF2005/Milkpuchasedbyfat.xls.

New Zealand Food Standards Agency (1993) Standard for Pasteurisation Heat Treatments MRD Standard 3 (Revision 2). Available at http://www.nzfsa.govt.nz/dairy/publications/standards/mrd03stan.htm.

OIE (2007) *Terrestrial Animal Health Code*, 16th edn, Appendix 3.6.2, World Organisation for Animal Health, Paris. Available at http://www.oie.int/eng/normes/mcode/en_chapitre_3.6.2.htm.

Randell, A. & Race, J. (1996) Regulatory and legal aspects of functional foods: an international perspective. *Nutrition Reviews*, **54**(11), 152–155.

Rysstad, G. & Kolstad, J. (2006) Extended shelf life milk-advances in technology. *International Journal of Dairy Technology*, **59**, 85–96.

Smillie, D.M. (1948) Pasteurisation – past, present and future. *Journal of the Society of Dairy Technology*, **1**, 147–154.

Staal, P. (1986) *IDF and the Development of Milk Pasteurization*. Pasteurized Milk, Document No. 200, pp. 4–6, International Dairy Federation, Brussels.

te Giffel, M.C., van Asselt, A.J. & de Jong, P. (2006) Shelf extension. *Dairy Industries International*, **71**(3), 44–49.

The Stationery Office (2006a) *The Food Hygiene (Scotland) Regulations 2006*, SSI 2006 No. 3. Her Majesty's Stationery Office, Edinburgh, UK.

The Stationery Office (2006b) *The Food Hygiene (Wales) Regulations 2006*, WSI 2006 No. 31 (W.5). Her Majesty's Stationery Office, Cardiff, UK.

The Stationery Office (2006c). *The Food Hygiene Regulations (Northern Ireland) 2006*, SR 2006 No. 3. Her Majesty's Stationery Office, Belfast, UK.

The Stationery Office Dublin (1973) *Merchandise Marks (Prepackaged Goods) (Marking and Quantities) Order*, SI No. 26 of 1973. The Stationery Sales Office, Dublin.

The Stationery Office Dublin (1981) *The Merchandise Marks (Prepacked Goods) (Marking and Quantities) (Amendment) Order*, SI No. 394 of 1981. The Stationery Sales Office, Dublin.

The Stationery Office Dublin (1990) *European Communities (Standards for Heat-Treated Milk in Intra-Community Trade) Regulations*, SI No. 124 of 1990. The Stationery Sales Office, Dublin.

The Stationery Office Dublin (1996) *European Communities (Hygienic Production and Placing on The Market of Raw Milk, Heat-Treated Milk and Milk-Based Products) Regulations*, SI No. 9 of 1996. The Stationery Sales Office, Dublin.

The Stationery Office Dublin (2005) *European Communities (Food and Feed Hygiene) Regulations*, SI No. 910 of 2005. The Stationery Sales Office, Dublin.

The Stationery Office Dublin (2006a) *European Communities (Food and Feed Hygiene) (Amendment) Regulations*, SI No. 387 of 2006. The Stationery Sales Office, Dublin.

The Stationery Office Dublin (2006b) *European Communities (Hygiene of Foodstuffs) Regulations*, SI No. 369 of 2006. The Stationery Sales Office, Dublin.

The Stationery Office Dublin. (2007) *European Communities (Food and Feed Hygiene) (Amendment) Regulations*, SI No. 56 of 2007. The Stationery Sales Office, Dublin.

US National Archives and Records Administration (2007a) *US Code of Federal Regulations*, Title 21: Part §131, pp. 295–313, Revision 1 April 2007. The Office of the Federal Register National Archives and Records Administration, Washington, DC.

US National Archives and Records Administration (2007b) *US Code of Federal Regulations*, Title 21: Part §101.62, pp. 105–115, Revision 1 April 2007. The Office of the Federal Register National Archives and Records Administration, Washington, DC.

US National Archives and Records Administration (2007c) *US Code of Federal Regulations*, Title 21: Part §113.3, pp. 230–231, Revision 1 April 2007. The Office of the Federal Register National Archives and Records Administration, Washington, DC.

US National Archives and Records Administration (2007d) *US Code of Federal Regulations*, Title 21: Part §1240.61, p. 692, Revision 1 April 2007. The Office of the Federal Register National Archives and Records Administration, Washington, DC.

US Public Health Service (2005) *Grade 'A' Pasteurized Milk Ordinance*, 2005 edn, pp. 1–357, US Department of Health and Human Services, Washington, DC. Available at http://www.foodscience.cornell.edu/cals/foodsci/extension/upload/PMO2005.pdf.

Westhoff, D.C. (1978) Heating milk for microbial destruction: a historical outline and update. *Journal of Food Protection*, **41**, 122–130.

Wilton, D.W. (1983) Some aspects of the law on pasteurisation. In: *Pasteurizing Plant Manual*, 3rd edn (ed. E. Geene), pp. 86–96, The Society of Dairy Technology, Huntingdon, UK.

6 The Safety of Raw Liquid Milk

M. O'Mahony, S. Fanning and P. Whyte

6.1 Introduction

Milk remains an integral dietary component of many societies. As a product of animal origin, milk has a relatively unique status as an inherently wholesome nutritious and safe product. As mammals, milk is deemed an appropriate dietary item for the young, and also other vulnerable sectors of our society. Even in a raw state, milk is rarely attributed the hygienic precautions accepted as necessary for other food products of animal origin, such as meat.

However, milk is the secretion of the mammary gland of an animal, and is never sterile (unlike, for example, meat which is a sterile product until contaminated at harvest by man). Milk is a food product which is harvested from animals on-farm, and temporarily stored as a food product on farm. Milk may contain pathogens, and indeed has historically been the vehicle of significant foodborne diseases, such as tuberculosis (Atkins, 1992). While milk has innate bacteriostatic constituents such as lactoferrin or lysosyme, the availability of nutrients and water, along with favourable tonicity and acidity, facilitate bacterial growth more readily than other products of animal origin.

The supply and production chain of milk has several unique characteristics. Dairy production is an intensive system with substantial reliance upon modern agronomic inputs, e.g. dry cow antimicrobials. There is a strong tradition of vertical integration of the various phases of the food production, from pre- to post-harvest. This has resulted in a significant awareness of food safety and quality issues even at primary production. Milk is a liquid and is, therefore, amenable to sampling, monitoring, and hence meaningful control. Milk is collected and processed in a manner which virtually guarantees cross-contamination of products of various animals, holdings and regions. Conversely, this mixing provides for significant dilution of any hazard present from a point source, e.g. low number of affected cows.

These, and other unique factors, have led to the widespread acceptance of the equally unique thermal treatment during processing prior to supply to the final consumer (Steele, 2000). Pasteurisation of milk represents one of the singularly most successful contributions to the safety of products of animal origin (Holsinger *et al.*, 1997)

6.2 Implication of milk in human disease

In the past, milk was regarded as a common source of disease in man (Galbraith *et al.*, 1982). However, widespread implementation of pre-harvest control measures and pasteurisation has resulted in a substantial reduction in the burden of milkborne disease in society (Djuretic *et al.*, 1997). Recent evidence would implicate milk and dairy products in a small but significant proportion (~2%) of outbreaks of foodborne infectious intestinal disease (Bonner *et al.*, 2001; Gillespie *et al.*, 2003). A recent estimation of food-specific risk has attributed 9% of cases of foodborne disease to milk consumption (Adak *et al.*, 2005). Intensification of the dairy industry with centralisation of processing facilities would explain the trend for outbreaks of milkborne disease to involve substantial morbidity (Gillespie *et al.*, 2003) and create a significant burden on healthcare systems (Adak *et al.*, 2005). The three pathogens, *Salmonella* spp., *Campylobacter* spp. and verocytotoxigenic *Escherichia coli*, have been reported as the most common aetiological agents implicated in milkborne disease (Gillespie *et al.*, 2003).

6.3 Microbial hazards in milk

Harvesting of products from animals in a commercial environment inevitably leads to transfer of microorganisms from animals to those products. Milk may be contaminated by bacteria or viruses. Viruses, which may be present in milk, do not generally pose a public health hazard, but may constitute a significant animal health risk. The presence of potentially pathogenic bacteria in raw milk is a well-documented phenomenon (Rea *et al.*, 1992; Jayarao & Henning, 2001). These microorganisms may enter the milk in the pre-harvest phase by extravasation from systemic circulation or by ingress from the external environment via the teat orifice. Intramammary, enteric, reproductive or systemic infections may result in intermittent presence of those bacteria in milk. Microorganisms may enter milk at harvest phase, contaminating the food product without infecting the gland. Contamination of the external surface of the teat with faecal and other environmental organisms is normal. Even with high standards of hygiene at harvest, some contamination of the product is difficult to avoid. The negative pressure associated with milking, further facilitates aspiration of aerosolised microorganisms from the animal environment. Contamination of milk may also arise at any stage in the post-harvest phase, e.g. on-farm storage, transport, pre-processing storage, processing or post-processing. Microbial contaminants of milk, thus, represent the organisms that exist in these animals, and their environment in which that food is produced and processed. Milk is an organic matrix that facilitates the survival and replication of bacterial species, so initial contamination may result in subsequently higher loads of bacteria depending on storage temperatures.

6.3.1 Salmonella *species*

Salmonella species may be found in milk, and have been implicated in milkborne disease. Some serovars, such as *Salmonella Dublin*, can achieve a carrier status in lactating ruminants, with sub-clinical enteric infections and intermittent faecal shedding, and may occasionally present as an animal disease. Other serovars, such as *Salmonella typhimurium*, tend to produce clinical disease in ruminants. In general terms, *Salmonella* contamination of milk represents faecal contamination of the product at harvest. In rare cases, sub-clinical mastitis due to *Salmonella* species has been implicated in milkborne disease. *Salmonella* species have a poor thermal tolerance, and are effectively removed by pasteurisation. *Salmonella* species have been implicated in outbreaks of foodborne disease associated with unpasteurised dairy products, with implication of raw milk contamination.

6.3.2 Campylobacter *species*

The *Campylobacter* species associated with disease in ruminants, such as *Campylobacter fetus*, are not of public health concern. Thermophillic campylobacters of public health concern may contaminate milk (Whyte *et al.*, 2004). Sub-clinical carriage and intermittent shedding of campylobacters relevant to public health, such as *Campylobacter jejuni* and *Campylobacter coli*, are common in clinically normal bovine animals (Minihan *et al.*, 2004). In general, *Campylobacter* contamination of milk indicates faecal contamination of the product at harvest. Occasional reports of sub-clinical mammary infections with pre-harvest entry to milk have been described. *Campylobacter* may also be shed by birds, and post-processing contamination of liquid milk with this organism has been well described in the UK with the pecking of foil-topped milk bottles on consumers' doorsteps by wild birds (Gillespie *et al.*, 2003). *Campylobacter* organisms are susceptible to desiccation and thermal treatment, including pasteurisation. Also due to the fastidious nature of these organisms and their suboptimal survival in the environment, including extremes of pH, they tend not to be associated with human illness resulting from the consumption of dairy products made from raw milk.

6.3.3 *Verocytotoxigenic* Escherichia coli

E. coli is normally a commensal organism found in the gastrointestinal tract of animals; however, a number of pathogenic strains can cause a range of illnesses in animals and humans.

The most significant of the pathogenic *E. coli* are frequently referred to as verocytotoxigenic *E. coli* (VTEC), shiga toxin-producing *E. coli* (STEC) or enterohaemorrhagic *E. coli* (EHEC). These organisms can produce a potent verotoxin which can cause severe damage to kidney cells in primates including humans. *E. coli* O157:H7 first emerged in the mid-1980s, and has been the VTEC strain most associated with human illness ever since. In more recent years, other non-O157:H7 VTEC serotypes have emerged as significant human enteropathogens. Infection can be induced in humans by as few as ten cells, and symptoms can range from mild to severe including non-bloody diarrhoea, bloody diarrhoea, haemolytic uraemic syndrome (HUS), thrombotic thrombocytopenic purpura (TTP) and in some instances death (Coia, 1998).

These organisms have been frequently isolated from the digestive tracts and faecal material of bovine, ovine and caprine animals (Cornick *et al.*, 2000; Cortes *et al.*, 2005; Schouten *et al.*, 2005). Shedding in faeces may be intermittent with seasonal peaks and animals can remain asymptomatically infected. Contamination of milk by these organisms has been frequently reported to date and is generally a result of direct exposure to faecal material or environmental contamination. VTEC *E. coli* serotypes have on occasion also been detected in bovine mastitic milk, suggesting that entry to milk supplies from cows with sub-clinical mammary infections is an additional route for these organisms.

Significant outbreaks of disease in humans associated with the consumption of raw milk, raw milk cheeses and yoghurt have been documented demonstrating the public health risks (Morgan *et al.*, 1993; Deschênes *et al.*, 1996; Trevena *et al.*, 1999; Vivegnis *et al.*, 1999; Allerberger *et al.*, 2001; McIntyre *et al.*, 2002). However, verocytotoxigenic *E. coli* are susceptible to thermal processes including pasteurisation; however, clear risks are associated with the consumption of raw milk or dairy products manufactured from raw milk. Raw milk has frequently been reported to contain VTEC *E. coli* organisms (Nam *et al.*, 2004; Fremaux *et al.*, 2006; Perelle *et al.*, 2007), while Murphy *et al.*, 2005 detected *E. coli* O157 in enriched in-line bulk tank milk filters, and reported that analysis of these samples was a sensitive means of assessing the VTEC contamination status of raw bulk tank milk. VTECs can also survive in low-pH foods including fermented dairy products. Its ability to survive during the ripening and storage of certain cheeses, in particular soft cheese varieties, has been reported previously (Maher *et al.*, 2001). Although VTEC *E. coli* cannot multiply in frozen ice cream, it can survive, and therefore, such products manufactured from raw milk may also pose a risk (Duffy *et al.*, 2002).

6.3.4 Enterotoxigenic *Staphylococcus aureus*

Staphylococci are gram-positive bacteria predominantly of animal origin, although environmental isolates have also been recovered. These organisms are recognised as part of the normal flora of both animals and humans. *S. aureus* causes serious diseases in humans and animals and is the most common aetiological agent of contagious bovine mastitis (Omoe *et al.*, 2002). In Japan, a study of 146 *S. aureus* strains from human and food samples linked to foodborne outbreaks along with healthy humans, mastitic bovine animals and raw milk (Omoe *et al.*, 2002) demonstrated the importance of assessing the links between enterotoxin production and food poisoning.

Staphylococcal food poisoning (or enterointoxication) is the result of the ingestion of thermotolerant toxins produced by these staphylococci during growth in foods. The staphylococcal enterotoxins (SE) form a heterogenic group, differing at the nucleotide and amino acid levels, and there are 11 recognised enterotoxins (denoted A, B, C_{1-3} and D–I) that represent the pyrogenic group. Further, toxin production may be affected by a number of factors including pH, a_w and temperature. Staphylococcal enterointoxication usually occurs 2–4 h after ingestion of contaminated food and lasts for a period of 24 h. Symptoms include nausea, vomiting and on occasion diarrhoea. Enterotoxin poisoning is usually self-limiting with rapid recovery.

A number of studies have reported the isolation of *S. aureus* and staphylococcal enterotoxins directly from milk (Jorgensen *et al.*, 2005), cheese production from cow's raw milk (Delbes *et al.*, 2006) and from the bulk milk tank (Muehlherr *et al.*, 2003). The presence of this organism in cow's and goat's milk may constitute a risk with respect to food poisoning following the consumption of raw product. In Japan, an extensive outbreak of staphylococcal enterointoxication was reported linked to milk and powdered skimmed milk (Asao *et al.*, 2003). An unusual feature of this outbreak was the fact that thermal processes had destroyed the bacteria in the milk; however, *Staphylococcus aureus* enterotoxin was present and retained sufficient activity to cause intoxication. Loncarevic *et al.* (2005) suggest that, when sampling raw milk and raw milk products for *S. aureus*, several isolates should be tested for enterotoxin production and that culturing may suppress or favour different strains. Modelling of *S. aureus* growth in milk may be a useful approach to control the production of bacterial toxins (Fujikawa & Morozumi, 2006).

Detection of staphylococci can be carried out using established conventional microbiological culture methods. The corresponding enterotoxins genes can be identified using modern molecular approaches including multiplex and 'real-time' polymerase chain reaction (PCR) (Cremonesi *et al.*, 2005). Subtyping approaches are useful to trace the routes of infection and provide additional epidemiological information on strains recovered from food, animals and humans (Katsuda *et al.*, 2005).

Staphylococcal organisms are susceptible to the temperatures used in pasteurisation; however, toxins are not eliminated by these thermal approaches. *S. aureus* is resistant to high-pressure processing and where this approach is used, it may be helpful to include nisin in the process.

Although it is impractical to consider the elimination of staphylococci from foods of animal origin, good hygienic practice should contribute to minimise contamination, and these along with other controls should be included in hazard appraisal (analysis) critical control points (HACCP) plans.

6.3.5 Listeria monocytogenes

L. monocytogenes is a gram-positive bacterium that has been recognised as a significant foodborne pathogen for several decades, with increasing recent evidence of a previously undescribed gastroenteric syndrome. The organism is widely distributed in the environment and on occasion, it can be detected in all raw foods. The bacterium is difficult to eliminate from food; nevertheless, control measures have been effective in reducing the incidences of associated infections.

Milk and products derived from milk can harbour a variety of microorganisms, and is an important source of foodborne pathogens (Oliver *et al.*, 2005). The presence of foodborne pathogens often arises from contamination resulting from direct contact in the dairy environment and from excretion through the udder of an infected animal. Identification of the pathogen reservoir is an important step towards control, and in the USA, Borucki *et al.* (2004) estimated that *L. monocytogenes* accounted for 28% of foodborne infections. Of the six recognised species, *L. monocytogenes* and *Listeria innocua* commonly occur in food. The former strain is a human pathogen, with serogroups 1a, 2a or 4b being those most often encountered in the majority of outbreaks. Approximately one-quarter of the human cases could be linked to on-farm sources using standardized pulsed-field gel electrophoresis (PFGE) subtyping (Borucki *et al.*, 2004). In 2004, of the varieties of foods analysed in Hungary, *L. monocytogenes* was frequently detected in milk and dairy products (Kiss *et al.*, 2006). Serotypes ½a and 4b were the most common. In Pennsylvania, *L. monocytogenes* was detected in 2.8% for the milk samples analysed (Jayarao *et al.*, 2006). In the latter study, farm families often consumed raw milk, suggesting that the risk of infection for these individuals may be high compared with those consuming pasteurised milk. More recently, it was established that the distribution of *L. monocytogenes* strains may not be constant, suggesting that cycles of elimination and recontamination are occurring along the production line (Chambel *et al.*, 2007). Once introduced into the milking parlour or other food processing environments, *L. monocytogenes* is difficult to eradicate. The organism can readily colonise moist/wet environments. Also, the fact that *L. monocytogenes* can grow at low temperatures (0–4°C) implies that even with correct cold-chain control, the organisms may not be completely eliminated. Implementation of appropriate cleaning and disinfection protocols are essential. During the production of certain types of cheeses from raw milk, this could be the difference between a safe and contaminated product.

Several strategies have appeared in the scientific literature aimed at controlling the dissemination of *L. monocytogenes* along the food chain. Inhibition of growth using bacteriocin-producing strains has been evaluated recently (Achemchem *et al.*, 2006; Azuma *et al.*, 2007). *Enterococcus faecium* F58 and *Carnobacterium piscicola* CS526 were shown to be effective in controlling the growth of *L. monocytogenes*. In certain types of cheese production, growth of *L. monocytogenes* may be restricted through the metabolism of the natural resident bacterial flora (Millet *et al.*, 2006). Further studies are required to elucidate the key contributors. Single-strand conformational polymorphism (SSCP) may be useful in defining the important bacterial constituents (Saubusse *et al.*, 2007). As an alternative to biological control, physical approaches have involved the use of irradiation and high pressure. Ultraviolet light has been shown to reduce *L. monocytogenes* numbers in goat's milk (Matak *et al.*, 2005), whilst the use of high pressures was shown to reduce *L. innocua* by between 1.5 and 1.8 $\log_{10}$ when homogenised at 300 MPa at 24°C (Picart *et al.*, 2006).

6.3.6 Mycobacterium bovis

Mycobacteria are aerobic non-spore-forming acid-fast bacilli. Their lipid-rich cell walls confer a measure of resistance to environmental influences. Some pathogenic mycobacteria exhibit a host preference, whilst others can occasionally infect a number of species. Mycobacterial disease in domestic animals is usually chronic and progressive. The closely related members of the *Mycobacterium tuberculosis* complex (MTC), including *M. tuberculosis*, *M. bovis* and *Mycobacterium africanum*, all cause tuberculosis (TB) in humans.

M. bovis has a broad host range and is the principal aetiological agent of TB in domestic and wild animals. Epidemiological investigations demonstrated that this organism can also infect humans, and it can be ingested or inhaled, causing zoonotic TB. In the UK, zoonotic TB was transmitted to humans following the consumption of unpasteurised milk (de la Rua-Domenech, 2006), and recent data suggested that cases can also arise from inhalation of infectious air droplets. Following the consumption of infected milk, extra pulmonary lesions may develop (Thoen *et al.*, 2006). However, this route of infection has become less significant following the introduction of pasteurisation of milk and national eradication programmes in cattle. Nevertheless, there are some rural populations that may be at increased risk of infection through consumption of unpasteurised milk and other dairy products, along with occupational exposures (de la Rua-Domenech, 2006). In a recent US study, two-thirds of dairy farmer staff and their families consumed raw milk, and more than half of them reacted positively when a tuberculin test was administered (Winthrop *et al.*, 2005). Cheese imported into the USA was identified as a likely source of infection, suggesting that the importation of dairy products from countries where the bacterium is common in the cow population may pose an increased risk to public health (Anonymous 2005). Similarly, in Brazil, animal products were analysed and found to contain mycobacteria, posing a public health risk (Leite *et al.*, 2003). According to EU standards, the Netherlands is regarded as having an official bovine TB-free status. However, in recent years, a small number of bovine outbreaks occurred and several surveillance strategies were evaluated as part of a national control programme. The application of these protocols may be useful in helping to eliminate this organism from the food chain (Fischer *et al.*, 2005).

M. bovis can be detected using bacteriological culture techniques. In Argentina, the frequency of *M. bovis* detection in milk samples was determined using a combination of bacteriological and molecular methods (Perez *et al.*, 2002). Culture-based protocols are known to be slow, taking several weeks to yield a diagnostic result. In contrast, molecular approaches based on PCR offer a more rapid interpretation. Several PCR-based approaches have been described in the literature, and may be useful as an adjunct to culture methods (Antognoli *et al.*, 2001; Perez *et al.*, 2002; Zumarraga *et al.*, 2005).

6.3.7 Brucella abortus *and* Brucella melitensis

Brucella species are the principal causative agents of brucellosis, a bacterial zoonosis that can affect a diverse range of mammals including livestock and humans (Gorvel & Moreno, 2002; Celli, 2006). They are intracellular pathogens, highly infectious and capable of causing disease in both animals and humans. The most pathogenic strains associated with disease in humans are *B. abortus* and *B. melitensis*. Whilst *B. abortus* is generally associated with chronic recurrent episodic disease in human, *B. melitensis* can have more serious life-threatening consequences. In relation to food animals, cattle are most commonly associated with *B. abortus*, while *B. melitensis* is more commonly found in sheep and goats. Transmission to humans can be via entry through a cut or skin abrasion, inhalation of aerosols containing the organism, direct contact with infected blood placenta, fetuses or uterine secretions or through the consumption of infected foods of animal origin, including milk and milk products (Gupta *et al.*, 2006). Most cases of foodborne brucellosis in humans are contracted through the consumption of raw milk and dairy products, including cheeses manufactured from unpasteurised milk (Leclerc *et al.*, 2002; Anonymous, 2007). Geographical differences in the numbers of human cases caused by either *B. abortus* or *B. melitensis* have been reported, and this reflects regional differences in the prevalence of these causative agents in food animal populations. For example, most reported human cases of brucellosis in the USA are *B. melitensis* while in Europe many states are free of *B. melitensis*, and most cases are reported in southern and eastern regions (Anonymous, 2006, 2007). As the likelihood of human exposure to this aetiological agent is dependent on prevalences in wildlife and livestock reservoirs, most human cases occur in countries with less well-developed disease control and prevention programmes. As a result, the consumption of raw milk or dairy products made from raw milk represents a substantial risk to public health in countries or regions where *Brucella* spp. has not been eradicated from livestock. The organism is not particularly resistant to thermal processing and normal pasteurisation of milk is sufficient to destroy these organisms.

Determining the herd status or infection at individual level can be carried out by detecting antibodies in raw milk. The *Brucella* ring test or enzyme-linked immunosorbent assays (ELISA) are most commonly used. Although most of these tests are highly specific, problems with sensitivity may arise following dilution in the bulk tank, but increased sensitivity has been reported with the ELISA-based detection techniques (Vanzini *et al.*, 2001).

6.4 Chemical contaminants and residues

Milk is the physiological secretion of the mammary gland, and its constituents reflect various chemical equilibria in the host body. Toxic chemicals present in the animals' bodeis may be shed into the milk, resulting in a hazard to public health. The dynamic daily synthesis of fat by the mammary gland facilitates the accumulation of endogenous lipophilic compounds in milk. Chemical residues are remnants of purposeful additions to the food chain, while chemical contaminants arise without purposeful addition. In most cases, chemical residues and contaminants are resistant to degradation, and will not be affected by thermal treatments, such as pasteurisation, or the pH decreases following fermentation.

6.4.1 *Industrially derived contaminants*

Lactating ruminants tend to consume extensively grown forage and be relatively long-lived, with resultant potential for bioaccumulation of any environmental contaminants present (Rabinowitz *et al.*, 2005). Examples include dioxins, furans, polychlorinated biphenols (PCBs), elemental 'heavy' metals and radionuclides. In general terms, the risk to public health arises through chronic exposure and build-up of contaminants partitioning into tissues harvested for food. In contrast, and somewhat non-intuitively, acute toxicity with overt clinical disease in the animal represents a lower risk to public health due to likelihood of detection and exclusion of those animals from food production (Sharpe & Livesey, 2006).

Dioxins are a group of chemical compounds (congeners) inadvertently produced by many anthropogenic industrial activities, including metallurgical works, incineration and paper mulch bleaching. PCBs are a group of molecules purposefully synthesised for incorporation into, for example, industrial coolants, or plastic compounds. Dioxins and PCBs demonstrate remarkable resilience to environmental degradation, along with potential for lipophilic bioaccumulation in animal tissues. Consumption of food of animal origin represents the principal route of human exposure to dioxins and PCBs (Furst *et al.*, 1992), and consumption of cow's milk has been implicated as a significant source of human exposure to lipophilic contaminants, such as dioxins. The risk of such contaminants bio-accumulating in lactating animals leading to high levels in milk is directly related to the potential for exposure from emitters in the geographic area of feed production. Contamination of feed after it has been harvested has also been described, e.g. contaminated storage containers (Bernard *et al.*, 2002) or inadvertent contaminant production in the manufacture of feed supplements. Contamination of animal feed with heavy metals has been linked to local industrial activity, such as mining, as well as farming practices, such as the application of sewage sludge to agricultural land. Radionuclide emissions present similar potential for bioaccumulation in foods derived from grazing animals, following fallout over a feed-producing area.

Potential emitting industries should engage in environmental impact assessment prior to beginning production. Ongoing emission monitoring should be supplemented with surveillance of contaminant concentrations in tissues of animal sentinels, to verify total environmental load, and assess the attendant risk to public health in high-risk areas.

6.4.2 *Biologically derived contaminants*

Some potential chemical contaminants of milk arise from biological processes in the feed of the lactating animals, with subsequent potential for secretion into milk.

Mycotoxin is a term used to refer to a group of secondary metabolites of fungi with toxic effects in animals or man. Fungi growing on plants may produce such toxins as the plant is growing, or after harvest of the plants during storage prior to utilisation as animal feed (Driehuis & Oude Elferink, 2000). Mycotoxin ingestion can result in animal disease with overt clinical syndromes, as is the case with aflatoxicoses of turkeys (turkey X disease) or may be associated with a chronic insidious loss of productivity in food animals. Various factors, particularly mycotoxin load and animal age, and synergism amongst mycotoxins, may permit apparently health animals to shed mycotoxins in their milk (Yiannikouris & Juany, 2002). Aflatoxins produced by *Aspergillus* species of fungi are well studied, with biological transformation prior to shedding of a specific milk derivative, Alfatoxin M1, with carcinogenic and hepatotoxic potential (Sweeney *et al.*, 2000). Several other mycotoxins, particularly Ochratoxin, are effectively detoxified in the ruminant fore stomach and, hence, are low risk in milk derived from ruminant animals. The risk of mycotoxin accumulation is higher when animals are consuming feed which has been stored following harvesting. Monitoring of mycotoxins in stored feed represents an important risk management strategy. Approaches to keeping fungal growth and toxin elaboration low include maintaining low a_w of feed, use of fungal inhibitors, such as organic acid, and cultivation of resistant plant varieties. Potential carcinogenicity of Aflatoxin M1 has resulted in zero-tolerance in many regulatory frameworks.

Phytotoxins are naturally expressed substances in plants. Toxicity in grazing animals is well described in cases, such as pyrrolizidine in ragworth, ptaquiloside in bracken, or glucosinolate in brassicas. Potential for these toxins to arise in milk of lactating animals consuming these plants represents a poorly understood risk (Panter & James, 1990; EFSA, 2007). One specific example is a hepatotoxicity syndrome of 'milk sickness' described in people in southern and mid-western USA, which is associated with milkborne tremetol and tremetone toxins due to the consumption of plants, such as white snakeroot or rayless goldenrod. The highest risk of phytotoxins of public health significance arises when animals are relatively resistant, as in the case of sheep and pyrrolizidine. Similarly, point-source milk supply and subsistence agriculture with one animal supplying all milk, for example to one family, presents opportunity for potentially high exposure for a small number of individuals. The grazing management and dilution effects involved in modern intensive dairying minimise phytotoxic milkborne risks.

6.4.3 *Pesticides and residues of plant health agrichemicals*

The development of a range of pesticides during the second half of the twentieth century resulted in dramatic increases in crop yields and has significantly reduced plant disease. They can be broadly divided into three sub-categories: insecticides, herbicides and fungicides. These chemicals can have harmful effects and cause acute or chronic illness in humans. The range of adverse effects includes damage to nervous tissue, lungs, reproductive organs, immune and endocrine systems, birth defects and cancer. Toxicities of the various classes of chemicals used as pesticides vary widely and have been well documented (Hura *et al.*, 1999; Mansour, 2004; Kan & Meijer, 2007). The use of the most toxic of these agents (e.g. organochlorines including DDT) have been banned in developed countries since the 1970s; however, they are still marketed and used in many developing countries. As some of the commonly used chemicals are stable, they can persist in the environment in some instances for years, and the main exposure route for humans to these substances is via the consumption of contaminated foods, including those of animal origin. Concentration of these substances can increase through the food chain as a result of bioaccumulation, and higher concentrations have frequently been found in animal-fat-rich foods, including milk (John *et al.*, 2001).

Many of these chemical contaminants are now almost ubiquitous in the environment, and as a result, they can also be present in animal feeds and depending on their physicochemical characteristics, may be metabolised into naturally occurring and generally harmless molecules or, as is more frequent with pesticides, may persist in the animal and animal products (Kan & Meijer, 2007). The more persistent and toxic pesticides have been detected in milk and dairy products, and clearly demonstrate the potential risk to public health (Mallatou *et al.*, 1997; Martinez *et al.*, 1997; Waliszewski *et al.*, 1997; Battu *et al.*, 2004). Furthermore, chemical hazards are extremely thermostable and remain unchanged in foods following cooking or other heat treatments (including pasteurisation, ultra-high temperature (UHT) or sterilisation of milk).

As agricultural chemicals are intentionally used in food production, their use can generally be controlled by legislation and voluntary codes of practice (Buncic, 2006). However, levels in feeds and foods can vary considerably based on geographical location and local regulations pertaining to their licensing and use. Effective control and risk management in foods of animal origin including milk is based around monitoring and surveillance of levels of these compounds in animal feeds, their raw constituents (particularly where imported from regions where highly toxic pesticides are still in use) and in food animals and in food products.

6.4.4 *Residues of animal remedies*

Pharmaceutical substances administered to lactating animals, may arise in milk in low concentrations, known as residues. Administration of medicines to food animals should be followed by a withdrawal period, during which food should not be harvested, in order to permit concentration in tissues to reach acceptable residue levels. Milk is a tissue harvested for human consumption for continuous periods of animals' lives. Administration of remedies to lactating animals will, therefore, result in a requirement to actively remove harvested food from human consumption, in contrast to, for example meat production where harvesting of food is merely postponed following treatment. Infectious mastitis remains a common indication for medicinal therapy of lactating animals (Hillerton & Berry, 2005). Dairying may, in many instances, involve intramammary administration of sustained-delivery antimicrobial formulations to all animals at drying-off, regardless of infection status. Blanket medicinal administration and variable depletion kinetics in diseased udders create potential for violative residues. Licensing regimens for veterinary pharmaceuticals require allocation of acceptable daily intake to various tissues, and consequently, frequent absence of licence for lactating animals. Extra-label utilisation of formulations not licensed in lactating animals creates definite potential for milkborne residues.

While perceptions of significant risk are generally attributed to residues of chemotherapeutic substances, the actual nature of the risks is difficult to quantify. Antimicrobials are compounds selected for remarkably low toxicity to mammals, so direct toxicity arising from their residues generally represents a negligible risk. However, low levels of antimicrobial products in milk may be associated with induction of hypersensitivity in consumers. Similarly, speculation exists on the residual antimicrobial activity in milk, even low levels with consequential risk of altering consumer microbial flora or selection for antimicrobial resistance. Milkborne antimicrobial residues create definite potential for interference with post-harvest biological processing such as fermentation in cheese manufacture. Other substances administered to animals for anthelmintic, analgesic, anti-inflammatory or theriogenologic reasons carry their specific risks to consumers if present in food. Much speculation and very little evidence exist on risks to public health arising from chronic low-level exposure due to levels below defined maximal residue limits.

Risks of residues in milk are managed through prudent use of medicines in lactating animals. Medicine usage should be minimised through non-chemotherapeutic approaches to disease management, e.g. vaccination. Active compliance with withdrawal periods is possible only with reliable animal identification and accurate treatment records. Withdrawal periods should be actively prolonged when depletion kinetics may be affected, e.g. intramammary administration into chronically infected glands with sequestration and fibrosis. Potential for contamination of animal feed with milk or urine from treated animals should also be addressed. Milk may be screened for 'inhibitory' antimicrobial residues using various microbial inhibition assays. Further chemical analysis, such as HPLC, is required to characterise the substance and quantity present.

6.5 Poorly understood and emerging hazards

Scientific understanding of milk safety is a dynamic and rapidly evolving area. Newly emergent hazards or increasing awareness of zoonotic potential associated with some milkborne hazards result in a group of hazards whose risk to public health is difficult to accurately characterise.

6.5.1 *Bovine spongiform encephalopathy (BSE)*

Transmissible spongiform encephalopathies (TSEs) are a group of fatal degenerative non-inflammatory neurological diseases, associated with proteinaceous agents, prions. The epidemic of BSE has been implicated in an analogous epidemic of a novel TSE in human, new variant Creutzfeldt-Jakob disease.

Ongoing questions remain regarding the nature of BSE (Capobianco *et al.*, 2007) and the role of prions in pathogenesis of TSEs (Manuelidis *et al.*, 2007) and ongoing zoonotic potential of BSE (Bishop *et al.*, 2006). However, there is widespread acceptance of a risk to public health arising from BSE, which was largely a disease of dairy cows, and prions have demonstrated remarkable resilience to thermal degradation, so would be unaffected by pasteurisation. There have been some indications of transmissibility associated with peripheral blood (Hunter *et al.*, 2002), raising concerns regarding peripheral leucocytes in milk.

Early in vitro studies demonstrated no evidence of transmission of BSE to mice following oral, intraperitoneal or intracerebral administration of milk from clinical cases of BSE (Taylor *et al.*, 1995). Whilst the UK dairy calf cohort study demonstrated some indications of enhanced maternal risk in offspring of dairy cattle with BSE, evidence was not consistent with milkborne transmission (Wilesmith *et al.*, 1997). Further evidence for the absence of a role for milk in transmission of diseases to offspring was found in a study of suckler (non-dairy) cows, none of whose calves developed BSE (Wilesmith & Ryan, 1997). More recent studies have also failed to find evidence of infectivity in affected cows milk (Everest *et al.*, 2006).

The safety of cows' milk with regard to BSE has been the subject of extensive reviews (Vetrugno *et al.*, 2004; TAFS, 2007; Tyshenko *et al.*, 2007) with a general conclusion of negligible risk to consumers. Also of relevance is the rapidly waning epidemic of BSE due to successful control measures, i.e. the slaughter of all BSE suspects out of food production, with very low residual incidence. Milk is harvested from live animals with non-existent risk of contamination of milk with specified risk material, such as neural tissues.

In summary, BSE is a disease whose transmissibility is not associated with milk.

6.5.2 Mycobacterium avium *subsp.* paratuberculosis

M. avium subsp. *paratuberculosis* (MAP) is the aetiological agent of paratuberculosis in ruminants and other animals. Paratuberculosis or Johne's disease in cows is an economically important disease especially in dairy cattle. The disease is of major global importance and the Office International des Epizooties (OIE) classifies it as a List B transmitted disease of socio-economic and/or public health importance. Johne's disease is a chronic infectious enteritis caused by MAP. For some time now, MAP has been suspected of involvement in chronic inflammatory changes in the human gastrointestinal tract leading to Crohn's disease, and more recently, there has been a suggestion of its involvement in Type 1 diabetes mellitus (Dow, 2006). Although considered to be autoimmune in origin, there is increasing evidence that Crohn's disease may have an infectious cause (Greenstein, 2003). The pathology of both diseases is similar; however, evidence remains inconclusive regarding a definitive link. If MAP has a role in the pathogenesis of Crohn's disease, then it is possible that infection may arise through food or water (Grant, 2005). In a recent case-control study in the UK, 218 Crohn's disease patients assessed, results could not conclusively support a role for water or dairy products contaminated with MAP in the aetiology of the disease (Abubakar *et al.*, 2007).

MAP has been cultured from the milk of clinically and sub-clinically infected bovines (O'Reilly *et al.*, 2004). The organism can also be cultured from the milk of other ruminants, including sheep and goats, affected by Johne's disease. The route of contamination can arise directly either from MAP infected milk in the udder or from faecal contamination during milking. The organism is not reproducibly eliminated by standard pasteurisation (Grant, 1998; Grant *et al.*, 1998). MAP is certainly more heat-resistant compared to other mycobacteria (including *M. bovis*) and on occasion low numbers may survive pasteurisation. Factors including the initial numbers of the organism together with its ability to clump may be important (Grant, 2005). Revised pasteurisation conditions required to eliminate MAP may lead to changes in the organoleptic properties of milk, leading to a reduction in consumer acceptance of the milk and derived products. The Food Standards Agency (FSA) in the UK has adopted a control strategy, on the basis of the available scientific data, to reduce human exposure to the organism (FSA, 2006).

In other dairy products, (e.g. cheese) made from raw or minimally treated milk, Donaghy *et al.* (2004) recovered numbers of MAP that were concentrated tenfold, and were culturable for up to 27 weeks during the ripening period. In a more recent study, Stephan *et al.* (2007) examined 143 Swiss cheeses from retail outlets, which were made using raw milk, for the presence of MAP by culture and molecular methods. No viable MAP was recovered, but 4.2% of the samples were found to be positive by molecular methods.

Reduced lactation performance occurs in animals infected with MAP (Raizman *et al.*, 2007). The negative economic impacts associated with infection may be a positive motivator towards the development and implementation of effective control programmers at herd level. It can be expected that without management approaches designed to reduce or eliminate MAP at farm level (Lombard *et al.*, 2005), the prevalence of Johne's disease will continue to impact milk production. Understanding the epidemiology of infection in dairy herds will also be important. The role of non-ruminant animals should also be considered (Palmer *et al.*, 2005; Judge *et al.*, 2006). Establishing useful testing strategies may form part of the management process (Nielsen & Ersboll 2006). Veterinarians are being encouraged (Lombard *et al.*, 2006) to implement environmental monitoring strategies as an aid to determine the infectious status of a herd.

6.5.3 Enterobacter sakazakii

E. sakazakii is an opportunistic pathogen, the aetiological agent in rare life-threatening cases of meningitis, necrotising enterocolitis and sepsis in infants (Forsythe, 2005; Drudy *et al.*, 2006; Mullane *et al.*, 2007a). Contaminated powdered infant formula (PIF) manufactured from milk is the recognised food vehicle, identified in several epidemiological studies wherein the bacterium kills an estimated 40–80% of infected infants (Bowen & Braden, 2006). Following a number of reported outbreaks, arising from neonates being fed rehydrated PIF the manufacturers and public health authorities alike are now exploring ways to reduce growth of the organism in PIF and limit transmission (Gurtler *et al.*, 2005). In health care institutions and the home, since PIF is not a sterile product, proper precautions should be taken during handling and reconstitution of formula prior to feeding in order to prevent contamination and proliferation of the bacterium. Foods other than PIF have rarely been examined for the presence of the organism; nevertheless *E. sakazakii* can be detected in various foods and there is a growing recognition of its presence in a diverse range of food matrices (Friedemann, 2007).

Edelson-Mammel *et al.* (2005) showed that *E. sakazakii* is well adapted to survive for long periods (>2 years) in PIF. Fifty-six *E. sakazakii* strains were examined (Lehner *et al.*, 2005) for characteristics normally associated with survival in high osmotic and dry conditions. Many of these isolates could form biofilms and possessed those biological characteristics that would contribute to their successful survival under the extreme conditions normally associated with milk-derived powders. In a study which examined thermotolerance (Breeuwer *et al.*, 2003), it was concluded that these organisms are not especially heat-resistant, when compared to other *Enterobacteriaceae*. However, stationary phase *E. sakazakii* was more adapted to osmotic and dry stress conditions, compared with *E. coli* and *Salmonella* spp. This feature was likely linked to the accumulation of trehalose thereby facilitating survival. Edelson-Mammel & Buchanan (2004) also examined the thermotolerance of this organism using kinetic approaches, and suggested that two distinct heat-resistant phenotypes exist. However, based on these and other data, it is difficult to get a clear view as to the ability of this organism to survive under the various temperatures tested, due in part to a lack of standardisation of approach and reported thermo-tolerant differences among strains at the protein level (Williams *et al.*, 2005).

Several studies have investigated the prevalence of *E. sakazakii* in PIF (Iversen & Forsythe, 2004). Muytjens *et al.* (1988) analysed 141 PIFs from 35 countries, and isolated *E. sakazakii* at a frequency of 14.2% representing 13 different countries. The organism was recovered from PIFs at levels ranging from 0.36 to 0.66 colony-forming units (cfu) 100 g^{-1} sample. More recently, the WHO reported a prevalence of $<1.8\%$ in 6 out of 12 documented studies (FAO-WHO, 2006), while a study of follow-on formula from five different South-East Asian manufacturers found *E. sakazakii* in 13.5% of powders analysed (Estuningsih *et al.*, 2006).

Development of rapid, sensitive and specific detection methods will facilitate the dairy industry's efforts to reduce the occurrence of *E. sakazakii* in added ingredients and the final powdered product. As a result of the growing recognition of *E. sakazakii* as a pathogen associated with PIF, the International Dairy Federation (IDF) and the International Organisation for Standardisation (ISO) standardised a reference method for the detection of the organism in milk powdered products and PIF (IDF, 2006; ISO, 2006). This method was recently evaluated using a number of differential and selective agars (Besse *et al.*, 2006; see also Iversen & Forsythe, 2007).

The manufacturing environment serves as a key route for sporadic contamination of PIF (Mullane *et al.*, 2007b). Molecular subtyping of *E. sakazakii* recovered as part of an environmental monitoring programme can facilitate the development of HACCP-based control strategies to limit transmission of the pathogen. In addition, others have demonstrated the effectiveness of high-pressure treatment (600 MPa for 1 min) to achieve 3 through 6.84 $\log_{10}$ reductions in cell numbers (Gonzalez *et al.*, 2006). Gamma irradiation may also be effective (Lee *et al.*, 2006). Caprylic acid is a naturally occurring constituent of human and bovine milk, and is approved for use by the US Food and Drug Administration (USFDA). Nair *et al.* (2004) showed that monocaprylin could be used to inactivate *E. sakazakii* in reconstituted PIF, no data were published showing whether or not the organoleptic properties of the food were altered. Evidence has also emerged suggesting that antibacterial peptides, derived from proteolysis of sodium caseinate, may be effective against planktonic *E. sakazakii* (Hayes *et al.*, 2006). However, the latter study was limited in that it used only an *E. sakazakii*-type strain to demonstrate the antibacterial activity and did not include environmental or clinical isolates.

6.5.4 *Antimicrobial resistance*

Antimicrobial resistance is a biological hazard resulting in increased morbidity and mortality. Use of antimicrobials to treat and control infectious diseases in animals is a common management tool; however, the indiscriminate use of these valuable agents gives rise to resistance among bacterial communities, including commensal organisms and pathogens alike. Consequently, the transfer of antimicrobial resistance from animals to humans remains a concern for both animal and public health medicine (Sischo, 2006). Milk and products derived from milk may contain zoonotic pathogens, some of which may be resistant to one or more antimicrobial agents. Antimicrobial susceptibility monitoring systems, such as the National Antimicrobial Resistance Monitoring System (NARMS) for enteric bacteria in the USA and DANMAP in Denmark, provide valuable information on resistance traits among foodborne bacteria. DANMAP is unique in that it collects data on both antimicrobial consumption and resistance in bacteria from animals, food and humans and provides a temporal relationship between usage and resistance (Bager, 2000).

The impact of antimicrobial use on resistance has been examined on dairy farm holdings, in milk and milk products (Makovec & Ruegg, 2003; Loch *et al.*, 2005; Munsch-Alatossava & Alatossava, 2007). In Washington State, the use of antimicrobials along with on-farm biosecurity measures was described in a recent study (Raymond *et al.*, 2006). One of its objectives was an attempt to limit the dependence on antimicrobials. Commonly used drugs to treat animal disease included penicillin, ceftiofur and oxytetracycline. In addition, 28% of dairy producers indicated that they used medicated milk replacer as a prophylactic measure. Similar drugs were also used in Pennsylvania when 113 dairy herds across 13 counties were studied (Sawant *et al.*, 2005). Results of this study indicated a significant dependence on antimicrobials for both therapeutic and prophylactic purposes. Variation on the protocols for drug use also existed.

Antimicrobial susceptibility patterns among *Campylobacter* spp. cultured from dairy animals on conventional and organic farms were studied (Halbert *et al.*, 2006a). Four hundred fifty and 1570 *Campylobacter* isolates cultured from organic and conventionally run farm operations, respectively, were reported to be susceptible to the antimicrobial agents tested. In that study, isolates from 128 conventional farms had a higher proportion of resistant isolates. Food-producing animals are known to play a role in the dissemination of antimicrobial resistance to humans and when *Campylobacter* isolates were investigated for genetic resistance markers, *tetO* was identified in isolates from both farm types (Halbert *et al.*, 2006b). Although *Campylobacter* spp. would be more commonly associated with poultry flocks, dairy animals are now becoming increasingly recognised as a risk to public health. In a large monitoring programme carried out in the USA, across 21 states, 735 *Campylobacter* isolates were cultured from the faeces of 1435 cows representing 96 dairy operations (Englen *et al.*, 2007). Generally, low levels of resistance were noted in isolates from dairy cattle; however, *C. coli* appeared to be more resistant compared to *C. jejuni*.

Other gram-negative organisms (e.g. *Salmonella* spp.) have also been cultured from milk and milk products (Cody *et al.*, 1999; Carneiro *et al.*, 2003; Berge *et al.*, 2004). In a study of dairy herds supplying liquid milk to the consumer market, 6% of the cow herds were positive for *Salmonella* spp. (Murphy *et al.*, 2008). Ten isolates were recovered following a herd-level surveillance programme. With a previously developed screening tool using in-line milk filters to recover organic matter from the milking line (Murphy *et al.*, 2005), woven cotton filters were enriched in appropriate culture media and *Salmonella* spp. were isolated; serotype typhimurium was the most common type recovered. These isolates demonstrated the characteristic penta-resistant phenotype, i.e. against ampicillin, chloramphenicol, spectinomycin, sulphonamide and tetracycline (ACSSuT), commonly associated with *S. typhimurium*. In a related study, cow and goat herds supplying milk for artisan cheese production were screened using the previous approach (Murphy *et al.*, 2007). A number of verocytotoxigenic *E. coli* (VTEC) strains including serotypes O157 and O26 were recovered, and interestingly, none of the cheese products manufactured from this milk supply were positive for VTEC. When these isolates were assessed for their susceptibility to antimicrobials, three of the *E. coli* O26 isolates were resistant to ampicillin, cephalothin and sulfamethoxazole/trimethoprim. All *E. coli* O157 isolates were susceptible. Interestingly, the prevalence of antimicrobial resistant bacteria is highest among younger animals and this situation may not be the direct result of selective pressure (Khachatryan *et al.*, 2004). Dairy calves are often fed milk from cows treated with antibiotics (Langford *et al.*, 2003). In a dose–response study, resistance to penicillin G increased with increasing concentrations of drug. It appears that drug-resistant traits linked to streptomycin, sulphonamide and tetracycline (SSuT) phenotype may involve environmental components independent of drug selective pressure (Khachatryan *et al.*, 2006). In studies that compared the duration of faecal *E. coli* O157 being shed by 1-week-old Holstein calves fed with a milk replacer containing or devoid of antibiotics (neomycin and oxytetracycline), supplementation of the replacer appeared to have a short impact of low magnitude (Alali *et al.*, 2004).

Mastitis is an economically important disease of dairy cattle, and *S. aureus* among others is a major pathogen. Antimicrobial resistance patterns of major mastitis causing pathogens cultured from milk samples were investigated over a number of years (Makovec & Ruegg, 2003). These data did not show any significant trend towards increased resistance among mastitis pathogens from dairy cows. In a more recent study (overlapping with the previous report by some 6 months), a nationwide study conducted in Norway and involving 3538 sets of 4 quarter milk samples taken from 2834 dairy cows during 2000 was reported (Osteras *et al.*, 2006). Milk culture data showed that *S. aureus* was the most prevalent isolate recovered, decreasing in prevalence throughout days in milk. However, a seasonal effect was noted which was related to resistance to penicillin G. Despite the apparent susceptibility of these strains to antimicrobials, the cure of diseased animals is rather disappointing (Melchior *et al.*, 2006), leading to recurrent infections of varying severity. This has been attributed to the growth of *S. aureus* in a biofilm, and under these conditions, this organism is highly resistant to antimicrobial agents. Widespread distribution of genes-mediating resistance to quaternary ammonium compounds (QAC) used in disinfectants have also been reported (Bjorland *et al.*, 2005) in staphylococci cultured from raw cow's and goat's milk. This may reflect the dissemination of these efflux-associated genes on resistance plasmids. When conventional and organic farms were compared in a Swiss study (Roesch *et al.*, 2006), there was no difference in the frequency of antibiotic resistance among mastitis-causing pathogens, suggesting that production factors may be responsible for these observations, and these need to be evaluated. Milk and dairy products are often contaminated with enterotoxigenic *S. aureus* strains. Anderson *et al.* (2006) investigated the relationship between resistant strains of *S. aureus* and their corresponding DNA profiles, generated by PFGE. Whilst resistance among mastitis-causing *S. aureus* isolated from milk samples was uncommon, a limited number of genotypes were identified, and these drug resistant strains were associated with particular PFGE types. In a 3-year study of meat and dairy products, 1634 samples were examined and 12.8% were contaminated with *S. aureus*. Over two-thirds were resistant to one or more antimicrobial agents. These data provide evidence of the emergence of drug-resistant toxin-producing strains, and it has been suggested that improvements in food controls should be devised to address this problem (Normanno *et al.*, 2007). Recently, methicillin-resistant *Staphylococcus aureus* (MRSA) has been identified in milk samples taken from Hungarian herds with sub-clinical mastitis, and compared to isolates contemporaneously cultured from those working in close contact with these animals (Juhasz-Kaszanyitzky *et al.*, 2007). Phenotypic and genotypic comparisons of the human and animal strains were indistinguishable. This is the first report of animal-to-human transmission of a Panton-Valentine leukocidin (PVL)-negative staphylococcal cassette chromosome (SCC)*mec*-type IVa strain, which was different from the PVL-positive SCC*mec*-type IVg cow's milk strains identified earlier in South Korea (Kwon *et al.*, 2005).

Enterococci are part of the dominant microflora of several dairy products. Occurrence of these strains in dairy produce is probably the result of faecal contamination during milking. A collection of *E. faecium* strains, recovered from dairy products, sheep's faeces and clinical sources, were examined for their resistance to a panel of antibiotics along with their associated virulence characteristics (Mannu *et al.*, 2003). Although similar susceptibility patterns were noted from both sources, isolates from cheese and sheep's milk were less pathogenic when compared with clinical strains. Of the faecal and bulk milk tank samples from 222 Murciano-Granadina dairy goats screened microbiologically, enterococci were recovered more often from goat kids (Cortes *et al.*, 2006) compared with older animals. A total of 17 of 134 isolates were resistant to at least 7 antibiotics, and 14 of these were vancomycin resistant. In Costa Rica, 8% of *Enterococcus* strains recovered from non-pasteurised milk samples were similarly resistant (Araya *et al.*, 2005).

Other gram-positive bacteria were cultured from pasteurised milk, and assessed for their resistance to a range of antibiotics (Perrin-Guyomard *et al.*, 2005). Two *Micrococcaceae* strains were found to contain the *ermC* gene encoding resistance to erythromycin, and of the corynebacteria analysed, none were erythromycin resistant. When specific genetic markers of resistance were investigated, the *ermX* gene was detected, but no *erm(C)* gene could be identified. These authors demonstrated the transfer from staphylococcal strains to *Enterococcus faecalis* and *S. aureus*.

Multi-drug-resistant zoonotic pathogens are a potential threat to public health, if these organisms gain access to the food chain. Some microorganisms, including MRSA are more generally recognised as nosocomial pathogens, but increasingly these are now being associated with animal reservoirs and steps to eliminate them from the food chain are now being reported. Increasing resistance to antimicrobials has led to a renewed search for novel antimicrobials. The use of naturally occurring antimicrobial agents elaborated by some gram-positive strains, such as the nisin-Z-producing *Lactococcus lactis* subsp. *lactis*, have been evaluated as a means of eliminating MRSA (Rilla *et al.*, 2004). Using defined culture conditions, *S. aureus* was undetectable in milk and cheese. Specific peptidoglycan hydrolases or endolysins have also been explored as a useful means of reducing numbers of *S. aureus* and *Streptococcus agalactiae* (Donovan *et al.*, 2006).Transgenic animals expressing these antimicrobials would be resistant to mastitis.

Naturally occurring antimicrobial peptides, bacteriocins, produced by some gram-positive organisms have also been used to target *Listeria* spp. often found in artisanal cheeses (Martinez *et al.*, 2005). A number of these inhibitory bacteriocins were assessed and pediocin PA-1 was the most active against the strains tested. However, some *L. monocytogenes* and *L. innocua* were highly resistant to PA-1. Stable nisin-resistant wild-type *L. monocytogenes* were detected and which exhibited cross-resistance to lysozyme, EDTA, NaCl at various pH levels. These strains could survive and grow in milk in the presence of nisin-producing *Lactococcus* strains. It was difficult to predict how nisin resistance would develop in a dairy environment. These data should be carefully evaluated as part of a risk assessment plan if bacteriocins are being used as a control measure against *L. monocytogenes* (Martinez *et al.*, 2005).

The dairy farm environment and animals are reservoirs of commensal and pathogenic bacteria that may carry antibiotic resistance determinants. These can gain access to the human food chain by several routes. Straley *et al.* (2006) reviewed the prevalence and role of previously unrecognised commensal gram-negative enteric organisms in the bulk milk tank. These organisms are subjected to the same selective pressures as are pathogens in these environments, and their contribution to the evolution of resistance is only now being evaluated. There is increasing evidence that these indicator bacteria are becoming a significant route of resistance gene transmission (Schlegelova *et al.*, 2002).

6.6 Risk management strategies to ensure safety of liquid milk

6.6.1 *Pre-harvest risk management*

Milk is a secretion of animals whose microbial and chemical content directly reflects the environment in which the animals are farmed. Controls in the pre-harvest phase of dairying provide significant potential for ensuring milk safety (Collins & Wall, 2004).

Feed safety assurance is a pivotal part of pre-harvest dairy assurance. Concentrates should be produced from high-quality ingredients and stored in a manner minimising contamination by wild animals or growth of deleterious fungi. Risk of recycling enteric pathogens through manure spreading onto grazed pasture should be minimised through passive or active manure treatment. Pasture preservation should be preformed with regard to safety and quality of feed. Only potable water should be provided for drinking by animals, and water troughs should be designed to minimise contamination and maintained in a hygienic state.

Herd biosecurity approaches should minimise ingress of pathogens, such as *Mycobacterium* or *Brucella* species. The prevalence of some enteric pathogens, such as *Salmonella* species, can be minimised by ensuring general overall animal health. Potential for contamination at harvest of milk by faeces containing asymptomatically shed enteric pathogens, such as *Campylobacter* species or VTECs, may be minimised by ensuring animal hygiene with adequate bedding and management of parasitic enteritis. Udder health should be monitored, e.g. by somatic cell counts, with control programmes for organisms, such as *Staphylococcus* species. Antimicrobials should only be used in a targeted therapeutic manner to minimise selection pressures for resistant organisms. All animal remedies should be used prudently, with identification and records of treated lactating animals and active withdrawal of food that might contain residues.

6.6.2 *Harvest phase risk management*

Harvest of milk can never succeed in producing sterile food. However, regard to hygiene can help minimise contamination from the general production environment, particularly faeces. Lactating animals should be maintained in a hygienic manner, with faecal soiling of the udder, tail and perineal area not allowed to accumulate. Visibly contaminated teats should be cleaned and dried prior to milking. Safe potable water should be provided in the dairy production area. Clusters should be fitted with high-flow vacuum cut-off valves to minimise aspiration of ground water if they fall off. Lactating animals treated with any products requiring withdrawal of milk from human consumption should be clearly identifiable, and milked last to prevent contamination with milk containing residues. Milk should be screened prior to storage to remove any particulate matter, such as faecal or hair material. Milk should be cooled rapidly and stored on farm in a manner minimising contamination.

6.6.3 *Post-harvest risk management*

Following harvest, it is necessary to control a number of parameters to ensure the safety and the quality of liquid milk. Milk must be stored hygienically on farm in bulk tanks until its collection and delivery to the processing plant. The cooling and maintenance of low temperature during storage, transport and holding at the plant are also essential to product safety and quality. This retards or prevents the proliferation of spoilage organisms and any pathogenic bacteria that might be present in the raw milk. For example within the EU, milk must be cooled immediately to 8°C or less if collected from farms daily, or 6°C or less if not collected daily. The temperature of milk must not be more than 10°C following transportation to the processing plant where on arrival it must be cooled to 6°C or less and stored at that temperature until processed (Anonymous, 2004). The microbiological quality of raw milk samples is also assessed from delivery vehicles and individual suppliers. Again in the EU, microbiological and other quality criteria have been established in the legislation, and must be enforced at national level by the relevant competent authorities. Within the EU, raw cow's milk must have a somatic cell counts of no more than 400 000 ml^{-1} with many dairy companies paying a price premium for milk supplies with a lower somatic cell counts (Small, 2006). In addition, the aerobic plate count (APC) at 30°C of raw cow's milk must not exceed 100 000 cfu ml^{-1}.

For most markets, liquid milk is then homogenised, which involves passing the liquid under high pressure through a very small aperture (typically in the region of 20.6 MPa, 210 kg cm^{-2} or 3000 lb $in.^{-2}$) (Small, 2006). This process irreversibly breaks up fat globules in the milk so that they are more uniformly dispersed throughout the liquid, and prevents subsequent formation of a layer of butterfat or cream on the top of the liquid during storage. This facilitates processing of a more uniform and palatable product, and is preferred by consumers in most developed countries.

Liquid milk is then normally heat treated to destroy any pathogenic microorganisms present. In many countries, legislation forbidding the sale of raw liquid milk has been introduced as a specific public health measure. The minimum heat treatment required to eliminate bacterial pathogens is pasteurisation, and under most commercial conditions, the milk is heated to a temperature of 71.7°C and held at that temperature for 15 s. The process is normally carried out using plate heat exchangers, which ensures rapid heat transfer between the separated heating/cooling medium and the milk. Rapid heating to the required temperature followed by rapid cooling once the holding period has elapsed ensures that the taste and other intrinsic organoleptic properties in the milk are preserved. Other more severe heat treatments can be used where extended shelf life of the product is required. These include UHT treatment where milk is 'flash' heated to 135°C before being cooled rapidly. Milk may also be sterilised, which involves in-container heating to over 100°C, usually between 115 and 120°C for 15–20 min. These more 'severe' heat treatments give significant increases in shelf life, and unopened products normally do not require refrigeration; however, they do cause a deterioration in taste with the development of 'cooked' flavours.

Once heat treated, it is imperative that post-process contamination is prevented, which could compromise either product quality or safety. Liquid milk should be cooled to below 10°C and aseptically filled into suitable containers along with record keeping of heat-processing parameters and other information to ensure food chain traceability. Testing to ensure that pasteurisation or other heat treatments have been performed satisfactorily should also be performed. This can be carried out by examining for the presence of phosphatase and peroxidase enzymes, which naturally occur in raw milk. Phosphatase is relatively heat sensitive, and is denatured by pasteurisation. Adequately pasteurised milk should, therefore, test negative for this enzyme. Peroxidase is a more heat-stable enzyme than phosphatase, and can be used to differentiate between raw or pasteurised milk and UHT or sterilised milk.

6.7 Summation

Milk is a safe and nutritious food. It should be harvested, processed and handled as an organic matrix capable of contamination and elaboration of hazards of public health significance.

References

Abubakar, I., Myhill, D.J., Hart, A.R., Lake, I.R., Harvey, I., Rhodes, J.M., Robinson, R., Lobo, A.J., Probert, C.S. & Hunter, P.R. (2007) A case-control study of drinking water and dairy products in Crohn's disease-further investigation of the possible role of *Mycobacterium avium paratuberculosis*. *American Journal of Epidemiology*, **165**, 776–783.

Achemchem, F., Abrini, J., Martinez-Bueno, M., Valdivia, E. & Maqueda, M. (2006) Control of *Listeria monocytogenes* in goat's milk and goat's jben by the bacteriocinogenic *Enterococcus faecium* F58 strain. *Journal of Food Protection*, **69**, 2370–2376.

Adak, G.K., Meakins, S.M., Yip, H., Loopman, B.J. & O'Brien, S.J. (2005) Disease risks from foods, England and Wales, 1996–2000. *Emerging Infectious Disease*, **11**, 365–372.

Alali, W.Q., Sargeant, J.M., Nagaraja, T.G. & DeBey, B.M. (2004) Effect of antibiotics in milk replacer on fecal shedding of *Escherichia coli* O157:H7 in calves. *Journal of Animal Science*, **82**, 2148–2152.

Allerberger, F., Wagner, M., Schweiger, P., Rammer, H-P., Resch, A., Dierich, M.P., Friedrich, A.W. & Karch, H. (2001) *Escherichia coli* O157 infections and unpasteurised milk. *Eurosurveillance*, **6**, 147–151.

Anderson, K.L., Lyman, R.L., Bodeis-Jones, S.M. & White, D.G. (2006) Genetic diversity and antimicrobial susceptibility profiles among mastitis-causing *Staphylococcus aureus* isolated from bovine milk samples. *American Journal of Veterinary Research*, **67**, 1185–1191.

Anonymous (2004) Corrigendum to Regulation (EC) No 853/2004 of the European Parliament and of the Council of 29 April 2004 laying down specific hygiene rules for the hygiene of foodstuffs. *Official Journal of the European Union*, **L226**, 22–82.

Anonymous (2005) Human tuberculosis caused by *Mycobacterium bovis*-New York city, 2001–2004. *MMWR Morbidity and Mortality Weekly Report*, **54**, 605–608.

Anonymous (2006) *Zoonoses in the European Union – 2005*, European Food Safety Authority. Available at http://www.efsa.europa.eu/etc/medialib/efsa/science/monitoring_zoonoses/reports/zoonoses_report_2005.Par.0001.File.dat/Zoonoses_report_2005.pdf.

Anonymous (2007) *Brucellosis*, Centres for Disease Control and Prevention. Available at http://www.cdc.gov/ncidod/dbmd/diseaseinfo/brucellosis_g.htm.

Antognoli, M.C., Salman, M.D., Triantis, J., Hernandez, J. & Keefe, T. (2001) A one-tube nested polymerase chain reaction for the detection of *Mycobacterium bovis* in spiked milk samples: an evaluation of concentration and lytic techniques. *Journal of Veterinary Diagnostic Investigation*, **13**, 111–116.

Araya, M., Davidovich, G., Arias, M.L. & Chaves, C. (2005) Identification of *Enterococcus* sp. isolated from raw milk samples coming from the metropolitan area of Costa Rica and evaluation of its antibiotic sensibility pattern. *Archives of Latinoamerican Nutrition*, **55**, 161–166.

Asao, T., Kumeda, Y., Kawai, T., Shibata, T., Oda, H., Haruki, K., Nakazawa, H. & Kozaki, S. (2003) An extensive outbreak of staphylococcal food poisoning due to low-fat milk in Japan: estimation of enterotoxin A in the incriminated milk and powdered skim milk. *Epidemiology and Infection*, **130**, 33–40.

Atkins, P.J. (1992) White poison? The social consequences of milk-borne infection. *Social History of Medicine*, **5**, 207–227

Azuma, T., Bagenda, D.K., Yamamoto, T., Kawai, Y. & Yamazaki, K. (2007) Inhibition of *Listeria monocytogenes* by freeze-dried piscicocin CS526 fermentate in food. *Letters in Applied Microbiology*, **44**, 138–144.

Bager, F. (2000) DANMAP: monitoring antimicrobial resistance in Denmark. *International Journal of Antimicrobial Agents*, **14**, 271–274.

Battu, R.S., Singh, B. & Kang, B.K. (2004) Contamination of liquid milk and butter with pesticide residues in the Ludhiana district of Punjab state, India. *Ecotoxicology and Environmental Safety*, **59**, 324–331.

Berge, A.C., Adaska, J.M. & Sischo, W.M. (2004) Use of antibiotic susceptibility patterns and pulsed-field gel electrophoresis to compare historic and contemporary isolates of multi-drug-resistant *Salmonella enterica* subsp. *enterica* serovar Newport. *Applied and Environmental Microbiology*, **70**, 318–323.

Bernard, A., Broeckaert, F., de Poorter, G., de Cock, A., Hermans, C., Saegerman, C. & Houins, G. (2002) The Belgian PCB/Dioxin incident: analysis of the food chain contamination and health risk evaluation. *Environmental Research Section A*, **88**, 1–18.

Besse, N.G., Leclercq, A., Maladen, V., Tyburski, C. & Bertrand, L. (2006) Evaluation of the International Organization for Standardization-International Dairy Federation (ISO-IDF) draft standard method for detection of *Enterobacter sakazakii* in powdered infant food formulas. *Journal of the American Organisation of Analytical Chemists International*, **89**, 1309–1316.

Bishop, M.T. Hart. P., Aitchinson, L., Baybutt, H.N., Plinston, C., Thomson, V., Tuzi, N.L., Head, M.W., Ironside, J.W., Will, R.G. & Manson, J.C. (2006) Predicting susceptibility and incubation time of human-to-human transmission of vCJD. *Lancet Neurology*, **5**, 393–398.

Bjorland, J., Steinum, T., Kvitle, B., Waage, S., Sunde, M. & Heir, E. (2005) Widespread distribution of disinfectant resistance genes among staphylococci of bovine and caprine origin in Norway. *Journal of Clinical Microbiology*, **43**, 4363–4368.

Bonner, C., Foley, B., Wall, P.G. & Fitzgerald M. (2001) Analysis of outbreaks of infectious intestinal disease in Ireland: 1998 and 1999. *Irish Medical Journal*, **94**, 140–142.

Borucki, M.K., Reynolds, J., Gay, C.C., McElwain, K.L., Kim, S.H., Knowles, D.P. & Hu, J. (2004) Dairy farm reservoir of *Listeria monocytogenes* sporadic and epidemic strains. *Journal of Food Protection*, **67**, 2496–2969.

Bowen, A.B. & Braden, C.R. (2006) Invasive *Enterobacter sakazakii* disease in infants. *Emerging Infectious Disease*, **12**, 1185–1189.

Breeuwer, P., Lardeau, A., Peterz, M. & Joosten, H.M. (2003) Desiccation and heat tolerance of *Enterobacter sakazakii*. *Journal of Applied Microbiology*, **95**, 967–973.

Buncic, S. (2006) On-farm phase in the context of the food chain. In: *Integrated Food Safety and Veterinary Public Health* (ed. S. Buncic), pp. 26–31, CABI, Oxford.

Capobianco, R., Casalone, C, Suardi, S., Mangieri, M., Miccolo, C., Limido, L., Catania, M., Rossi, G., Di Fede, G., Giaccone, G., Grazia Bruzzone, M., Minati, L., Corona, C., Acutis, P., Gelmetti, D., Lombardi, G., Groschup, M.H., Buschmann, A., Zanusso, G., Monaco, S., Caramelli, M. & Tagliavini, F. (2007) Conversion of the BASE prion strain into the BSE strain: the origin of BSE? *PLos Pathogens*, **3**(e31), 1–8.

Carneiro, L.A., Silva, A.P., Merquior, V.L. & Queiroz, M.L. (2003) Antimicrobial resistance in Gram-negative bacilli isolated from infant formulas. *FEMS Microbiology Letters*, **228**, 175–179.

Celli, J. (2006) Surviving inside a macrophage: the many ways of *Brucella*. *Research in Microbiology*, **157**, 93–98.

Chambel, L., Sol, M., Fernandes, I., Barbosa, M., Zilhao, I., Barata, B., Jordan, S., Perni, S., Shama, G., Adriao, A., Faleiro, L., Requena, T., Pelaez, C., Andrew, P.W. & Tenreiro, R. (2007) Occurrence and persistence of *Listeria* spp. in the environment of ewe and cow's milk cheese dairies in Portugal unveiled by an integrated analysis of identification, typing and spatial-temporal mapping along production cycle. *International Journal of Food Microbiology*, **116**, 384–390.

Cody, S.H., Abbott, S.L., Marfin, A.A., Schulz, B., Wagner, P., Robbins, K., Mohle-Boetani, J.C. & Vugia, D.J. (1999) Two outbreaks of multidrug-resistant *Salmonella* serotype *typhimurium* DT104 infections linked to raw-milk cheese in Northern California. *Journal of the American Medical Association*, **281**, 1805–1810.

Coia, J.E. (1998) Clinical, microbiological and epidemiological aspects of *Escherichia coli* O157 infection. *FEMS Immunology and Medical Microbiology*, **20**, 1–9.

Collins, J.D. & Wall, P.G. (2004) Food safety and animal production systems: controlling zoonoses at farm level. *Scientific and Technical Review of the OIE*, **23**, 685–700.

Cornick, N.A., Booher, S.L., Casey, T.A. & Moon, H.W. (2000) Persistent colonisation of sheep by *Escherichia coli* O157:H7 and other *E. coli* pathotypes. *Applied and Environmental Microbiology*, **66**, 4926–4934.

Cortes, C., De la Fuente, R., Blanco, J., Blanco, M., Blanco, J.E., Dhabi, G., Mora, A., Justel, P., Contreras, A., Sanchez, A., Corrales, J.C. & Orden, J.A. (2005) Serotypes, virulence genes and intimin types of verotoxin-producing *Escherichia coli* and enteropathogenic *E. coli* isolated from healthy dairy goats in Spain. *Veterinary Microbiology*, **110**, 67–76.

Cortes, C., De la Fuente, R., Contreras, A., Sanchez, A., Corrales, J.C., Ruiz-Santa-Quiteria, J.A. & Orden, J.A. (2006) Occurrence and preliminary study of antimicrobial resistance of enterococci isolated from dairy goats in Spain. *International Journal of Food Microbiology*, **110**, 100–103.

Cremonesi, P., Luzzana, M., Brasca, M., Morandi, S., Lodi, R., Vimercati, C., Agnellini, D., Caramenti, G., Moroni, P. & Castiglioni, B. (2005) Development of a multiplex PCR assay for the identification of *Staphylococcus aureus* enterotoxigenic strains isolated from milk and dairy products. *Molecular and Cell Probes*, **19**, 299–305.

de la Rua-Domenech, R. (2006) Human *Mycobacterium bovis* infection in the United Kingdom: Incidence, risks, control measures and review of the zoonotic aspects of bovine tuberculosis. *Tuberculosis (Edinburgh)*, **86**, 77–109.

Delbes, C., Alomar, J., Chougui, N., Martin, J.F. & Montel, M.C. (2006) *Staphylococcus aureus* growth and enterotoxin production during the manufacture of uncooked, semihard cheese from cows' raw milk. *Journal of Food Protection*, **69**, 2161–2167.

Deschênes, G., Casenave, C., Grimont, F., Desenclos, J.C., Benoit, S., Collin, M., Baron, S., Mariani, P., Grimont, P.A-D. & Nivet, H. (1996) Cluster of cases of haemolytic uraemic syndrome due to unpasteurised cheese. *Pediatric Nephrology*, **10**, 203–204.

Djuretic, T., Wall, P.G. & Nicholas, T. (1997) General outbreaks of infectious intestinal disease associated with milk and dairy products in England and Wales: 1992 to 1996. *Communicable Disease Reports CDR Review*, **7**, R41–R45.

Donaghy, J.A., Totton, N.L. & Rowe, M. (2004) Persistence of *Mycobacterium paratuberculosis* during manufacture and ripening of Cheddar cheese. *Applied and Environmental Microbiology*, **70**, 4899–4905.

Donovan, D.M., Dong, S., Garrett, W., Rousseau, G.M., Moineau, S. & Pritchard, D.G. (2006) Peptidoglycan hydrolase fusions maintain their parental specificities. *Applied and Environmental Microbiology*, **72**, 2988–2996.

Dow, C.T. (2006) Paratuberculosis and type I diabetes: is this the trigger? *Medical Hypotheses*, **67**, 782–785.

Driehuis, F. & Oude Elferink S.J. (2000) The impact of the quality of silage on animal health and food safety: a review. *Veterinary Quarterly*, **22**, 212–216.

Drudy, D., Mullane, N.R., Quinn, T., Wall, P.G. & Fanning, S. (2006) *Enterobacter sakazakii*: an emerging pathogen in powdered infant formula. *Clinical Infectious Disease*, **42**, 996–1002.

Duffy, G., Garvey, P. & Sheridan, J.J. (2002) *A European Study on Animal Food and Biomedical Aspects of* E. coli *O157 : H7*. Final Report of an EU FAIR Agriculture and Fisheries Fourth Framework programme, Teagasc, The National Food Centre, Ireland.

Edelson-Mammel, S.G. & Buchanan, R.L. (2004) Thermal inactivation of *Enterobacter sakazakii* in rehydrated infant formula. *Journal of Food Protection*, **67**, 60–63.

Edelson-Mammel, S.G., Porteous, M.K. & Buchanan, R.L. (2005) Survival of *Enterobacter sakazakii* in a dehydrated powdered infant formula. *Journal of Food Protection*, **68**, 1900–1902.

EFSA (2007) Opinion of the Scientific Panel CONTAM related to pyrrolizidine alkaloids as undesirable substances in animal feed. *The EFSA Journal*, **447**, 1–51. Available at http://www.efsa.europa.eu/etc/medialib/efsa/science/contam/contam_opinions/ej447_pyrrolizidine.Par.0001.File.dat/contam_ej447_op_pyrrolizidine%20alkaloids%20in%20feed_en.pdf.

Englen, M.D., Hill, A.E., Dargatz, D.A., Ladely, S.R. & Fedorka-Cray, P.J. (2007) Prevalence and antimicrobial resistance of *Campylobacter* in US dairy cattle. *Journal of Applied Microbiology*, **In press**.

Estuningsih, S., Kress, C., Hassan, A.A., Akineden, O., Schneider, E. & Usleber, E. (2006) *Enterobacteriaceae* in dehydrated powdered infant formula manufactured in Indonesia and Malaysia. *Journal of Food Protection*, **69**, 3013–3017.

Everest, S.J., Thorne, L.T., Hawthorn, J.A., Jenkins, R., Hammersley, C., Ramsay, A.M., Hawkins, S.A., Venables, L., Flynn, L., Sayers, R., Kilpatrick, J., Sach, A., Hope, J. & Jackman, R. (2006) No abnormal prion protein detected in the milk of cattle infected with the bovine spongiform encephalopathy agent. *Journal of General Virology*, **87**, 2433–2441.

FAO-WHO (2006) *Enterobacter sakazakii* and *Salmonella* in powdered infant Formula. Second Risk Assessment Workshop, Food and Agriculture Organization of the United Nations, Rome.

Fischer, E.A., van Roermund, H.J., Hemerik, L., van Asseldonk, M.A. & de Jong, M.C. (2005) Evaluation of surveillance strategies for bovine tuberculosis (*Mycobacterium bovis*) using an individual based epidemiological model. *Preventative Veterinary Medicine*, **67**, 283–301.

Forsythe, S.J. (2005) *Enterobacter sakazakii* and other bacteria in powdered infant milk formula. *Maternal and Child Nutrition*, **1**, 44–50.

Fremaux, B., Raynaud, S., Beutin, L. & Vernozy Rozand, C. (2006) Dissemination and persistence of Shiga toxin-producing *Escherichia coli* (STEC) strains on French dairy farms. *Veterinary Microbiology*, **117**, 180–191.

Friedemann, M. (2007) *Enterobacter sakazakii* in food and beverages (other than infant formula and milk powder). *International Journal of Food Microbiology*, **116**, 1–10.

FSA (Food Standards Agency) (2006) Strategy for the control of *Mycobacterium avium* subspecies *paratuberculosis* (MAP) in cow's milk. Available at http://www.food.gov.uk/multimedia/pdfs/map_strategy.pdf.

Fujikawa, H. & Morozumi, S. (2006) Modelling *Staphylococcus aureus* growth and enterotoxin production in milk. *Food Microbiology*, **23**, 260–267.

Furst, P., Beck, H. & Theelan, R.M.C. (1992) Assessment of human intake of PCDDs and PCDFs from different environmental sources. *Toxic Substances Journal*, **12**, 133–150.

Galbraith, N.S., Forbes, P. & Clifford, C. (1982) Communicable diseases associated with milk and dairy products in England and Wales 1951–80. *British Medical Journal (Clinical Research Edition)*, **284**, 1761–1765.

Gillespie, I.A., Adak, G.K., O'Nrien, S.J. & Bolton, F.J. (2003) Milkborne general outbreaks of infectious intestinal disease in England and Wales, 1992–2000. *Epidemiology and Infection*, **130**, 461–468.

Gonzalez, S., Flick, G.J., Arritt, F.M., Holliman, D. & Meadows, B. (2006) Effect of high-pressure processing on strains of *Enterobacter sakazakii*. *Journal of Food Protection*, **69**, 935–937.

Gorvel, J.P. & Moreno, E. (2002) *Brucella* intracellular life: from invasion to intracellular replication. *Veterinary Microbiology*, **90**, 281–297.

Grant, I.R. (1998) Does *Mycobacterium paratuberculosis* survive current pasteurisation conditions? *Applied and Environmental Microbiology*, **64**, 2760–2761.

Grant, I.R. (2005) Zoonotic potential of *Mycobacterium avium* ssp. *paratuberculosis*: the current position. *Journal of Applied Microbiology*, **98**, 1282–1293.

Grant, I.R., Ball, H.J. & Rowe, M. (1998) Isolation of *Mycobacterium paratuberculosis* from milk by immunomagnetic separation. *Applied and Environmental Microbiology*, **64**, 3153–3158.

Greenstein, R.J. (2003) Is Crohn's disease caused by a mycobacterium? Comparisons with leprosy, tuberculosis and Johne's disease. *The Lancet*, **3**, 507–514.

Gupta, V.K., Verma, D.K., Rout, P.K., Singh, S.V. & Vihan, V.S. (2006) Polymerase chain reaction (PCR) for detection of *Brucella melitensis* in goat milk. *Small Ruminant Research*, **65**, 79–84.

Gurtler, J.B., Kornacki, J.L. & Beuchat, L.R. (2005) *Enterobacter sakazakii*: a coliform of increased concern to infant health. *International Journal of Food Microbiology*, **104**, 1–34.

Halbert, L.W., Kaneene, J.B., Linz, J., Mansfield, L.S., Wilson, D., Ruegg, P.L., Warnick, L.D., Wells, S.J., Fossler, C.P., Campbell, A.M. & Geiger-Zwald, A.M. (2006a) Genetic mechanisms contributing to reduced tetracycline susceptibility of *Campylobacter* isolated from organic and conventional dairy farms in the midwestern and northeastern United States. *Journal of Food Protection*, **69**, 482–488.

Halbert, L.W., Kaneene, J B., Ruegg, P L., Warnick, L.D., Wells, S.J., Mansfield, L.S., Fossler, C.P., Campbell, A.M. & Geiger-Zwald, A.M. (2006b) Evaluation of antimicrobial susceptibility patterns in *Campylobacter* spp isolated from dairy cattle and farms managed organically and conventionally in the midwestern and northeastern United States. *Journal of the American Veterinary Medical Association*, **228**, 1074–1081.

Hayes, M., Ross, R.P., Fitzgerald, G.F., Hill, C. & Stanton, C. (2006) Casein-derived antimicrobial peptides generated by *Lactobacillus acidophilus* DPC6026. *Applied and Environmental Microbiology*, **72**, 2260–2264.

Hillerton, J.E. & Berry, E.A. (2005) Treating mastitis in the cow – a tradition or an archaism. *Journal of Applied Microbiology*, **98**, 1250–1255.

Holsinger, V.H., Rajkowski, K.T. & Stabel J.R. (1997) Milk Pasteurisation: a brief history and update. *Scientific and Technical Review of the OIE*, **16**, 441–451.

Hunter, N., Foster, J., Chong, A., McCutcheon, S., Parnham, D., Eaton, S., MacKenzie, C. & Houston, F. (2002) Transmission of prion diseases by blood transfusion. *Journal of General Virology*, **83**, 2897–2905.

Hura, C., Leanca, M., Rusu, L. & Hura, B.A. (1999) Risk assessment of pollution with pesticides in food in the Eastern Romania area (1996–1997). *Toxicology Letters*, **107**, 103–107.

IDF (2006) *Milk and Milk Products – Detection of* Enterobacter sakazakii, Standard No. 210, International Dairy Federation, Boulevard Auguste Reyers 80, Brussles.

ISO (2006) *Milk and Milk Products – Detection of* Enterobacter sakazakii, ISO / DTS 22964, ISO Case Postal 56, CH-1211 Geneva 20.

Iversen, C. & Forsythe, S. (2004) Isolation of *Enterobacter sakazakii* and other *Enterobacteriaceae* from powdered infant formula and related products. *Food Microbiology*, **21**, 771–776.

Iversen, C. & Forsythe, S. (2007) Comparison of media for the isolation of *Enterobacter sakazakii*. *Applied and Environmental Microbiology*, **73**, 48–52.

Jayarao, B.M., Donaldson, S.C., Straley, B.A., Sawant, A.A., Hegde, N.V. & Brown, J.L. (2006) A survey of foodborne pathogens in bulk tank milk and raw milk consumption among farm families in Pennsylvania. *Journal of Dairy Science*, **89**, 2451–2458.

Jayarao, B.M. & Henning, B.R. (2001) Prevalence of foodborne pathogens in bulk tank milk. *Journal of Dairy Science* **84**, 2157–2162.

John, P.J., Bakore, N. & Bhatnagar, P. (2001) Assessment of organochlorine pesticide residue levels in dairy milk and buffalo milk from Jaipur city, Rajasthan, India. *Environmental International*, **26**, 231–236.

Jorgensen, H.J., Mork, T., Caugant, D.A., Kearns, A. & Rorvik, L.M. (2005) Genetic variation among *Staphylococcus aureus* strains from Norwegian bulk milk. *Applied and Environmental Microbiology*, **71**, 8352–8361.

Judge, J., Kyriazakis, I., Greig, A., Davidson, R.S. & Hutchings, M.R. (2006) Routes of intraspecies transmission of *Mycobacterium avium* subsp. *paratuberculosis* in rabbits (*Oryctolagus cuniculus*): a field study. *Applied and Environmental Microbiology*, **72**, 398–403.

Juhasz-Kaszanyitzky, E., Janosi, S., Somogyi, P., Dan, A., Van Der Graff-van Bloos, L., van Duijkeren, E. & Wagnaar, J.A. (2007) MRSA transmission between cows and humans. *Emerging Infectious Disease*, **14**, 630–632.

Kan, C.A. & Meijer, G.A.L. (2007) The risk of contamination of food with toxic substances present in animal feed. *Animal Feed Science and Technology*, **133**, 84–108.

Katsuda, K., Hata, E., Kobayashi, H., Kohmoto, M., Kawashima, K., Tsunemitsu, H. & Eguchi, M. (2005) Molecular typing of *Staphylococcus aureus* isolated from bovine mastitic milk on the basis of toxin genes and coagulase gene polymorphisms. *Veterinary Microbiology*, **105**, 301–305.

Khachatryan, A.R., Besser, T.E., Hancock, D.D. & Call, D.R. (2006) Use of a nonmedicated dietary supplement correlates with increased prevalence of streptomycin-sulfa-tetracycline-resistant *Escherichia coli* on a dairy farm. *Applied and Environmental Microbiology*, **72**, 4583–4588.

Khachatryan, A.R., Hancock, D.D., Besser, T.E. & Call, D.R. (2004) Role of calf-adapted *Escherichia coli* in maintenance of antimicrobial drug resistance in dairy calves. *Applied and Environmental Microbiology*, **70**, 752–757.

Kiss, R., Tirczka, T., Szita, G., Bernath, S. & Csiko, G. (2006) *Listeria monocytogenes* food monitoring data and incidence of human listeriosis in Hungary, 2004. *International Journal of Food Microbiology*, **112**, 71–74.

Kwon, N.H., Park, K.T., Moon, J.S., Jung, W.K., Kim, S.H., Kim, J.M., Hong, S.K., Koo, H.C., Joo, Y.S. & Park, Y.H. (2005) Staphylococcal cassette chromosome mec (SCCmec) characterization and molecular analysis for methicillin-resistant *Staphylococcus aureus* and novel SCCmec subtype IVg isolated from bovine milk in Korea. *Journal of Antimicrobial Chemotherapy*, **56**, 624–632.

Langford, F.M., Weary, D.M. & Fisher, L. (2003) Antibiotic resistance in gut bacteria from dairy calves: a dose response to the level of antibiotics fed in milk. *Journal of Dairy Science*, **86**, 3963–3966.

Leclerc, V., Dufour, B., Lombard, B., Gauchard, F., Garin-Bastuji, B., Salvat, G., Brisabois, A., Poumeyrol, M., De Buyser, M.-L., Gnanou-Besse, N. & Lahellec, C. (2002) Pathogens in meat and milk products: surveillance and impact on human health in France. *Livestock Production Science*, **76**, 195–202.

Lee, J.W., Oh, S.H., Kim, J.H., Yook, H.S. & Byun, M.W. (2006) Gamma radiation sensitivity of *Enterobacter sakazakii* in dehydrated powdered infant formula. *Journal of Food Protection*, **69**, 1434–1437.

Lehner, A., Riedel, K., Eberl, L., Breeuwer, P., Diep, B. & Stephan, R. (2005) Biofilm formation, extracellular polysaccharide production, and cell-to-cell signaling in various *Enterobacter sakazakii* strains: aspects promoting environmental persistence. *Journal of Food Protection*, **68**, 2287–2294.

Leite, C.Q., Anno, I.S., Leite, S R., Roxo, E., Morlock, G.P. & Cooksey, R.C. (2003) Isolation and identification of mycobacteria from livestock specimens and milk obtained in Brazil. *Memórias do Instituto Oswaldo Cruz*, **98**, 319–323.

Loch, I.M., Glenn, K. & Zadoks, R.N. (2005) Macrolide and lincosamide resistance genes of environmental streptococci from bovine milk. *Veterinary Microbiology*, **111**, 133–138.

Lombard, J.E., Garry, F.B., McCluskey, B.J. & Wagner, B.A. (2005) Risk of removal and effects on milk production associated with paratuberculosis status in dairy cows. *Journal of the American Veterinary Medical Association*, **227**, 1975–1981.

Lombard, J.E., Wagner, B.A., Smith, R.L., McCluskey, B.J., Harris, B.N., Payeur, J.B., Garry, F.B. & Salman, M.D. (2006) Evaluation of environmental sampling and culture to determine *Mycobacterium avium* subspecies *paratuberculosis* distribution and herd infection status on US dairy operations. *Journal of Dairy Science*, **89**, 4163–4171.

Loncarevic, S., Jorgensen, H.J., Lovseth, A., Mathisen, T. & Rorvik, L.M. (2005) Diversity of *Staphylococcus aureus* enterotoxin types within single samples of raw milk and raw milk products. *Journal of Applied Microbiology*, **98**, 344–350.

Maher, M.M., Jordan, K.N., Upton, M.E. & Coffey, A. (2001) Growth and survival of *E. coli* O157:H7 during manufacture and ripening of a smear-ripened cheese produced from raw milk. *Journal of Applied Microbiology*, **90**, 201–207.

Makovec, J.A. & Ruegg, P.L. (2003) Antimicrobial resistance of bacteria isolated from dairy cow milk samples submitted for bacterial culture: 8905 samples (1994–2001). *Journal of the American Veterinary Medical Association*, **222**, 1582–1589.

Mallatou, H., Pappas, C.P., kondyli, E. & Albanis, T.A. (1997) Pesticide residues in milk and cheese from Greece. *The Science of the Total Environment*, **196**, 111–117.

Mannu, L., Paba, A., Daga, E., Comunian, R., Zanetti, S., Dupre, I. & Sechi, L.A. (2003) Comparison of the incidence of virulence determinants and antibiotic resistance between *Enterococcus faecium* strains of dairy, animal and clinical origin. *International Journal of Food Microbiology*, **88**, 291–304.

Mansour, S.A. (2004) Pesticide exposure – Egyptian scene. *Toxicology*, **198**, 91–115.

Manuelidis, L., Yu, Z.-X., Banquero, N. & Mullins, B. (2007) Cells infected with scrapie and Creutzfeldt-Jakob disease agents produce intracellular. *Proceedings of National Academy of Sciences*, **104**, 1965–1970.

Martinez, M.P., Angulo, R., Pozo, R. & Jodral, M. (1997). Organochlorine pesticides in pasteurised milk and associated health risks. *Food and Chemical Toxicology*, **35**, 621–624.

Martinez, B., Bravo, D. & Rodriguez, A. (2005) Consequences of the development of nisin-resistant *Listeria monocytogenes* in fermented dairy products. *Journal of Food Protection*, **68**, 2383–2388.

Matak, K.E., Churey, J.J., Worobo, R.W., Sumner, S.S., Hovingh, E., Hackney, C.R. & Pierson, M.D. (2005) Efficacy of UV light for the reduction of *Listeria monocytogenes* in goat's milk. *Journal of Food Protection*, **68**, 2212–2216.

McIntyre, L., Fung, J., Paccagnella, A., Isaac-Renton, J., Rockwell, F., Emerson, B. & Preston, T. (2002) *Escherichia coli* O157 outbreak associated with the ingestion of unpasteurised goat's milk in British Columbia, 2001. *Canada Communicable Disease Report*, **28**(1), 6–8.

Melchior, M.B., Fink-Gremmels, J. & Gaastra, W. (2006) Comparative assessment of the antimicrobial susceptibility of *Staphylococcus aureus* isolates from bovine mastitis in biofilm versus planktonic culture. *Journal of Veterinary Medicine B Infectious Disease and Veterinary Public Health*, **53**, 326–332.

Millet, L., Saubusse, M., Didienne, R., Tessier, L. & Montel, M.C. (2006) Control of *Listeria monocytogenes* in raw-milk cheeses. *International Journal of Food Microbiology*, **108**, 105–114.

Minihan, D., Whyte, P., O' Mahony, M., McGill, K., Fanning, S. & Collins, J.D. (2004) *Campylobacter* spp. in Irish feedlot cattle: a longitudinal study involving pre-harvest and harvest phases of the food chain. *Journal of Veterinary Medicine Series B*, **51**, 28–33.

Morgan, D., Nawman C.P., Hutchinson, D.N., Walkes, A.M., Rowe, B. & Maijd, F. (1993) Verotoxin-producing *Escherichia coli* O157:H7 infections associated with the consumption of yoghurt. *Epidemiology and Infection*, **111**, 181–187.

Muehlherr, J.E., Zweifel, C., Corti, S., Blanco, J.E. & Stephan, R. (2003) Microbiological quality of raw goat's and ewe's bulk-tank milk in Switzerland. *Journal of Dairy Science*, **86**, 3849–3856.

Mullane, N.R., Iversen, C., Healy, B., Walsh, C., Whyte, P., Wall, P.G., Quinn, T. & Fanning, S. (2007a) *Enterobacter sakazakii*: an emerging bacterial pathogen with implications for infant health. *Minerva Pediatrica*, **59**, 137–148.

Mullane, N.R., Whyte, P., Wall, P.G., Quinn, T. & Fanning, S. (2007b) Application of pulsed-field gel electrophoresis to characterise and trace the prevalence of *Enterobacter sakazakii* in an infant formula processing facility. *International Journal of Food Microbiology*, **116**, 73–81.

Munsch-Alatossava, P. & Alatossava, T. (2007) Antibiotic resistance of raw-milk-associated psychrotrophic bacteria. *Microbiology Research*, **162**, 115–123.

Murphy, B.P., Buckley, J.F., O'Connor, E.M., Gilroy, D. & Fanning, S. (2008) Comparison of *Salmonella* species recovered from Irish liquid milk production holdings with temporal clinical veterinary isolates. *International Journal of Hygiene and Environmental Health*, **211**, 283–291.

Murphy, M., Buckley, J.F., Whyte, P., O'Mahony, M., Anderson, A. & Fanning S. (2007) Surveillance of dairy production holdings supplying raw milk to the farmhouse cheese sector for verocytotoxin producing *Escherichia coli* O157 and other VTEC serotypes. *Zoonoses and Public Health*, **54**, 358–365.

Murphy, B.P., Murphy, M. Buckley, J. Gilroy, D. Rowe, M. McCleery, D. & Fanning S. (2005) In-line milk filter analysis: *Escherichia coli* O157 surveillance of milk production holdings. *International Journal of Hygiene and Environmental Health*, **208**, 407–414.

Muytjens, H.L., Roelofs-Willemse, H. & Jaspar, G.H. (1988) Quality of powdered substitutes for breast milk with regard to members of the family *Enterobacteriaceae*. *Journal of Clinical Microbiology*, **26**, 743–746.

Nair, M.K., Joy, J. & Venkitanarayanan, K.S. (2004) Inactivation of *Enterobacter sakazakii* in reconstituted infant formula by monocaprylin. *Journal of Food Protection*, **67**, 2815–2819.

Nam, H.M., Murinda, S.E., Nguyen, L.T. & Oliver, S.P. (2004) Evaluation of universal pre-enrichment broth for isolation of *Salmonella* spp., *Escherichia coli* O157:H7, and *Listeria monocytogenes* from dairy farm environmental samples. *Foodborne Pathogens and Disease*, **1**, 37–44.

Nielsen, S.S. & Ersboll, A.K. (2006) Age at occurrence of *Mycobacterium avium* subspecies *paratuberculosis* in naturally infected dairy cows. *Journal of Dairy Science*, **89**, 4557–4566.

Normanno, G., La Salandra, G., Dambrosio, A., Quaglia, N.C., Corrente, M., Parisi, A., Santagada, G., Firinu, A., Crisetti, E. & Celano, G.V. (2007) Occurrence, characterization and antimicrobial resistance of enterotoxigenic *Staphylococcus aureus* isolated from meat and dairy products. *International Journal of Food Microbiology*, **115**, 290–296.

Oliver, S.P., Jayarao, B.M. & Almeida, R.A. (2005) Food borne pathogens in milk and the dairy farm environment: food safety and public health implications. *Foodborne Pathogens and Disease*, **2**, 115–129.

Omoe, K., Ishikawa, M., Shimoda, Y., Hu, D.L., Ueda, S. & Shinagawa, K. (2002) Detection of *seg*, *seh*, and *sei* genes in *Staphylococcus aureus* isolates and determination of the enterotoxin productivities of *S. aureus* isolates Harboring *seg*, *seh*, or *sei* genes. *Journal of Clinical Microbiology*, **40**, 857–862.

O'Reilly, C.E., O'Connor, L., Anderson, W., Harvey, P., Grant, I.R., Donaghy, J., Rowe, M. & O'Mahony, P. (2004) Surveillance of bulk raw and commercially pasteurised cows' milk from approved Irish liquid milk pasteurising plants for *Mycobacterium paratuberculosis*. *Applied and Environmental Microbiology*, **70**, 5138–5144.

Osteras, O., Solverod, L. & Reksen, O. (2006) Milk culture results in a large Norwegian survey–effects of season, parity, days in milk, resistance, and clustering. *Journal of Dairy Science*, **89**, 1010–1023.

Palmer, M.V., Stoffregen, W.C., Carpenter, J.G. & Stabel, J.R. (2005) Isolation of *Mycobacterium avium* subsp *paratuberculosis* (Map) from feral cats on a dairy farm with Map-infected cattle. *Journal of Wildlife Disease*, **41**, 629–635.

Panter, K.E. & James, L.F. (1990) Natural plant toxicants in milk: a review. *Journal of Animal Science*, **68**, 892–904.

Perelle, S., Dilasser, F., Grout, J. & Fach, P. (2007) Screening food raw materials for the presence of the world's most frequent clinical cases of Shiga toxin-encoding *Escherichia coli* O26, O103, O111, O145 and O157. *International Journal of Food Microbiology*, **113**, 284–288.

Perez, A., Reniero, A., Forteis, A., Meregalli, S., Lopez, B. & Ritacco, V. (2002) Study of *Mycobacterium bovis* in milk using bacteriological methods and the polymerase chain reaction. *Reviews of Argentinian Microbiology*, **34**, 45–51.

Perrin-Guyomard, A., Soumet, C., Leclercq, R., Doucet-Populaire, F. & Sanders, P. (2005) Antibiotic susceptibility of bacteria isolated from pasteurized milk and characterization of macrolide-lincosamide-streptogramin resistance genes. *Journal of Food Protection*, **68**, 347–352.

Picart, L., Thiebaud, M., Rene, M., Pierre Guiraud, J., Cheftel, J.C. & Dumay, E. (2006) Effects of high pressure homogenisation of raw bovine milk on alkaline phosphatase and microbial inactivation – a comparison with continuous short-time thermal treatments. *Journal of Dairy Research*, **73**, 454–463.

Rabinowitz, P.M., Gordon, Z., Holmes, R., Taylow, B., Wilcox, M., Chudnov, D., Nadkarni, P. & Dein, F.J. (2005) Animals as sentinels of human environmental health hazards: an evidence-based analysis. *Eco-Health*, **2**, 26–37.

Raizman, E.A., Fetrow, J., Wells, S.J., Godden, S.M., Oakes, M.J. & Vazquez, G. (2007) The association between *Mycobacterium avium* subsp. *paratuberculosis* fecal shedding or clinical Johne's disease and lactation performance on two Minnesota, USA dairy farms. *Preventative Veterinary Medicine*, **78**, 179–195.

Raymond, M.J., Wohrle, R.D. & Call, D.R. (2006) Assessment and promotion of judicious antibiotic use on dairy farms in Washington State. *Journal of Dairy Science*, **89**, 3228–3240.

Rea, M.C., Cogan, T.M. & Tobin S. (1992) Prevalence of pathogenic bacteria in raw milk in Ireland. *Journal of Applied Bacteriology*, **73**, 331–336.

Rilla, N., Martinez, B. & Rodriguez, A. (2004) Inhibition of a methicillin-resistant *Staphylococcus aureus* strain in Afuega'l Pitu cheese by the nisin Z-producing strain *Lactococcus lactis* subsp. *lactis* IPLA 729. *Journal of Food Protection*, **67**, 928–933.

Roesch, M., Perreten, V., Doherr, M.G., Schaeren, W., Schallibaum, M. & Blum, J.W. (2006) Comparison of antibiotic resistance of udder pathogens in dairy cows kept on organic and on conventional farms. *Journal of Dairy Science*, **89**, 989–997.

Saubusse, M., Millet, L., Delbes, C., Callon, C. & Montel, M.C. (2007) Application of single strand conformation polymorphism – PCR method for distinguishing cheese bacterial communities that inhibit *Listeria monocytogenes*. *International Journal of Food Microbiology*, **116**, 126–135.

Sawant, A.A., Sordillo, L.M. & Jayarao, B.M. (2005) A survey on antibiotic usage in dairy herds in Pennsylvania. *Journal of Dairy Science*, **88**, 2991–2999.

Schlegelova, J., Babak, V., Klimova, E., Lukasova, J., Navratilova, P., Sustackova, A., Sediva, I. & Rysanek, D. (2002) Prevalence of and resistance to anti-microbial drugs in selected microbial species isolated from bulk milk samples. *Journal of Veterinary Medicine B Infectious Disease and Veterinary Public Health*, **49**, 216–225.

Schouten, J.M., Graat, E.A.M., Frankena, K., ven de Giessen, A.W., Van Der Zwaluw, W.K. & de Jong, M.C.M. (2005) A longitudinal study of *Escherichia coli* O157 in cattle of a Dutch dairy farm and in the farm environment. *Veterinary Microbiology*, **107**, 193–204.

Sharpe, R.T. & Livesey, C.T. (2006) Lead poisoning in cattle and its implications for food safety. *Veterinary Record*. **159**, 71–74.

Sischo, W.M. (2006) Stakeholder position paper: dairy producer. *Preventative Veterinary Medicine*, **73**, 203–208.

Small, A. (2006). Hygiene of production-processing of other foods. In: *Integrated Food Safety and Veterinary Public Health* (ed. S. Buncic), pp. 283–303, CABI, Oxford.

Steele, J.H. (2000) History, trends, and extent of pasteurisation. *Journal of the American Veterinary Medical Association*, **217**, 175–178.

Stephan, R., Schumacher, S., Tasara, T. & Grant, I.R. (2007) Prevalence of *Mycobacterium avium* subsp. *paratuberculosis* in Swiss raw milk cheeses collected at retail level. *Journal of Dairy Science*. **90**, 3590–3595

Straley, B.A., Donaldson, S.C., Hedge, N.V., Sawant, A.A., Srinivasan, V., Oliver, S.P. & Jayarao, B.M. (2006) Public health significance of antimicrobial-resistant gram-negative bacteria in raw bulk tank milk. *Foodborne Pathogens and Disease*, **3**, 222–233.

Sweeney, M.J., White, S. & Dobson, A.J. (2000) Mycotoxins in Agriculture and food safety. *Irish Journal of Agricultural and Food Research*, **39**, 235–244.

TAFS (2007) *International Forum for Transmissible Animal Diseases and Food Safety*, Position paper on the safety of bovine milk and bovine milk products. Available at http://www.tseandfoodsafety.org/position_papers/TAFS_POSITION_PAPER_ON_MILK.pdf.

Taylor, D.M., Ferguson, C.E., Bostock, C.J. & Dawson, M. (1995) Absence of disease in mice receiving milk from cows with bovine spongiform encephalopathy. *Veterinary Record*, **136**, 592.

Thoen, C., Lobue, P. & de Kantor, I. (2006) The importance of *Mycobacterium bovis* as a zoonosis. *Veterinary Microbiology*, **112**, 339–345.

Trevena, W.B., Willshaw, G.A., Cheasty, T., Domingue, G. & Wray, C. (1999) Transmission of vero cytotoxin producing *Escherichia coli* O157 infection from farm animals to humans in Cornwall and West Devon. *Communicable Disease and Public Health*, **2**, 263–268.

Tyshenko, M.G. (2007) Bovine spongiform encephalopathy and the safety of milk from Canadian dairy cattle. *Veterinary Record*, **160**, 215–218.

Vanzini, V.R., Aguirre, N.P., Valentini, B.S., Torioni de Echaide, S., Lugaresi, C.I., Marchesino, M.D. & Nielsen, K. (2001) Comparison of an indirect ELISA with the Brucella milk ring test for detection of antibodies to *Brucella abortus* in bulk milk samples. *Veterinary Microbiology*, **82**, 55–60.

Vetrugno, V. (2004) Safety of milk and milk derivatives in relation to BSE: the lactoferrin example. *Biometals*, **17**, 353–356.

Vivegnis, J., El Lioui, M., Leclercq, A., Lambert, B. & Decallonne, J. (1999) Detection of shiga-like toxin producing *Escherichia coli* from raw milk cheeses produced in Wallonia. *Biotechnology Agronomy Society and Environment*, **3**, 159–164.

Waliszewski, S.M., Pardio, V.T., Waliszewski, K.N., Chantiri, J.N., Aguirre, A.A., Infanzon, R.M. & Rivera, J. (1997) Organochlorine pesticide residues in cow's milk and butter in Mexico. *The Science of the Total Environment*, **208**, 127–132.

Whyte, P., McGill, K., Cowley, D., Madden, R.H., Moran, L., Scates, P., Carroll, C., O'Leary, A., Fanning, S., Collins, J.D., McNamara, E., Moore, J.E & Cormican, M. (2004) Occurrence of Campylobacter in retail foods in Ireland. *Journal of Food Microbiology*, **95**, 111–118.

Wilesmith J.W. & Ryan, J.B.M. (1997) Absence of BSE in the offspring of pedigree suckler cows affected by BSE in Great Britain. *Veterinary Record*, **141**, 250–251.

Wilesmith, J.W., Wells, G.A., Ryan, J.B., Gavier-Widen, D. & Simmons, M.M. (1997) A cohort study to examine maternally-associated risk factors for bovine spongiform encephalopathy. *Veterinary Record*, **141**, 239–243.

Williams, T.L., Monday, S.R., Edelson-Mammel, S., Buchanan, R. & Musser, S.M. (2005) A top-down proteomics approach for differentiating thermal resistant strains of *Enterobacter sakazakii*. *Proteomics*, **5**, 4161–4169.

Winthrop, K.L., Scott, J., Brown, D., Jay, M.T., Rios, R., Mase, S., Richardson, D., Edmonson, A., MacLean, M. & Flood, J. (2005) Investigation of human contacts: a *Mycobacterium bovis* outbreak among cattle at a California dairy. *International Journal of Tuberculosis and Lung Disease*, **9**, 809–813.

Yiannikouris, A. & Juany, J.-P. (2002) Mycotoxins in feeds and their fate in animals: a review. *Animal Research*, **51**, 81–99.

Zumarraga, M.J., Meikle, V., Bernardelli, A., Abdala, A., Tarabla, H., Romano, M.I. & Cataldi, A. (2005) Use of touch-down polymerase chain reaction to enhance the sensitivity of *Mycobacterium bovis* detection. *Journal of Veterinary Diagnostic Investigation*, **17**, 232–238.

7 Heat Treatment of Milk

M.J. Lewis and H.C. Deeth

7.1 Introduction

The aim of this chapter is to communicate the underlying principles for producing heat-treated milk which is safe and of high quality, based on many years of experience which has been gained from teaching, research and troubleshooting.

Although a small amount of milk is still sold as 'untreated' or raw, the two main heat treatments used for milk sold in the retail sector are pasteurisation and sterilisation. Treatments somewhere between these are also being used for extending shelf life. The main aims of heat treatment are to reduce the microbial population, both pathogenic and spoilage, in raw milk to inactivate enzymes and to minimise chemical reactions and physical changes. An overview of the changes taking place when milk is heated is given by Walstra & Jenness (1984). Some important ones are a decrease in pH, precipitation of calcium phosphate, denaturation of whey proteins and interaction with casein, lactose isomerisation, Maillard browning and modifications to the casein micelle. The overall effect is to alter the sensory characteristics, i.e. overall appearance, colour, flavour and texture and the nutritional value as well as making it safe and improving its keeping quality.

Most of the milk destined for processing into dairy products is also heat treated in some form, although some cheeses are made from raw milk. One industrial process is thermisation, which involves heating milk at temperatures between 57 and 68°C for about 15 s. Another is preheating or forewarming, applied to milks prior to evaporation and powder production. Conditions can be as low as 72°C for low-heat powders and up to 90–95°C for 5–10 min for high-heat powders, although temperatures above 100°C for shorter times are also used.

Patterns of consumption and preferences for milk vary from one country to another. For example, in Great Britain in 2003, 92.9% of heat-treated milk for drinking was pasteurised, 1.4% (in-container) sterilised and 5.7% ultra-high temperature (UHT) treated (Anonymous, 2003a). In Australia, the corresponding figures for white milk are 91.9%, 0% and 8.1%, respectively. The balance is totally different in other European countries, and in some, such as France and Germany, UHT milk is the main milk product.

Going back about 50 years, only a limited number of products were available. These were mainly white milk, limited flavoured milks and some creams. Domestic refrigeration was not widespread and pasteurised milk stored in the larder would keep for only 24–48 h. Sterilised milk with its characteristic flavour and slightly brown colour was quite popular in the UK and powdered, evaporated and sweetened condensed milks were available. UHT milk was in its infancy and fermented milks were for the future. Heat treatment was on a small to medium scale.

Jumping forward to the twenty-first century, one big change is the scale of operation, with a move towards highly automated large-scale, energy-saving continuous processes. In terms of product availability, we are now spoilt for choice. Milk is standardised for both protein and fat: semi-skimmed milk at 1.5–1.8 g fat 100 g^{-1} is now popular, accounting for 57.3% of all liquid milk drunk in Great Britain in 2003, and skimmed milk at 0.1 g fat 100 g^{-1} provides a low-fat version. For those who prefer a richer fuller flavour, milk from Jersey and Guernsey cows and other rare breeds can be purchased; these no longer are said to contain 5 g fat 100 g^{-1}, but are now '95% fat-free'. There is a wide range of flavoured milks, and creams ranging from 12 to 55 g fat 100 g^{-1}, with textures ranging from thin coffee creams to very viscous spoonable dessert creams. Competing with cow's milk are milks from goat, sheep and buffalo, as well as specialty milks for cats and dogs.

The industry is rightly promoting the health benefits of components naturally present in milk, i.e. specific fat fractions, bioactive peptides, beneficial minerals, especially calcium and magnesium, or producing milks with a healthy image, for example fermented milks, lactose-reduced milk, melatonin milk and milks containing probiotic microorganisms, prebiotic compounds, plant sterols or fish oils. In many parts of the world, conditions are not conducive to producing fresh milk; hence reconstituted and recombined milk are produced. Some dairy companies are now also processing 'vegetable milks', based on extracts from soy, rice, sweet corn and other vegetables.

7.2 Milk composition

Although there are many sources of data for composition of milk and milk products (McCance & Widdowson (2002), these report average values, giving no indication of the variability of raw milk due to breed, diet, climate and stage of lactation. This should not be forgotten when processing milk or when using data from the literature. The complexity and changing composition of raw milk pose key challenges.

Milk is an emulsion, containing fat globules in the range 1–10 microns in diameter, dispersed in an aqueous phase. Above 45°C, all this fat will be in the liquid phase; below this temperature, it will start to crystallise. This is not an instantaneous process and during crystallisation, latent heat will be released. The proteins in milk are divided into two fractions: (a) the casein fraction and (b) the whey protein fraction. The casein fraction is complex and exists in micelles with a size range of 30–300 nm. In the context of heat treatment, heat stability is very important and is influenced by several factors, particularly pH and ionic calcium.

The casein micelle is remarkably stable to heat, and good quality milk can withstand temperatures of 140°C for at least 10 min and often longer without coagulating. However, if the milk is not properly handled, its stability can deteriorate drastically. Some manifestations of poor heat stability are fouling or deposit formation on heat exchangers, sediment in milk and heat-induced thickening and coagulation. These problems tend to increase as the processing temperature increases, but are also dependent on raw milk quality.

In this context, a very important property is milk acidity, measured as pH or titratable acidity. Whereas pH is a direct measure of H^+activity, titratable acidity is a measure of buffering capacity between its own pH and that of the colour change (from colourless to red) in phenolphthalein, which is about 8.3. The pH of milk may influence many other aspects related to quality, in particular, the colloidal stability of milk and other heat-induced reactions, such as Maillard browning and lactulose formation. Both microbial activity and microbial inactivation are also influenced by pH, as is enzyme activity. The pH of raw milk is usually between 6.6 and 6.7, but it can be outside this range – for example 6.40–6.89 (Tsioulpas *et al.*, 2007a). Its exact value is influenced by its protein, mineral and acid contents. The pH of milk falls during heat treatment, but this is largely reversible on cooling. Walstra & Jenness (1984) illustrated that pH could fall to below 6.0, when the temperature exceeds 100°C. Another important determinant is ionic calcium, but its influence on heat stability is less well established, mainly because of difficulties in measuring it. Thus, the pH and ionic calcium in raw milk may be useful heat stability indicators, especially in sterilisation and UHT treatment of normal milk and in situations where milk is fortified with calcium or acidified.

Milk is a bland fluid with a characteristic creamy flavour. Because of this blandness, it is very susceptible to off-flavours; for example, a rancid flavour may develop due to excessive agitation of raw milk (Deeth & Fitz-Gerald, 2006). Raw milk from healthy animals has a very low microbial count, but it easily becomes contaminated with spoilage bacteria and perhaps some pathogenic microorganisms. These need to be inactivated and this is readily achieved by heat treatment. From the standpoint of the milk processor, raw milk composition and its microbial loading will vary from day to day.

7.3 Reaction kinetics

All thermal processes involve three distinct periods: a heating period, a holding period and a cooling period. In theory, all three periods may contribute to the reactions taking place, although in situations where heating and cooling are rapid, the holding period is the most significant. However, procedures are needed to evaluate each of these periods individually to determine the overall effect. One such example of this approach is offered by Browning *et al.* (2001).

The two reaction kinetic parameters of interest are the rate of reaction or inactivation at a constant temperature (e.g. D and k values), and the effect of temperature change on reaction rate (z and E values).

For pasteurisation processes, the range of interest is 60–80°C, and for sterilisation, from 100 to over 150°C. Chemical reaction rates are less temperature-sensitive than microbial inactivation rates. Thus, using heat treatment at higher temperatures for shorter times will result in less chemical damage occurring for an equivalent level of microbial inactivation. In practice, deviations from first order reaction kinetics are often encountered (Gould, 1989), as are deviations from the log–linear relationships between processing time and temperature discussed recently by Peleg (2006).

7.4 Principles of heat transfer

In thermal processing, the aim is to maximise the rate of heat transfer (Js^{-1} (W) or British thermal units (BTU) h^{-1}), i.e. to heat and then cool the product down as quickly as possible. This will improve the economics of the process and in many cases also lead to an improvement in product quality. Heating processes can be classified as direct or indirect. The most widely used is indirect heating, where the heat transfer fluid and the milk are separated by a barrier; for in-container sterilisation this will be the wall of the bottle and for continuous processes, the heat exchanger plate or tube wall. In direct processes, steam is the heating medium and the steam comes into direct contact with the milk. Indirect heating also implies that the two fluids will not come into direct contact. It is important to ensure that this is the case, and the integrity of the barrier is a very important safety consideration, especially in the regeneration section where the heating medium is the hot heat-treated milk. The heating medium is usually saturated steam but hot water and hot air are sometimes used. At temperatures above 100°C, the steam and the hot water are above atmospheric pressure. For steam, there is a fixed relationship between its pressure and temperature, given by the steam tables (Lewis, 1990; Holdsworth, 1997). Thus, a steam pressure gauge will act indirectly as a second temperature-monitoring device. Discrepancies between temperature and pressure readings suggest that there may be some air in the steam or that the instruments are incorrect (Lewis & Heppell, 2000). Cooling is achieved using mains water, chilled water or glycol. Regeneration is used in continuous processes to further reduce energy utilisation. Heating can be by either batch or continuous processing. Section 7.13 gives a review of the main physical properties of fluid that influence heat transfer as well as some of the basic heat transfer equations.

7.5 Thermisation and tyndallisation

Thermisation is the mildest heat treatment given to milk. It is used to extend the keeping quality of raw milk when it is known that raw milk may be held chilled for some time, prior to being further processed. The aim is to reduce the growth of psychrotrophic bacteria, which may release heat-resistant proteases and lipases into the milk. These enzymes will not be totally inactivated during pasteurisation and may give rise to off-flavours if the milk is used for cheese or milk powders. Conditions used for thermisation are 57–68°C for 15 s, followed by refrigeration. Thermised raw milk can be stored at a maximum of 8°C for up to 3 days (IDF, 1984). The milk should also be phosphatase-positive in order to distinguish it from pasteurised milk, which is phosphatase-negative. It is usually followed later by pasteurisation or a more severe heat treatment.

Another thermal process, which has been investigated, is tyndallisation; it involves successive heat treatments in order to inactivate spores. According to Wilbey (2002), Tyndall in 1877 suggested that if a medium was heated at 100°C for 3 min on 3 successive days, first the vegetative cells would be killed and the spores would germinate, and then be killed on either the second or third days. In practice, such double heat treatments are rarely encountered, and the process is not successful in totally inactivating spores because of the unpredictability of the spore germination process. The same applies to double pasteurisation processes, which have not been found to be effective (Brown *et al.*, 1979).

7.6 Pasteurisation

7.6.1 *Introduction and principles*

In terms of historical perspective, two key references still worth consulting are Cronshaw (1947) and Davis (1955). The first holder pasteurisation system was introduced in Germany in 1895 and in the USA in 1907. Thus by 1895, it was well recognised what was required for an effective pasteurisation process: 'we know that this process (pasteurisation) if properly carried out will destroy all disease germs' and 'a thoroughly satisfactory product can only be secured where a definite quantity of milk is heated for a definite period of time at a definite temperature. Then too, an apparatus to be efficient must be arranged so that the milk will be uniformly heated throughout the whole mass. Only when all particles of milk are actually raised to the proper temperature for the requisite length of time is the pasteurisation process complete.' This sound advice has withstood the test of time and forms the main thrust of current milk heat treatment regulations, reviewed recently by Komorowski (2006).

High-temperature, short-time (HTST) continuous processes were developed between 1920 and 1927, and for some time, the ability of this process to produce safe milk was questioned. In 1927, North and Park established 15 combinations of temperature and time, which inactivated the tuberculosis bacillus (Cronshaw, 1947). These experiments were performed by heating milk heavily infected with bacilli under different conditions and injecting the treated bacilli into guinea pigs. Successful temperature–time combination heat treatments, i.e. those where the animals survived, ranged from 54.4°C (130°F) for 60 min up to 100°C (212°F) for 10 s. Others were 71.1°C (160°F) for 20 s or 60°C (140°F) for 10 min. Further developments were made in the classification of tests for evaluating the pasteurisation process. These included tests for raw milk quality: (a) as visual inspection and detection and removal of faulty milk on arrival at the dairy (i.e. the platform test, Davis, 1955), (b) assessing pasteurisability by the survival of thermodurics (c) measuring the efficiency of pasteurisation by measuring pathogen inactivation and phosphatase activation, first described in 1935 on the basis of the finding that conditions required to inactivate *Mycobacterium tuberculosis* were just slightly less than those required to inactivate alkaline phosphatase, (d) assessing recontamination in terms of thermophilic and coliform bacteria, and the methylene blue test and (e) measuring general bacterial quality, including organisms surviving pasteurisation together with contaminating organisms (plate count). The methylene blue test is now little used, but the detection of alkaline phosphatase activity is still used as a statutory test in many countries.

By this time, the bacteria in pasteurised milk were being identified and the detrimental effects of thermoduric bacteria were being recognised. Factors affecting keeping quality were being investigated as well as conditions that induced a cooked flavour and resulted in the loss of the cream line. The role of phosphatase as an indicator was introduced and there were interesting comparisons between the holder or batch process and the HTST process. Problems with milkstone had also been recognised.

By the early 1950s, HTST processing accounted for about 75% of all milk pasteurised in the UK, and was the favoured process by the larger dairies (Davis, 1955). Energy saving was not an important consideration. Plant capacities were typically 1818 L h^{-1} (400 gal h^{-1}) in 1947, but a few years later had increased to 9092 L h^{-1} (2000 gal h^{-1}), with a range from 2273 to 22 730 L h^{-1} (50–5000 gal h^{-1}). Run times of 4–5 h were the norm; maintaining the cream line was important, and there were excellent descriptions of different equipment set-ups for conducting the more traditional holder process. One then current concern was that some coliform bacteria were surviving HTST pasteurisation (Davis, 1955).

Moving toward the present, pasteurisation has been defined by the International Dairy Federation (IDF, 1986) as a process applied with the aim of avoiding public health hazards arising from pathogenic microorganisms associated with milk, by heat treatment which is consistent with minimal chemical, physical and organoleptic changes in the product. According to Codex Alimentarius (Anonymous, 2003b),

> *Pasteurisation is a heat treatment aimed at reducing the number of any harmful micro-organisms in milk and liquid milk products, if present, to a level at which they do not constitute a significant health hazard. In addition, it results in prolonging the keepability of milk or the liquid milk product and in only minimal chemical, physical and organoleptic changes. Pasteurisation conditions are designed to effectively destroy the organisms (*Mycobacterium tuberculosis *and* Coxiella burnetti*). Pasteurisation of milk and cream results in a negative alkaline phosphatase reaction immediately after the treatment. For milk, the minimum pasteurisation conditions are those having bactericidal effects equivalent to heating every particle of the milk to 72°C for 15 s (continuous flow pasteurisation) or 63°C for 30 min (batch pasteurisation). Other equivalent conditions can be obtained by plotting a line joining these points on a log time versus temperature graph.*

Pasteurisation causes little change to the colour, flavour and appearance of the milk, (although devotees of raw milk will contest this), and no significant reduction in nutritional value. It causes minimal whey protein denaturation (5–15%) and does not alter enzymatic coagulation properties during the manufacture of cheeses.

Pasteurised products should last for up to 48 h without refrigeration (at, say, 20°C), and for several days when stored refrigerated. However, longer keeping qualities and between 10 and 16 days at 4°C are now achievable, when produced from high-quality raw milk, under optimum technical and hygienic conditions. Milk can still be pasteurised by the holder or batch process at 63°C for 30 min, but as discussed earlier, the HTST process now predominates, with capacities over 50 000 l h^{-1}, and running times of up to 20 h. Minimal conditions are at 72°C for 15 s, but the actual conditions will vary from country to country. A recent survey of the conditions used in Australian factories revealed a range from 72°C for 15 s to 78–80°C for 25 s (Juffs & Deeth, 2007). The more severe heating conditions are being used as a precautionary measure for the presence of any heat-resistant *Mycobacterium avium* subsp. *paratuberculosis* (MAP). As described elsewhere, the holding tube temperature and time is not the whole story, and the heating and cooling periods provide an extra margin of safety.

The original phosphatase test for assessing the adequacy of pasteurisation was based on the reaction of phosphatase with disodium phenyl phosphate. If phosphatase is present, it will release phenol, which is determined colorimetrically (Davis, 1955). It was claimed to be able to detect the presence of about 0.2% raw milk (addition) in pasteurised milk, as well as under processing, for example 62°C instead of 62.8°C for 30 min or 70°C rather than 72°C for 15 s. Since then, a more automated test based on fluorescence measurement (e.g. Fluorophos) has increased the sensitivity of the method further, being able to detect the presence of 0.006% added raw milk. This is a very useful quality assurance test procedure, and its introduction should further help detect low levels of post-pasteurisation contamination, which should also reduce the incidence of pathogens in pasteurised milk. Tests for detecting post-pasteurisation contamination are reviewed in IDF (1993).

In some regulations, it is required that pasteurised milk should show a positive lactoperoxidase activity, to prevent the milk being over processed (Statutory Instruments – SI, 1995). In Europe, milks which show a negative lactoperoxidase activity are designated high temperature pasteurised (European Union – EU, 1992). The EU regulations require that freshly pasteurised milk should be deemed to pass a coliform test, and to have a plate count of less than 50 000 mL^{-1} after incubation for 5 days at 6°C, although these are in the process of revision.

The heat resistance of a wide range of other enzymes found in raw milk in the pasteurisation range has also been reviewed by Griffiths (1986) and Andrews *et al.* (1987). Lactoperoxidase activity, determined on a plate heat exchanger (PHE) for 15 s, was generally lower than expected from the laboratory data. Using a PHE, enzyme activity was almost destroyed at 78°C for 15 s, and completely destroyed at 80°C for 5 s. The enzyme appeared sensitive to temperatures above 75°C, with a z-value of 5.4°C. Griffiths (1986) determined the heat resistance of several other indigenous milk enzymes; these have also been summarised by Lewis & Heppell (2000).

Enzymes in raw milk may cause some other problems in pasteurised milk. For example, indigenous lipases may give rise to soapy off-flavours, especially if raw milk is subjected to excessive agitation at temperatures up to 40°C, e.g. during pumping or when mixing flavoured milk or other similar products. However, it is unlikely that bacterial lipases and proteases, which are very heat resistant, will cause problems in pasteurised milks because of their relatively short shelf life and refrigerated storage conditions.

7.6.2 *Methods of pasteurisation*

Holder or batch heating

Cronshaw (1947) and Davis (1955) both provide excellent descriptions of equipment for the holder or batch process – individual vessels (heated internally) and externally heated systems with one or more holding tanks. The batch heating time and factors affecting it can be predicted from equations given in Section 7.13. It is still being used in many countries in some small-scale pasteurisation processes. In answer to the question – Does HTST pasteurisation result in as good a bottle of milk as does the holder process, Yale in 1933 concluded that one method of pasteurisation produces as good a bottle of pasteurised milk as does the other when sound methods are used and when conditions are comparable. The authors have not seen anything of late to contradict this.

Continuous heating

The main types of indirect heat exchanger for milk are the PHE and the tubular heat exchanger (THE). PHEs are widely used for pasteurisation processes; they have a high overall heat transfer coefficient (OHTC), and are generally more compact than THEs. Their main limitation is pressure, with an upper limit of about 2 MPa (20 bar). The normal gap width between the plates is between 2.5 and 5 mm, but wider gaps are available for viscous

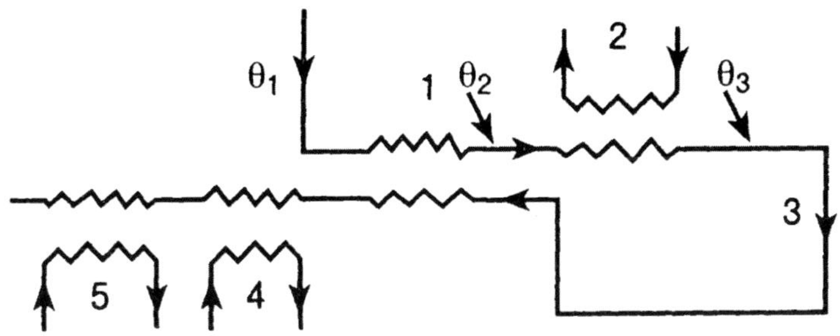

Fig. 7.1 Heat exchange sections for HTST pasteuriser. Note: 1, regeneration; 2, hot water section; 3, holding tube; 4, mains water cooling section; 5, chilled water cooling section. Reprinted from *Modern Dairy Technology*, Volume 1, 1994, R.K. Robinson, with kind permission of Springer Science and Business Media.

liquids to prevent large pressure drops. In general, PHEs are the cheapest option, and the one most widely used for low-viscosity fluids. Maintenance costs may be higher, as gaskets may need replacing, and the integrity of the plates also needs evaluating regularly. This is especially so for plates in the regeneration section, where a cracked or leaking plate may allow raw milk to contaminate already pasteurised milk. They are also more prone to fouling-related problems.

THEs have a lower OHTC than plates, and generally occupy a larger space. They have slower heating and cooling rates with a longer transit time through the heat exchanger. In general, they have fewer seals, and provide a smoother flow passage for the fluid. A variety of tube designs are available to suit different product characteristics. Most tubular plants use a multi-tube design. They can withstand higher pressures than PHEs. Although they are still susceptible to fouling, high pumping pressures can be used to overcome the flow restrictions. THEs give longer processing times than PHEs with viscous materials and with products, which are more susceptible to fouling.

The viscosity of the product is one major factor, which affects the choice of the most appropriate heat exchanger and the selection of pumps. Viscosity will influence the pressure drop causing a problem in the cooling section and when phase transition may take place, for example if coagulation or crystallisation takes place. For more viscous or particulate products, e.g. starch-based desserts or yoghurt with fruit pieces, a scraped surface heat exchanger may be required.

One of the main advantages of continuous systems over batch systems is that energy can be recovered in terms of regeneration. The layout for a typical regeneration section is shown in Figure 7.1. The hot fluid can be used to heat the incoming fluid, thereby saving on heating and cooling costs. Regeneration efficiencies over 90% can be obtained (see equation for regeneration efficiency in Section 7.13). Although high-regeneration results in considerable savings in energy, it necessitates the use of higher surface areas, resulting from the lower-temperature driving force, and a slightly higher capital cost for the heat exchanger. This also means that the heating and cooling rates are also slower, and the transit times longer, which may affect the quality.

For milk and cream, homogenisation must be incorporated to prevent fat separation. As homogenisation of raw milk is a very effective way of initiating lipolysis (Deeth & Fitz-Gerald, 2006), it must be carried out immediately before or after pasteurisation, which

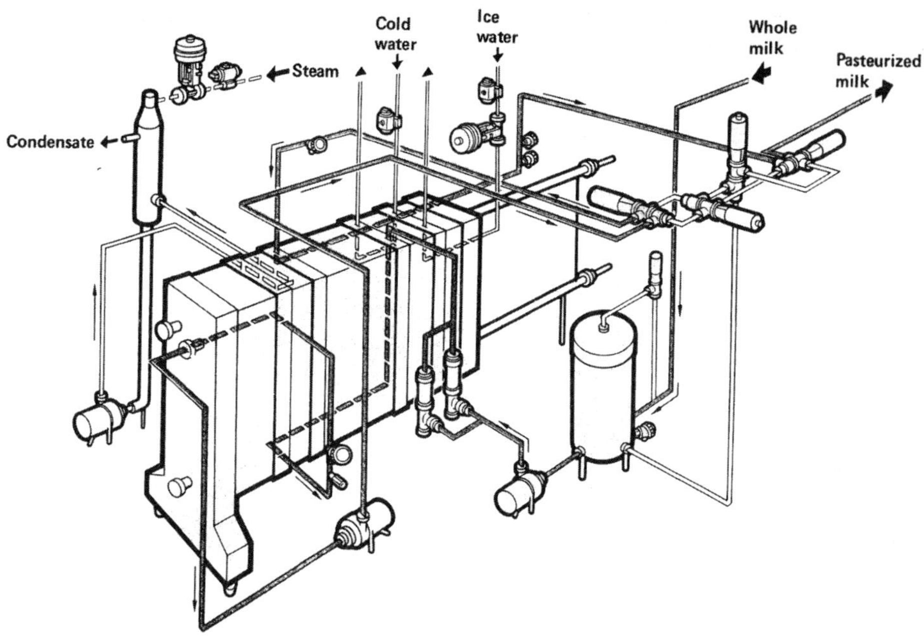

Fig. 7.2 The layout of a typical HTST pasteuriser. With permission of Tetra Pak, Lund, Sweden.

inactivates the native lipase. Homogenisation before pasteurisation is preferable as homogenisers can introduce post-pasteurisation contamination if used after pasteurisation. While pre-pasteurisation homogenisation is simple in a continuous flow system, it is more difficult to link with batch pasteurisation as the time delay between homogenisation, and when the milk reaches pasteurisation temperature, can result in an unacceptable amount of lipolysis. However, this problem can be largely overcome by homogenising the milk at $\geq$50°C (Deeth, 2002).

The layout of a typical HTST pasteuriser and its accessory services is shown in Figure 7.2. The holding time is controlled either by using a positive displacement pump or by a centrifugal pump linked to a flow controller, and the temperature is usually controlled and recorded. Note that a booster pump can be incorporated to ensure that the pasteurised milk is at a higher pressure than the raw milk in the regeneration section, i.e. to eliminate post-processing contamination in this section. A flow diversion valve diverts underprocessed fluid back to the feed tank. In continuous processing operations, there will be a distribution of residence times, and it is vital to ensure that the minimum residence time (i.e. the time for the fastest element of the fluid to pass through the holding tube) is greater than the stipulated time, to avoid underprocessing. In a fully developed turbulent flow, the minimum residence time is about $0.83t_{av}$, whereas in streamline flow, it is $0.5t_{av}$, where t_{av} is the average transit time through the holding tube (see Section 7.13).

Most HTST pasteurisers are of the plate type, and these should be tested for leaks periodically. Consideration should be given to ensuring that if leaks do occur, they do so in a safe fashion; i.e. pasteurised milk is not contaminated with cooling water or raw milk in the regeneration section. The control instrumentation, diversion valves and other valves should be checked regularly.

7.6.3 *Factors affecting the quality of pasteurised milk*

The main control points for ensuring good quality pasteurised milk products are

- raw milk quality,
- processing conditions: temperature and holding time,
- post-processing contamination (PPC) and
- storage temperature.

Raw material quality

Raw milk may contain pathogenic microorganisms from the farm environment, including vegetative bacteria, such as *Staphylococcus aureus*, *Campylobacter jejuni*, *Salmonella* spp., *Escherichia coli*, *Yerisinia enterocolitica*, and spore formers, such as *Bacillus* and *Clostridium* species. These major vegetative pathogens can be effectively controlled by pasteurisation, and are not the main determinants of keeping quality. The main interest is in what survives pasteurisation or mild heat treatments. Thermoduric bacteria are defined as those which survive pasteurisation conditions, e.g. 63°C for 30 min or 72°C for 15 s, whereas spores produced by spore-forming bacteria survive 80°C for 10 min. *Bacillus cereus* spores are relevant here, being the main pathogen which will survive pasteurisation and grow at low temperature. *Bacillus* can cause defects in heat-treated milk, for example bitty cream, and produce an intense bitter flavour, but it rarely causes food poisoning because infected products are so unacceptable.

Processing times and temperatures

Normal HTST conditions for milk are 72°C for 15 s. One interesting question relates to the use of higher temperatures (up to 90°C) for pasteurisation. In general, milk treated at such a temperature has a reduced keeping quality compared to milk heated at 72°C for 15 s. This was first recognised by Kessler & Horak (1984), and then by Schroder & Bland (1984), Schmidt *et al.* (1989), Gomez Barroso (1997) and Barrett *et al.* (1999). It is a question that should be often revisited, since it would be logical to expect a more severe heating process to result in improved keeping quality. Another drawback is that a cooked flavour will start to be noticeable between 85 and 90°C. Thus, those using or considering using more stringent pasteurisation conditions than the minimum conditions should be aware of these disadvantages. The usual explanation for this unexpected phenomenon is that the more severe conditions cause heat shocking of the *Bacillus* spp. spores which can then germinate, grow and reduce the keeping quality of the milk. However, recent evidence suggests that the lactoperoxidase system (LPS) also plays a role. The LPS involves the enzyme lactoperoxidase, hydrogen peroxide and thiocyanate, all of which are present in raw milk. The oxidation products, e.g. hypothiocyanite, exhibit strong anti-microbial activity by oxidising sulphydryl groups of bacterial cell walls (Reiter & Harnulv, 1982). The LPS can be further activated in raw milk by small additions of thiocyanate and hydrogen peroxide, and can be used to keep raw milk longer in countries where refrigeration is not widespread (IDF, 1988). Lactoperoxidase inactivation is very temperature-sensitive, and as described earlier, some heat treatment regulations now require that pasteurised milk should show a positive lactoperoxidase activity. Marks *et al.* (2001) showed that pasteurisation conditions of 72°C for 15 s, resulting in an active LPS, greatly increased the keeping quality of milks inoculated with *Pseudomonas aeruginosa*, *S. aureus* and *Streptococcus thermophilus*, when compared to heating at 80°C for 15 s. However, pasteurisation had no effect on the keeping quality of milks challenged with *B. cereus* spores. It may be possible to exploit some other natural anti-microbial systems in raw milk. These have been described in more detail by the IDF (1994). Double pasteurisation processes have been found not to be effective (Brown *et al.*, 1979) and, as such, are rarely used.

Problems arising from a build-up of thermophilic bacteria in the heating and cooling sections associated with long operating times in continuous heat exchangers have been recognised for some considerable time (Cronshaw, 1947). Bacterial numbers in pasteurised milk have been found to increase slowly over the initial 8–9 h, and then more rapidly over the remaining period of operation. The main growth occurs in the regenerative section. *Bacillus licheniformis* and *S. thermophilus* have been implicated (Lehmann *et al.*, 1992; Lehmann, 1996).

There has been much interest recently in *M. avium* subsp. *paratuberculosis* (MAP), and whether it would survive pasteurisation. MAP levels found in raw milk appear to be low, but there is no real indication of true levels because of the decontamination procedures used to remove the other bacteria in raw milk and its extremely slow growth rate. MAP levels found in milks subjected to pasteurisation are also low. There are many inconsistencies in the experimental results. These are discussed in several publications (Hammer *et al.*, 1998; Grant *et al.*, 2001; IDF, 2004; Grant, 2006).

Results from surveys on raw milks and pasteurised milks are also inconclusive in that MAP was found in 2% of both raw and pasteurised milk samples tested. This again suggests that pasteurisation has no significant effect. Clearly, the heat resistance data generated to date for MAP are inconclusive and do not permit an accurate assessment of the efficacy of the pasteurisation process with regard to MAP. Collated information has been published by the IDF (1998, 2004). In the UK, it has been recommended that HTST pasteurisation conditions should be increased to 72°C for 25 s as part of a strategy for controlling MAP in cows' milk.

One parameter which has been used for some considerable time to compare pasteurisation processes is the pasteurisation unit (PU). One PU results from a heat treatment at 60°C for 1 min, and the equivalent z-value is high (10°C); thus, the number of PUs is given by:

$$PU = 10^{\frac{T-60}{10}} t,$$

where t = time (min)

Thus, a treatment at 63°C for 30 min would have a value of ~60, whereas HTST conditions would give a value of about 4. Clearly, there is an inconsistency here, no doubt derived from the high z-value chosen for this calculation.

Kessler (1989) introduced a different parameter (p^*) for characterising and comparing pasteurisation processes, especially at temperatures higher than 72°C. According to him, 72°C for 15 s corresponds to a $p* = 1$. It is noteworthy that 63°C for 30 min corresponds to a p^* of approximately 9, suggesting from this approach that such milk would be considerably overprocessed. Since this is known not to be the case, it demonstrates that practical experience should always dictate what conditions to use.

However, pasteurisation conditions do vary from one country to another. In the USA, a wide range of conditions are used including 63°C for 30 min, 77°C for 15 s, 90°C for 0.5 s and 100°C for 0.01 s (Busse, 1981)

Other products pasteurised are creams and ice cream mix. In the UK, minimum temperature–time conditions are 72°C for 15 s and 79°C for 15 s, respectively, although conditions for these products are more severe in other countries. Codex Alimentarius (Anonymous, 2003b) states that the fat content of cream makes it is necessary to apply minimum pasteurisation condition of 75°C for 15 s.

Post-pasteurisation contamination

Post-pasteurisation contamination (PPC) was recognized as a problem in the 1930s, and is now considered to be a very important determinant of keeping quality. Muir (1996a,b) describes how this was recognised both for milk and for cream in the early 1980s. PPC encompasses the recontamination of the product anywhere downstream of the end of the holding tube. It can occur in the regeneration or cooling sections, in storage tanks and in the final packaging of the product, due to poor hygienic practices. It can greatly be reduced by ensuring that all internal plant surfaces in contact with the product are heated at 95°C for 30 min. It can only be completely eliminated by employing aseptic techniques downstream of the holding tube. One of the main safety concerns is recontamination of the product with pathogens from raw milk, and this could occur due to bypassing of the holding tube by a number of possible routes, including pinhole leaks in plates and through pipelines that have been set up for cleaning and disinfecting. In terms of reducing keeping quality, recontamination with Gram-negative psychrotrophic bacteria is very important.

The presence in a pasteurised product with a high count of microorganisms (e.g. coliform bacteria), which should have been inactivated by pasteurisation, is indicative of PPC. IDF (1993) catalogued a large number of tests, which can be used to determine the extent of the problem. In practical situations where the keeping quality of milk starts to deteriorate or is below expectations, the most likely explanation is an increase in PPC and this should be the first factor to be investigated.

Storage temperature and time

In general, the lower the storage temperature, the better will be the keeping quality, bearing in mind the costs and practical problems of ensuring low temperatures throughout the cold chain and in domestic refrigerators. Before domestic refrigeration was commonplace, Cronshaw (1947) reported that the shelf life of pasteurised milk was about 24 h. Household refrigeration helped to improve this considerably, and in the UK by 1957, 10% of households had a refrigerator, increasing to 30% by 1962 and up to 90% by 1979. Raw milk is stored at typically 4°C; temperatures in the cold chain are slightly higher, and are likely to be higher still in domestic refrigerators. Many of our results have confirmed that pasteurised milk produced from good quality raw milk could be stored for up to 18 days at 8°C and between 25 and 40 days at 4°C (Ravanis & Lewis, 1995; Gomez Barroso, 1997). However, it must be emphasised that these experiments were performed with good quality raw milk, i.e. the counts immediately after pasteurisation were never above 10^3 colony-forming units (cfu) mL^{-1}, even for raw milk stored for up to 8 days at 4°C prior to pasteurisation. Also, care was taken to minimise PPC. These results also illustrate that good keeping quality can be achieved by eliminating PPC, and can be further enhanced by using low storage temperatures.

From our experience, there are some other interesting questions relating to pasteurised milk. For example, why does pasteurised skimmed milk have a shorter shelf life than pasteurised whole milk? This observation has been reported by several authors (Janzen *et al.*, 1982; Brown *et al.*, 1984; Deeth *et al.*, 2002). The latter authors reported that the rates of growth of psychrotrophic bacteria were not significantly different in the two milks and the bacterial types, all pseudomonads, present at spoilage were also similar. The different spoilage behaviours were attributed to greater proteolysis in skimmed milk than in whole milk, caused by higher production of protease and greater susceptibility of the protein to protease attack. Lipolysis in the whole milk also contributed to the spoilage flavours of the product, but not skimmed milk. Some other interesting questions relate to differences in the keeping qualities of pasteurised cow's and goat's milk and whether organic milk has a better keeping quality than non-organic milk. To answer these questions in a scientific manner is not straightforward, as one would likely find significant variations between different cow's and goat's milks, and between different organic milks for reasons outlined earlier in the chapter.

7.7 Sterilisation – safety and spoilage considerations

Sterilisation of milk to enable it to be kept at room temperature for several months became a commercial proposition in 1894. Milk can either be sterilised in bottles or other sealed containers, or by continuous UHT processing followed by aseptic packaging (see below). Very good accounts of the procedures for producing in-container sterilised milk and problems associated with it have been provided by Cronshaw (1947) and Davis (1955).

From a safety standpoint, the primary objective is the production of commercially sterile products with an extended shelf life. The main concern is inactivation of the most heat-resistant pathogenic spore, namely *Clostridium botulinum*. Since milk is a low-acid food ($pH > 4.5$), the main criterion is to achieve 12 decimal reductions of *Cl. botulinum*. This occurs when a product is heated at 121°C for 3 min, at its slowest heating point (Anonymous, 1991). The microbial severity of an in-container sterilisation process is traditionally expressed in terms of its Fo value. This takes into account the contributions of the heating, holding and cooling periods to the total lethality and is expressed in terms of minutes at 121°C. It provides a useful means of comparing processes. The minimum Fo value for any low-acid food should be 3.

Cl. botulinum is rarely found in raw milk. More common sporeformers are *Bacillus* species of which some, such as *B. stearothermophilus* (now known as *Geobacillus stearothermophilus*) and *B. sporothermodurans* (Hammer *et al.*, 1996), form highly heat-resistant spores, which are not destroyed by a process with an Fo of 3. These bacteria may cause spoilage, but they are not pathogenic. Thus, a minimum 'botulinum cook' will produce a product which is safe, but not necessarily sterile. For foods which may contain highly heat-resistant spores, a heat treatment achieving two or more decimal reductions is recommended, corresponding to an Fo value of 8. Target contamination rates should be less than 1 in every 10 000 containers.

Spore counts in raw milk have been rarely reported to exceed 10^3 cfu mL^{-1}, although Bramley & McKinnon (1990) reported that they may reach 5000 cfu mL^{-1}. Spores are mainly derived from surfaces of teats in contact with bedding materials. The most common *Bacillus* spores isolated from teat surfaces are *B. licheniformis*, *B. subtilis* and *B. pumilis* with lower numbers of *B. cereus*, *B. firmus* and *B. circulans.* Most common in raw milk are *B. licheniformis*, *G. stearothermophilus* and *B. cereus.* Very heat-resistant spores, such as *G. stearothermophilus*, are usually only a small proportion of the total.

7.8 In-container sterilisation

Foods have been sterilised in sealed containers, such as cans, for over 200 years. Milk was originally sterilised in glass bottles sealed with a crown cork, but more recently, plastic bottles have been used. Milk sterilisation really developed after 1930 with the advent of the crown cork, which helped with the mechanisation of the bottle-filling process, and the reuse of bottles. In general, the basic principles have remained the same.

The main aim is to inactivate heat-resistant spores, thereby producing a product which is 'commercially sterile', with an extended shelf life. Practical drawbacks of in-container sterilisation processes are that the product heats and cools relatively slowly, and that temperatures are limited by the internal pressure generated. However, many dairy products are still produced this way worldwide, including sterilised milk, evaporated milk and canned desserts such as custard and rice pudding.

Sterilised milk is still produced in many countries, and in essence, the manufacturing procedure is not too far removed from that used over 50 years ago. Milk is clarified using a centrifuge, e.g. a Bactofuge™, with claimed spore removal of greater than 99% (2 $\log_{10}$ reductions). It is heated using similar equipment to that used for pasteurisation. It is then homogenised at 63–82°C, for example at a single-stage pressure of about 20 MPa or double stage at about 17 and 3.5 MPa. It is then filled into glass bottles between 74 and 80°C under conditions which give minimal frothing, and sealed using a crown cork. Plastic bottles are sealed at a lower temperature of 54–55°C. Care should be taken to avoid conditions in balance tanks, which may be conducive to growth of thermophiles. Ashton & Romney (1981) cite sterilisation processing conditions of 110–116°C for 20–30 min, depending on the extent of cooked flavour and colour preferred by the consumer. Batch or continuous retorting processes may be used (Davis, 1955). Other processing details are outlined by Ashton & Romney (1981); these include more detail on continuous retorts, such as hydrostatic or rotary valve sealed sterilisers, which are capable of higher temperatures and shorter times

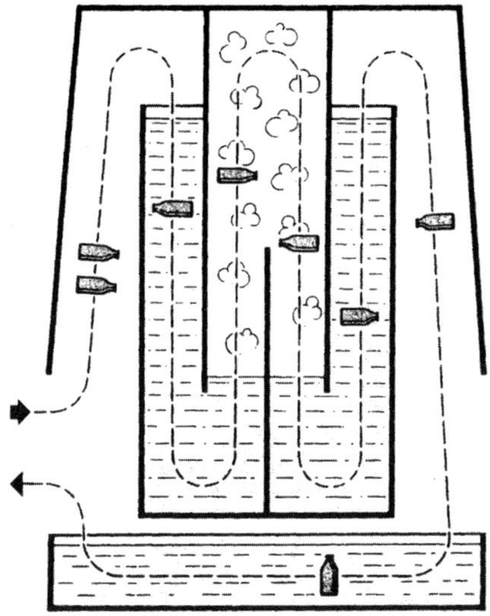

Fig. 7.3 A continuous retort for sterilisation of glass or plastic bottles. With permission of Tetra Pak, Lund, Sweden.

(132–140°C for 12 min), and the use of steam for glass bottles or steam/air mixtures for plastic bottles (see Figure 7.3).

Davis (1955) recognised the need for ensuring that raw milk to be used for sterilisation was not heavily contaminated with bacterial spores. Today, this remains an important control variable. Sweet curdling was the chief bacterial fault, due to highly resistant spores of *B. subtilis* and *B. cereus*. Bacterial growth was found to produce other taints, such as carbolic, bad (e.g. oxidised) or cardboard taints. Ashton & Romney (1981) reported that the failure level of well-produced sterilised milks is of the order of 1 in 1000 units, although it may be as high as 5–10% in situations where there are large numbers of thermotolerant spores in the raw material or other contamination arising in the process.

A recognised test for ensuring adequate sterilisation is the turbidity test, developed by Aschaffenburg in 1950. This test measures whey protein denaturation and it is an indirect test (similar to phosphatase), as complete denaturation would indicate that the milk was adequately sterilised. Milk (20 mL) is mixed with ammonium sulphate (4 g), which causes casein and any associated denatured whey protein to precipitate. The mixture is filtered, producing a clear filtrate, which contains any undenatured whey protein present in the milk sample. The filtrate is then boiled, which causes any undenatured whey protein to be denatured, thereby producing a turbid solution, the amount of turbidity being proportional to the amount of undenatured whey protein in the milk. Some regulations state that sterilised milk should produce a negative turbidity result; i.e. it should be heated for such a time as to fully denature the whey proteins. In principle, UHT heating of milk should result in some undenatured whey protein, but UHT processes with extensive heating and cooling profiles are more severe and will also give a negative turbidity result. Thus, the turbidity test will not always distinguish UHT milk from sterilised milk. In fact, many UHT milk samples show a negative turbidity. Methods for distinguishing between sterilised and UHT milk have been discussed in detail by Burton (1988). However, this is made difficult by the wide range of times and temperatures which are permitted. This includes a combined process, which involves the production of milk under UHT conditions, e.g. 137°C for 4 s, and is filled into bottles which are then sealed and passed through a conventional retorting process. Although the retorting is much reduced, it is generally just sufficient to ensure a negative turbidity result. In terms of determining the sterilisation effect, if this is to be treated as a single process, the critical point is to ensure that the milk does not become recontaminated in the intermediate filling process, especially with bacterial spores. This process was found to reduce the incidence of spoilage due to spore survivors (Ashton & Romney, 1981). There is also plenty of opportunity for spore inactivation due to high temperatures being maintained for some considerable time.

Sterilised milk has a rich creamy appearance, perhaps helped by Maillard browning components, a distinct cooked flavour (rich, nutty, caramelised), which once acquired, makes other heat-treated products taste insipid. It is considerably browner than raw milk, the extent of browning depending on the severity of the heat treatment. Thus, Maillard browning contributes to product quality in terms of its colour and flavour, although not everybody will find sterilised milk to be as palatable as pasteurised milk.

Sterilisation causes more loss of nutrients than any other heat treatment. For example, losses of the water-soluble vitamins B1 and B12 have been reported to be 30–40% and 80–100%, respectively (Schaafsma, 1989). Furthermore, it cannot be coagulated with rennet, unless calcium chloride is added (Kessler, 1981, 1989).

To summarise, sterilised milk is still produced in quantity in some countries, with much of it now produced in retortable disposable plastic bottles with a metal foil cap, rather than in returnable glass bottles.

7.9 UHT processing

7.9.1 *Introduction and principles*

UHT processing of milk combined with aseptic packaging was introduced to produce a shelf-stable product with minimal chemical damage compared with in-container sterilised milk. UHT milk may have a shelf life of up to 12 months, although in practice, it is usually consumed much earlier than this. In countries where it is a minor segment of the milk market, it is often used as a convenience product, and used when pasteurised milk is not available; whilst in countries where it is the major type of milk available, it is used regularly. In the former situation, UHT milk may need to be stable over a long period of time, while in the latter case, the desired shelf life may be $\leq$3 months.

UHT treatment is normally in the range 135–150°C in combination with appropriate holding times necessary to achieve 'commercial sterility'; i.e. microorganisms are unlikely to grow in the product under the normal conditions of storage (Burton, 1988; Lewis & Heppell 2000; Anonymous, 2003b). In practice, the products are checked for sterility by incubating at 55°C for 7 days or at 30°C for 15 days, and testing for bacterial growth (Anonymous, 2003b).

Some useful bacteriological and chemical indices have been developed to describe the effects of a particular heating regime on the bacteria and chemical components of milk, respectively. The major ones are B^* and Fo (bacterial) and C^* (chemical). B^* is a measure of the bacteriological effect of a heat treatment relevant to treatment at a reference temperature of 135°C. A process with $B^* = 1$ produces a nine-decimal reduction of thermophilic spores assuming a z value of 10.5°C, and is equivalent to holding the product at 135°C for 10.1 s; this is the recommended minimal value of B^* for a UHT process. This differs from the other bacterial index Fo, which is commonly used for in-container sterilisation, where the reference temperature is 121.1°C (250°F) (see Section 7.7). While it is a good measure of the lethality of the heat treatment, it is more appropriate to heating at temperatures around 120°C than around 140°C. As indicated above, sterilisation processes should have an Fo $\geq$ 3 to ensure bacteriological safety. Although a direct correlation is not strictly correct, an Fo of 3 corresponds roughly to a B^* of about 0.85.

C^* is a chemical index of heat treatment relevant to a reference temperature of 135°C based on the kinetics of destruction of the vitamin thiamine; a $C^* = 1$, which equates to a 3% loss of thiamine, is considered to be the desired upper limit for a UHT process to avoid excess damage. While thiamine is seldom measured in milk, the kinetics of its destruction by heat are deemed to provide a reasonable indication of the chemical effect of heat, and allow the effect of different heating conditions to be compared. However, other chemical components change with temperature and time differently. For example, heating at 90°C for 30 s, common UHT preheating conditions, destroys less than 0.1% of thiamine, but denatures up to 75% of the major whey protein, i.e. β-lactoglobulin.

There are two basic principles of UHT processing, which distinguish it from in-container sterilisation. First, for the same bactericidal effect, a high-temperature-short-time treatment (as in UHT) results in less chemical change than a low-temperature – long-time treatment (as in in-container sterilisation). This is because the Q_{10}, the relative change in reaction rate with a 10°C change in temperature, is much lower for chemical change (typically ~3) than for bacterial kill (typically ~10 for spore destruction), or alternatively, the z values for chemical reactions are higher (see Section 7.3). Based on the Q_{10} values of 3 and 10, the chemical change at 145°C is only about 2.7% of that at 115°C, for the same bactericidal effect.

The second principle is that the need to inactivate thermophilic bacterial spores dictates the minimum times and temperatures which can be used, while the need to minimise undesirable chemical alterations, such as undesirable flavour and colour changes, and vitamin destruction dictate the maximum times and temperatures. In terms of the indices discussed above, the recommended minimum conditions are those with a B^* of 1, and the maximum conditions with a C^* of 1.

It is worth noting that a B^* of 1 refers to 9 $\log_{10}$ reduction of thermophilic spores, which represents a more intense heat treatment than 9 $\log_{10}$ reduction of mesophilic spores. However, since the discovery of the extremely heat-resistant mesophilic sporeformer, *B. sporothermodurans* in UHT milk (IDF, 2000), a higher level heat treatment has been recommended.

In order to inactivate highly heat-resistant spores, higher temperatures ($\geq$150°C) for very short times (<1 s) have been proposed. The use of such extreme treatments is generally limited by the UHT plant's physical configuration. However, an innovative steam injection (ISI) process has been developed in the Netherlands to heat milk at 150–200°C for less than 0.1 s. It was shown to destroy heat-resistant spores (Huijs *et al.*, 2004).

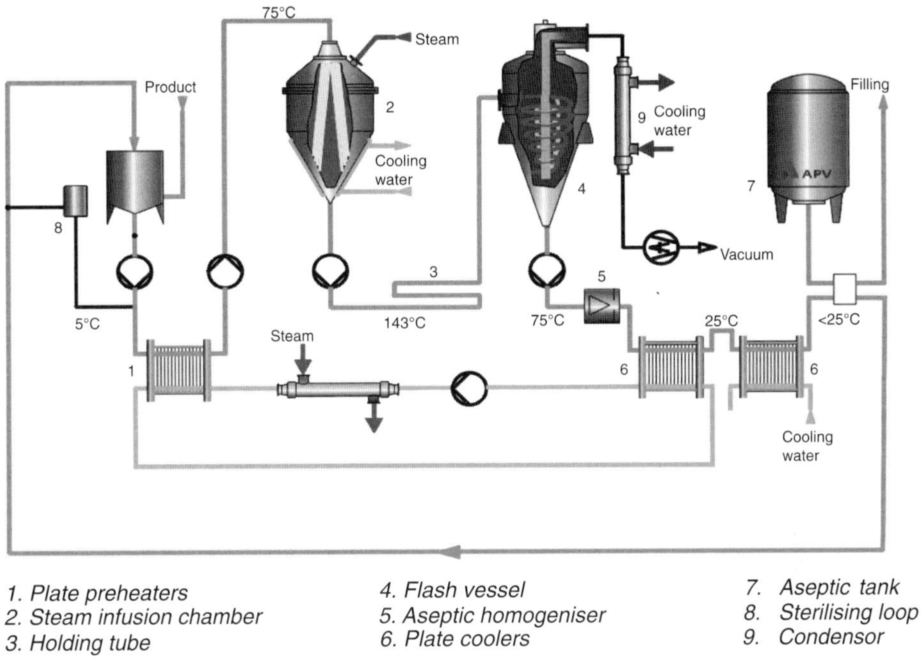

Fig. 7.4 The layout of a direct (infusion) UHT plant. With permission of APV Company.

7.9.2 *Methods of UHT processing*

Background

UHT heating can be either 'direct' or 'indirect'. In direct heating, superheated steam is mixed with milk while, in indirect heating, a heat exchanger transfers heat across a partition between the milk and the heating medium, either steam or hot water (Mehta, 1980; Burton, 1988). In THEs, the partition is the wall of the stainless steel tube, and in PHEs, it is the stainless steel plate.

The UHT process involves the following stages: preheating with heat regeneration, holding at preheat temperature, heating to sterilisation temperature, holding at sterilisation temperature, cooling and aseptic packaging (see Figure 7.4). In addition, a homogenisation step is usually included either before or after the high-heat holding section. In commercial processing of UHT milk, preheating (to ~80–95°C) is usually achieved by using the hot processed milk to heat the incoming cold raw milk. This enables much of the heat used in the process to be regenerated and the cooling water requirement to be reduced (Lewis & Heppell, 2000). The preheated milk is often held for a short time (15 s to a few min) to denature the whey proteins, principally β-lactoglobulin, to reduce their ability to foul, or deposit on, the hot surfaces of the high-temperature heating section (Burton, 1988). This processing approach, which is known as a 'protein stabilisation' step, is not usually employed in direct plants because they are much less likely to foul in the high-heat section than the indirect plants due to the reduced access of the milk to high-temperature surfaces.

The preheating and final cooling steps are always performed in indirect heat exchangers; however, the heating to sterilisation temperature after preheating can be either direct or indirect. This is the stage that characterises a plant as either 'direct' or 'indirect'. The cooling section immediately after the high-heat section also differs in the two systems: (a) direct plants use expansion cooling in a vacuum chamber and (b) indirect plants use THEs or PHEs (Burton, 1988).

All UHT processes involve aseptic packaging of the product into cartons, plastic bottles or laminated plastic cartons (Robertson, 2006). This is an essential part of UHT processing as it ensures that the sterile product is not contaminated during packaging, thus enabling the product to be stored at room temperature for several months without spoilage by bacterial growth.

Direct heating

There are two major types of direct heating UHT systems known as *steam injection* (steam into milk) and *steam infusion* (milk into steam) (see Figure 7.4). In the former, superheated steam is 'injected' into a stream of milk, while in the latter, milk is sprayed into or allowed to fall as a thin film or fine streams, through a chamber of superheated steam. A major feature of these two systems is the almost instantaneous rise in the temperature of the milk from preheat to sterilisation temperature through the transfer of the latent heat of vaporisation of the steam to the milk. During this heating stage, steam is condensed, and the milk is diluted with water. The degree of dilution depends on the temperature rise of the milk, but for an increase in temperature of 60°C (e.g. from 80°C to 140°C), it is approximately 11% (Lewis & Heppell, 2000). After the milk passes through the high-heat-holding tube, the entrained water is removed in the vacuum chamber, which also rapidly cools the milk to approximately the same temperature as that of the preheated milk. To ensure neither dilution nor concentration of the milk occurs, the total solids of the incoming milk and the processed milk are monitored, and the temperature in the vacuum chamber adjusted if necessary.

The steam used in direct UHT plants must be of high-quality, culinary grade. Poor-quality steam can lead to carry over of off-flavours into the milk (Burton, 1988).

Indirect heating

In indirect heating using PTEs or THEs, the heating medium is either steam or superheated water. When water is used, it flows in the reverse direction to that of the milk. The reverse flow minimises the temperature differential between the two liquids and, in turn, minimises the amount of burn-on. Hot water is a significantly better heating medium than steam with respect to burn-on and flavour of the product as it enables a smaller temperature differential between the milk and the heating medium (Dentener, 1984).

Heating from the preheat temperature to sterilisation temperatures and initial cooling of the sterilised milk is much slower in indirect systems than in direct systems. For example, Biziak *et al.* (1985) reported less than 1 s for each heat transfer operation in direct heating compared with greater than 10 s with indirect methods. In practice, the heating times from preheat to final UHT temperature can be up to 100 s (Tran *et al.*, 2008). Therefore, for a given temperature increase, indirect processes are more severe than direct processes. In

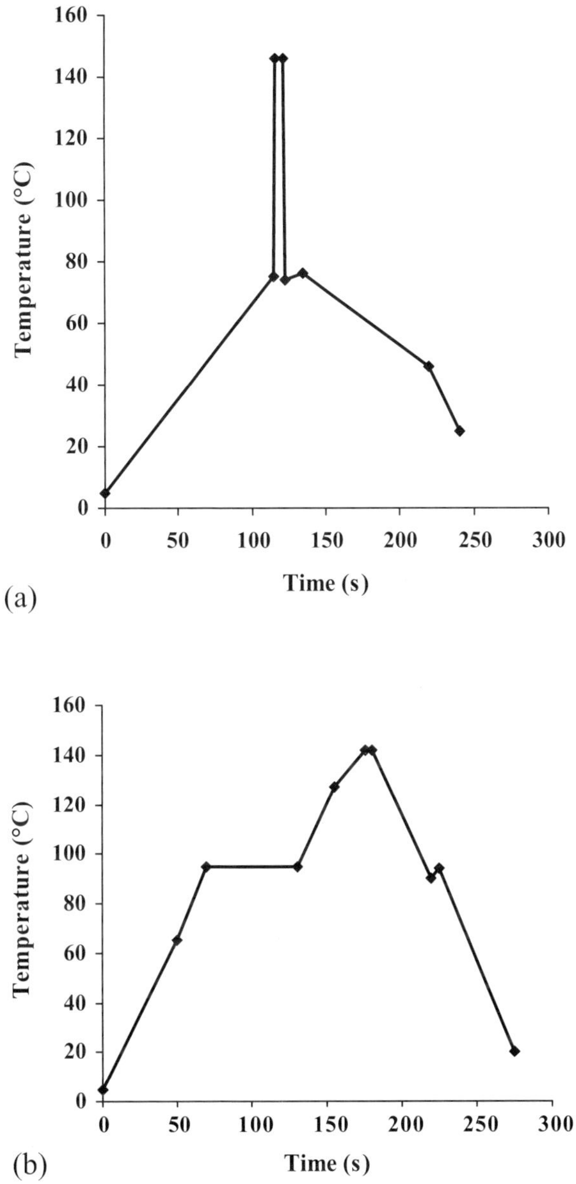

Fig. 7.5 Temperature–time profiles of (a) a direct and (b) an indirect UHT plant.

other words, milk produced in an indirect plant is subjected to a greater heat load than milk processed in a direct plant, with equivalent bactericidal effectiveness. A comparison of direct and indirect temperature–time profiles of two commercial UHT plants is shown in Figure 7.5.

The different heating processes and temperature–time profiles of direct and indirect systems give rise to several differences in the processing characteristics and parameters.

Some key differences are listed in Table 7.1. Of particular note is the higher propensity of indirect plants to foul because of the large area of hot surfaces. In the first stages of the plant where the temperature reached is <100°C, the main reaction is denaturation of whey proteins, mainly β-lactoglobulin, with subsequent deposition of the denatured protein. In the higher-temperature (>100°C) sections of the plant, the major reaction is deposition of calcium phosphate, which has reduced solubility at high temperatures. Consequently, the deposit in the high-temperature sections is predominantly mineral, while the deposit in the lower-temperature section is predominantly protein. The build-up of fouling deposit causes a reduction in heat transfer and an increase in pressure. In commercial plants, the product temperature is maintained by increasing the temperature of the heating medium. This exacerbates the fouling because of the increased temperature differential between the heating medium and the product, and eventually, the plant has to be shut down for cleaning (Ansari *et al.*, 2006).

The degree of heat energy recovery is another difference between the two systems. In direct heating systems, less regeneration of heat is possible since the steam flashed-off in the vacuum chamber is condensed, and the useful heat is lost from the system. In indirect systems, all the heat in the hot sterile product at the sterilisation temperature (135–150°C) can be used in the regeneration section. Thus, heat regeneration in indirect systems is usually >90%, while it is only about 50% in direct systems (Lewis & Heppell, 2000).

Aseptic packaging

Following sterilisation and subsequent cooling, the sterile product is filled into a sterile container in an aseptic environment and hermetically sealed to ensure sterility is maintained throughout the handling and distribution processes. Two main aseptic packaging systems are used commercially. First, the type that uses pre-formed containers, and second, the type that forms, fill and seals the containers in the aseptic packaging system (von Bockelmann & von Bockelmann, 1998). For both systems, the containers can be either plastic or paperboard. In the first system, plastic bottles are pre-blown while in the second they are blown online. For paperboard cartons, in the first system, pre-cut and folded individual packages are assembled online, while for the second system, the cartons are formed, filled and sealed from a continuous roll of paperboard.

Whatever system is employed, the packaging material or container must be sterilised before being filled with the sterile product. The major sterilisation technique used is a combination of hydrogen peroxide and heat, which is a very effective sterilant for the surface of packaging material. A typical procedure is to treat the packaging material with 35% H_2O_2 at 70°C for 6 s followed by hot air treatment at 125°C to evaporate the residual H_2O_2. An alternative to H_2O_2 alone is a mixture of peracetic acid and H_2O_2 (4%) (Carlson, 1998). Gamma radiation is used to sterilise heat-sensitive packaging materials, such as plastics and laminates; an example is the flexible bags used in Intasept^{TM} aseptic packaging machines.

It is important that a sterile environment is maintained during aseptic packaging to ensure the product remains sterile during transfer from the processing line to the sterile container. Following sterilisation of the filling machine with gaseous hydrogen peroxide, the air entering the filling machine is sterilised and filtered, and maintained at a positive

Table 7.1 Summary comparison of direct and indirect UHT heating systems.

Parameter	Direct systems	Indirect systems
Processing characteristics and parameters		
Preheat hold (at ~90°C) 'protein stabilisation' step	Uncommon	Common
Sterilising temperature for equal sterilisation effect	3–4°C higher than in indirect systems	
Homogeniser placement	Generally downstream of high-heat section (requires aseptic homogeniser)	Upstream or downstream of high-heat section
Heating rate from preheat to high heat	Fast (<0.5 s)	Slow (~30–120 s)
Ability to process viscous product	Reasonable, especially with infusion	Little with plate but good capability with tubular heat exchangers
Fouling/burn-on	Usually minimal	A major problem. Tubular better than plate heat exchangers
Run time	Long	Short (tubular longer than plate type)
Heat regeneration	~50%	≥90%
Steam quality requirement	Very high	No specific requirement
Energy requirement	Higher than indirect	
Ability to reach very high temperature (i.e. >145°C)	Capable	Limited
Ability to destroy heat-resistant sporeformers without excessive chemical damage	Better than indirect	
Process control issues	Careful control of water removal after high heat treatment required to prevent concentration or dilution	Need to control pressure increase and temperature differential between product and heating tube or plate as fouling layer builds up
Possibility of contamination from heating medium through pinholes	Nil for sterilising section Possible in regeneration and other indirect heating and cooling	Significant especially with plate heat exchanger
Water requirement	Greater (~1500 L water per 1000 L product) than for indirect system	
Other process features	Steam injection causes some homogenisation	Tubular is most common UHT heating system; corrugated tubes are used to increase turbulence

(Continued)

Table 7.1 *(Continued)*

Parameter	Direct systems	Indirect systems
Product (UHT milk) characteristics		
Flavour (assuming same sterilisation effect)	Mild cooked flavour; chalky if homogenised before high-heat section	Strong cooked flavour
Oxygen level (assuming, no headspace in package, no use of aseptic tank, package not permeable to O_2)	Low (<1 μmL mL^{-1})	High (7–9 μmL mL^{-1})
Sediment formation during storage	Higher than for indirect	
Susceptibility to age gelation	Higher than for indirect	
Plasmin and plasminogen level	Neither completely inactivated	Plasmin generally inactivated but some residual plasminogen may remain
Fat separation	Low, especially for steam injection	More than for direct
Heat indices – HMF, lactulose, furosine	Low	Higher than direct
Undenatured β-lactoglobulin (mg L^{-1})	>700	<200
Folic acid and vitamin C retention	Higher than indirect due to lower oxygen level	Low

Adapted from Datta *et al.* (2002).

pressure of about 0.05 MPa in the filling chamber. Such precautions are necessary to prevent post-sterilisation contamination of the product with airborne bacteria and moulds.

7.9.3 *Factors affecting the quality of UHT milk*

Raw milk quality is affected by: (a) growth of psychrotrophic bacteria and (b) heat-resistant spore-forming bacteria.

Growth of psychrotrophic bacteria

Good quality raw milk is essential for producing UHT milk with a long shelf life. In general, milk destined for UHT processing should be stored refrigerated (<5°C) for no more than 48 h. Storage at higher temperatures and/or for longer times promotes the growth of psychrotrophic bacteria, which cause the production of lactic acid, reduction of the pH of the milk and also production of enzymes, notably proteases and lipases that can have considerable heat stability.

When the pH is reduced to ≤6.5, milk becomes unstable to heat. In UHT processing, such milk readily causes fouling of the heat exchangers, and the final product will show considerable sediment formation (see Section 7.9.4)

The UHT process destroys all vegetative bacteria and most sporeformers but does not inactivate some of the enzymes produced by psychrotrophic bacteria, such as *Pseudomonas* species, i.e. the most common bacterial contaminants of raw milk. Such enzymes are typically produced when the bacterial count exceeds $\sim 10^6$ cfu mL^{-1}. If milk with such bacterial counts is UHT processed, these enzymes, particularly proteinases and lipases, can remain active in the UHT milk. Since UHT milk is usually kept at room temperature and may be stored for several months, even traces of these enzymes can produce noticeable changes, and result in bitter flavour and gelation (due to proteinases) and rancid flavours (due to lipases).

Heat-resistant spore-forming bacteria

Bacterial spores in raw milk present the fundamental challenge for UHT processors. Without such organisms, there would be no need for the heating intensity of UHT to produce a shelf-stable product. The heat-resistant thermophiles, such as *G. stearothermophilus* and *B. licheniformis*, are the most commonly encountered. They can cause the 'flat sour' defect in UHT milk, which is characterised by acid production, but no gas production. However, these thermophiles do not grow in milk under 'normal' storage conditions (≤30°C), and have a growth optimum of ~55°C. They have been known to cause problems if the milk temperature reaches high levels during transportation.

A concern related to raw milk quality is the occurrence of the bacterium *B. sporothermodurans* (IDF, 2000), which produces highly heat-resistant spores. This organism is mesophilic, which means that it can grow at room temperature. Fortunately, it does not appear to cause spoilage other than a slight discolouration of the milk, and seldom reaches counts of greater than 10^5 cfu mL^{-1}. However, it is extremely difficult to remove from contaminated equipment, and has caused the closure of some UHT plants. Because of its heat resistance, its spores may be present in UHT milk and be of little concern. However, the practice of reprocessing out-of-date UHT milk has been shown to enrich the UHT milk with this organism, and cause levels which are of concern.

Scheldeman *et al.* (2004) investigated a case of 'obstinate contamination' of UHT milk from a company, and found two very heat-resistant mesophilic organisms. One was *B. sporothermodurans* and the other was identified as *Paenibacillus lactis*. This was the first time *Paenibacillus* had been isolated from UHT milk, although *Paenibacillus* spores have been previously reported to survive heating at 120°C. *Paenibacillus* spores have been isolated from silage and feed concentrates, which may be the origin of the organism in milk.

Other aspects that can affect the quality of UHT milk are given in the next three sections.

Processing times and temperatures

The major effects of processing on the quality of UHT milk are due to the higher heat load in indirect systems and, to a lesser extent, the lower dissolved oxygen content in directly processed milk, caused by the concomitant extraction of air and water in the vacuum cooling chamber. Table 7.1 contains a summary of the effects of direct and indirect systems on product characteristics.

The two major differences in product characteristics are in flavour and susceptibility to gelation during storage, i.e. age gelation. Indirectly processed milk has a more cooked flavour, and may develop more oxidised or stale flavours due to the higher dissolved oxygen level. However, directly processed milk is more prone to gelation. A major cause of age gelation is proteolysis catalysed by either the native milk plasmin or bacterial proteases resulting from growth of psychrotrophic bacteria, principally pseudomonads, in the milk before processing. Both plasmin and the bacterial enzymes have considerable heat stability; the higher heat load of indirect heating inactivates the proteases much more than does direct heating. While both plasmin and bacterial proteases cause gelation and also bitterness, their modes of action on the caseins are quite different. Bacterial proteases preferentially attack κ-casein in much the same way as chymosin in rennet, while plasmin preferentially attacks β-casein. Thus, the two proteases release different peptides, and this difference can be exploited to determine the cause of age gelation (Datta & Deeth, 2003).

Post-sterilisation contamination

A major consideration in the handling of milk after the high-temperature sterilisation is contamination.This may result from several sources, but two important ones are the seals in the homogeniser (if downstream) and the air supply to the aseptic packaging unit. Kessler (1994) showed that spores trapped under seals had enhanced heat stability, largely attributable to a very low water activity in their microenvironment, and could act as a reservoir of contaminating spores. Flat sour defects due to contamination by *G. stearothermophilus* can arise in this way. Frequent seal changes have been found to be an effective, although expensive, way of minimising such contamination.

Another microbial problem, which has caused problems in several companies in recent years, is the filamentous fungus, *Fusarium oxysporum*. This organism can cause an off-flavour similar to blue-vein cheese in UHT milk within a few weeks, and it also produces gas. It is often detected when packages become swollen or 'blown'. It is a common fungus of plants and soils, and can enter UHT milk packages through contaminated air in the filling machine. Negative air pressures in aseptic filling areas may facilitate contamination of the packaging equipment if there is a source of the fungus nearby. Once the fungus has contaminated a filling machine, it is difficult to eliminate (K. Scrimshaw, personal communication, 2004).

Storage temperature and time

UHT and in-container sterilised milks are subject to considerable chemical and physical changes during storage, although they are more noticeable in the latter (Burton, 1988; Lewis & Heppell, 2000). Each of the changes is dependent on both time and temperature of storage. For example, non-enzymatic browning develops during storage, and proceeds at a faster rate at higher temperatures. It is particularly noticeable above 30°C, and is accelerated in lactose-hydrolysed and sugar-sweetened products.

Age gelation, which is often the limiting factor for the shelf life of UHT milk, is also temperature and time dependent (Datta & Deeth, 2001). However, the temperature dependence is only true up to about 30°C after which gelation is retarded (Kocak & Zadow, 1985) (see Figure 7.6). The exact reason for the retardation at higher temperatures is not known, but may be due to autolysis of the protease involved, excessive proteolysis at the

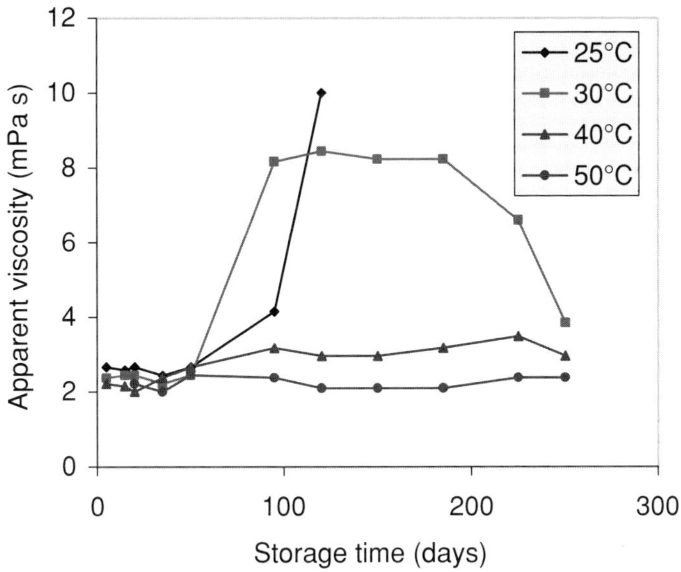

Fig. 7.6 Thickening of UHT milk during storage at different temperatures. Adapted from Kocak & Zadow (1985).

caseins preventing the protein network get from forming or the formation of non-disulphide covalent cross links between the caseins preventing the gel formation (McMahon, 1996).

Oxidation reactions occur, starting with oxidation of sulphydryl compounds produced during heating, then vitamin C and folic acid followed by fats. All these reactions are dependent on the level of dissolved oxygen. This is close to saturation in most products processed by indirect heating, but is reduced in products which have been subject to de-aeration, and in directly treated products subjected to vacuum flash cooling. However, the levels of dissolved oxygen can increase if the sterile milk is stored for some time in an aseptic tank before packaging, if the packaging material is permeable to oxygen or if there is a headspace in the filled package. The headspace volume can vary considerably with the type of package. Perkins *et al.* (2005) found that 1-L Tetra Brik packages to contain an average of ~8 mL, a 1-L Combiloc package ~34 mL and a 1-L plastic bottle ~58 mL. Directly processed milk packaged in the plastic bottles exhibited quite high dissolved oxygen levels (5.8–7.2 mg L^{-1}). The headspace volume can have a major influence on dissolved oxygen levels, and development of volatile carbonyl compounds caused by oxidation (Perkins *et al.*, 2005, 2006).

7.9.4 *Heat stability, sediment formation and fouling*

A major concern with UHT processing is the heat stability of the milk. Milk having poor heat stability can give rise to fouling of heat exchangers and sediment formation, and it would be best to avoid UHT treating such milk. However, there is no test for quickly assessing the heat stability of milk to UHT processing conditions. The much researched heat coagulation test for UHT processing, which measures the time required to coagulate milk at 140°C, is not easy to perform, and it has not been established that it is useful for predicting susceptibility to sediment or fouling in UHT milk.

Most UHT milk contains a slight amount of sediment, which is not usually sufficient to be a problem (Burton, 1988). It has been found that sediment increases with severity of the treatment, and is present in greater quantities following direct processes (Ramsey & Swartzel, 1984). Zadow (1978) found that little sediment was formed in UHT cow's milk if the pH was kept above 6.62; below this value, sedimentation increased rapidly. In contrast, sedimentation was severe in goat's milk when the pH was below 6.9. Similar trends were observed for concentrated skimmed milk, which were found to be stable above pH 6.55, but below this value, severe sedimentation occurred (Zadow & Hardham, 1981). However, on some occasions, a more voluminous sediment appears, and it is more of a problem in goat's than in cow's milk (Zadow *et al.*, 1983; Montilla & Calvo, 1997).

Zadow *et al.* (1983) suggested that ionic calcium may play a role in sediment formation in UHT goat's milk. Up until recently, the only detailed study on ionic calcium in cow's milk at its natural pH was that of White & Davies (1958a,b). A good correlation was found between ethanol stability and ionic calcium for 132 individual cows. Ionic calcium was measured on ultrafiltration permeates derived from each individual milk, using a chemical method. Since then, ion-selective electrodes have been introduced to measure ionic calcium in milk; the most significant of early studies being that of Geerts *et al.* (1983) and more recently by Chavez *et al.* (2004), Lin *et al.* (2006) and Tsioulpas *et al.* (2007a). Thus, there is considerable evidence that ionic calcium varies considerably both in milk from individual cows and in bulk milk, and that this may influence heat stability. However, despite this, its measurement is not routinely practiced as a quality assurance procedure.

One alternative test is the ethanol stability test. By adding equal volumes of different strength alcohol solutions to milk, one can establish the concentration which just fails to cause the milk to coagulate – this is termed the ethanol stability. Pioneering work on understanding factors affecting ethanol stability, and the mechanisms was conducted by Horne and co-workers starting in the 1980s, has been recently summarised by Horne (2003). Shew (1981) recommended that milk should be stable in 74% ethanol to be suitable for UHT processing.

One key question is whether there is a correlation between ethanol stability and stability to UHT processing conditions. The authors' experience with pilot plant and laboratory experiments on cow's and goat's milk suggests that reducing ionic calcium is beneficial in terms of reducing fouling of heat exchangers and sediment formation (Prakash *et al.*, 2006; Boumpa *et al.*, 2008). Also, reducing ionic calcium was found to increase ethanol stability.

Therefore, in situations where sediment formation or fouling is a problem, the following suggestions are offered: routinely monitor pH, ethanol stability and, if possible, ionic calcium in raw milk to establish their contribution to fouling and sediment-related problems. Over time, this should provide data to be able to assess, understand and eventually reduce the problem.

As intimated by Shew (1981), raw milk with an ethanol stability below 74% is likely to be problematic. There are two main reasons why ethanol stability may be low. The first and most likely is a high microbial count, and the second is due to a salt imbalance. The former situation is likely to arise with milk of poor hygienic quality or poor refrigeration. As raw milk quality deteriorates, i.e. as its bacterial count increases, its pH will fall, which in turn will increase ionic calcium and reduce ethanol stability.

However, as Horne (2003) cautions, milk with a low ethanol stability may still not be of poor microbial quality; it may just have a salt imbalance. In this context, salt imbalance refers to a combination of circumstances that leads to the micelle being made more susceptible to coagulation. For example, any factors which reduce the negative charge, such as concentrations of H^+, Ca^{++}, Mg^{++}, Na^+ and K^+ ions, as well as the proportions of different casein fractions in the micelle. One example is goat's milk, which has a high concentration of ionic calcium compared to cow's milk. Similar detrimental changes may occur to milk when calcium supplements are added, or when the pH is reduced by whatever means. Where permitted, stabilisers, such as disodium hydrogen phosphate and trisodium citrate, may help improve stability; their effect is to reduce ionic calcium, increase pH slightly and increase ethanol stability. Levels used are between 0.05 and 0.2 g 100 g^{-1}, but practical experience should dictate. One possible drawback is more browning, brought about by the increase in pH. This is not likely to show immediately after UHT processing, but may manifest itself during storage, especially when the product is stored at above 30°C.

Overall, it is recommended that ionic calcium, pH and ethanol stability are measured to develop a better understanding of how this relates to UHT plant performance and indicators of fouling and sediment.

It has also been suggested that factors that give rise to sediment formation are also responsible for fouling (Burton, 1988). Much of the research on the mechanism of fouling has focused on the role of β-lactoglobulin denaturation, whereas sediment research has revolved around casein micelle aggregation, in terms of pH and more recently ionic calcium. Of course, denatured β-lactoglobulin is involved through its aggregation with the casein micelle on heating. The mechanism of fouling in UHT plants is usually explained in terms of two types of fouling, A and B (Burton, 1988). Type A deposit is soft and voluminous and forms between ~75 and ~110°C in the preheat section; while type B deposit is hard and granular, and forms in the high-temperature sections of the plant. The two deposits differ considerably in composition, with type A being predominantly protein and type B being predominantly mineral, largely calcium phosphate. This mechanism, therefore, focuses on whey protein denaturation and mineral deposition, but does not consider the role of casein. A comprehensive mechanism of fouling and sediment formation involving whey protein denaturation, casein destabilisation and mineral insolubilisation is, therefore, still required.

7.10 High-temperature processing (extended shelf life)

There is a requirement to further increase the shelf life of pasteurised products, both for the convenience of the consumer and to provide additional protection against temperature abuse. However, it is important to avoid the onset of cooked flavour, which results from severe heating conditions. High-temperature pasteurisation has been introduced to meet this requirement. It is a continuous heat treatment between HTST pasteurisation and UHT sterilisation, and is used to produce what has become known as extended shelf life (ESL) milk. It is also known as 'ultrapasteurisation', although this term has a different meaning in some countries. ESL milk can also be produced using non-thermal technologies, such as

Table 7.2 Comparison of HTST pasteurisation, higher pasteurisation and UHT treatment.

Characteristic	HTST pasteurisation	Higher pasteurisation (ESL)	UHT
Heating temp, time	72°C for 15 s	120–135°C for 4–1 s	135–145°C for 10–2 s
Heat index enzyme inactivation	Phosphatase-negative, lactoperoxidase-positive	Phosphatase-negative, lactoperoxidase-negative	Phosphatase-negative, lactoperoxidase-negative
Storage conditions	Refrigerated	Refrigerated	Room temperature
Packaging	Clean	Ultraclean (or aseptic)	Aseptic
Shelf life	10–14 days	30–60 days	>6 months
Flavour	Little heated flavour	Mild heated flavour	Definite heated flavour
Lactulose (mg L^{-1})	~0	20 to ≤40 (Brandes, 2000; Gallmann, 2000; Kjærulff, 2000; Ranjith, 2000)	80–500 (Gallmann, 2000); Ranjith, 2000)
Furosine (mg g^{-1} protein)	~0	200	400–1200 (Gallmann, 2000)
α-Lactalbumin denaturation (%)[a]	~5 (Fredsted *et al.*, 1996)	~5 (Fredsted *et al.*, 1996)	~30–80 (Elliott *et al.*, 2005)
β-Lactoglobulin denaturation (%)[b]	~13 (Fredsted *et al.*, 1996)	~22 (Fredsted *et al.*, 1996)	~60–100 (Andreini *et al.*, 1990; Elliott *et al.*, 2005)
Immunoglobulin denaturation (%)	~67 (Fredsted *et al.*, 1996)	~100 (Fredsted *et al.*, 1996)	~100

[a] Assuming concentration in raw milk = 1200 mg L^{-1}.
[b] Assuming concentration in raw milk = 3000 mg L^{-1}.

microfiltration (Larsen, 1996; Kjærulff, 2000; Hoffmann *et al.*, 2006) and bactofugation, but its production by heating processes only is discussed here.

There is no single definition for ESL milk; however, the EU legislation (EU, 1992) states that pasteurised milk, which shows a negative reaction to the peroxidase test, must be labelled 'high temperature pasteurised milk'. In a recent review on ESL milk, Rysstad & Kolstad (2006) used the following definition: *ESL products are products that have been treated in a manner to reduce the microbial count beyond normal pasteurisation, packaged under extreme hygienic conditions, and which have a defined prolonged shelf life under refrigeration conditions* (see also Rysstad & Spikkestad, 2005). These indicate some of the features of ESL milk, and how ESL milk differs from HTST pasteurised and UHT milks. A more detailed comparison is given in Table 7.2.

A range of different temperature–time combinations has been suggested for producing ESL milk. HTST pasteurisation (72°C for 15 s) does not destroy sporeformers or thermoduric non-spore-forming bacteria, such as coryneforms, micrococci and thermoduric streptococci. As discussed above, increasing the temperature of pasteurisation from 72°C to ~90°C destroys some of the thermoduric non-spore-forming bacteria, but has a detrimental effect on shelf life (Schroder & Bland, 1984; Schmidt *et al.*, 1989). Thus, this temperature range is unfavourable for ESL processing. Therefore, an approach is to use temperatures >100°C for very short times. Wirjantoro & Lewis (1996) showed that milk heated to 115°C for 2 s had a much better keeping quality than milks heated at both 72°C for 15 s and 90°C for 15 s. Ranjith (2000) reported that treatment of milk at temperatures $\leq$117.5°C resulted in high total counts (>10^6 cfu mL^{-1}) after 13 days, whereas milks treated at temperatures $\geq$120°C showed counts of <10^2 cfu mL^{-1} after storage at 7°C for >40 days. It appears that heating at $\geq$120°C is required to inactivate psychrotrophic spore-forming organisms, such as *B. cereus* and *B. circulans.* The upper temperature limit, which Blake *et al.* (1995) concluded was 134°C, is governed by the heat-induced chemical changes, and can cause flavour impairment. Thus, the most common conditions for ESL heat treatment are in the 120–130°C range for a short time (<1 to ~4 s).

Another approach to produce ESL milk is to use small amounts of a bacteriocin. The addition of small amounts of nisin (40 IU mL^{-1}) was also effective in reducing microbial growth following heat treatment at 72°C for 5 s, and even more effective at 90°C for 15 s. It was particularly effective at inhibiting *Lactobacillus* spp. at both temperatures. Results for milk heat treated at 117°C for 2 s with 150 IU mL^{-1} nisin were even more spectacular. Such milks have been successfully stored for over 150 days at 30°C with only very low levels of spoilage (Wirjantoro *et al.*, 2001). Local regulations would need to be checked to establish whether nisin is a permitted additive in milk and milk-based beverages.

A key aspect of the heating conditions to produce ESL milk with a flavour similar to that of pasteurised milks is short heating, holding and cooling times. This is most easily achieved by direct steam heating, either steam injection or steam infusion, for example in the APV Pure-Lac™ system (Fredsted *et al.*, 1996).

For ESL milk, the lowest advisable packaging level is ultraclean, although aseptic packaging is recommended by some authors (e.g. Gallmann, 2000). There is no doubt that the longest shelf lives are obtained with aseptic packaging, but this is achieved at a cost, i.e. the cost of the aseptic packaging equipment installation and maintenance. For this reason, ESL milk is usually packaged in ultraclean rather than aseptic systems.

7.11 Reconstituted and recombined milk products

The issues for reconstituted milk products are similar to those posed by liquid milk products. Most milk powder is now produced by spray drying, which has largely superseded the earlier process of roller drying. The production of milk powder involves a number of processes, including centrifugal separation (for skimmed powder), forewarming, preconcentration and drying. A good account of technology is provided by Kelly (2006).

One of the main characteristics for distinguishing between powders is to the degree of heat treatment, the basis of the whey protein nitrogen index. Low- or medium-heat powders are usually used for reconstituted milk destined for consumption as liquid milk. It is essential to ensure that a high quality milk powder with no off-flavours is used. In terms of minimising problems during heat treatment, it is important to ensure that the powder is well mixed and properly dissolved. There are a number of important properties of milk powders, such as wettability, ability to sink, dispersibility and solubility, which influence this. Usually mixing is achieved at 40–50°C to fully rehydrate the powder. After mixing for about 15–20 min, the milk is often left for another 20 min to remove occluded air, since milk powder may contain up to about 40 mL 100 mL^{-1} of occluded air. Mixing can be done at 5°C, but it will take longer as the powder solubility is lower. Also, oxygen solubility is higher at low temperatures, and the air is less easily removed. Air can also be removed by vacuum deaeration, which can be incorporated as part of the heat treatment, normally after regeneration and before the homogeniser and the holding tube. Too much air will result in fouling of the heat exchanger, cavitation of the homogeniser and excessive oxidation of the product. Poorly dispersed powder can result in blockage of homogeniser valves and narrow gaps between the plates of the PHE.

Water quality is important – it should be good drinking quality, free of pathogenic microorganisms and an acceptably low hardness, less than 100 $\mu g\, g^{-1}$ of calcium carbonate. Excessive mineral content will jeopardise the salt balance, which may cause problems related to heat stability. Copper and iron contents are important as they promote oxidation and may have an adverse effect on flavour.

A very important property of powders in certain applications is their heat stability. This is especially relevant when they are to be used for UHT milk or concentrated milks. It is related to the variability in the composition and types of protein and minerals, and interactions that take place during processing. Of special interest are stability issues occurring due to changes of season, eating patterns and diet. The underlying causes are not well understood.

Some other important properties of milk powders are bulk density, sensory characteristics, nutritional value and microbial population. The source and composition of the fat used in recombined or filled milks may also influence the sensory characteristics of the milk. In countries where fresh milk and reconstituted milks are available, there is an interest in being able to distinguish between them. This is not a straightforward analytical problem to solve.

7.12 Conclusions

Heating is the major processing treatment applied to milk. In most countries, all market milk and most, if not all, milk used for manufacture undergoes some form of heat treatment. The types of heating vary from the mildest treatment, thermisation, through pasteurisation, high-temperature heating or ESL treatment, to UHT and in-container sterilisation, the most severe treatment. All processes reduce bacterial spoilage, and all but thermisation destroy most pathogenic organisms. The technologies used for heating are very mature and have served and continue to serve the consumer and the dairy industry very well. Considerable research has been carried out on the effects of the different methods and severity of heating on the bacterial flora and the chemical constituents of milk, and these are now reasonably well understood. This understanding has enabled the dairy industry to meet challenges related to heating. For example, there has been increasing consumer demand for minimal processing and natural flavours and, hence, UHT milks with very little cooked flavours. However, the discovery of very heat-resistant bacterial spores in UHT milk some years ago suggested a need for more intense UHT treatment to ensure sterility. These apparently conflicting demands were met by development of technologies capable of very rapid direct heating to high temperatures followed by rapid vacuum cooling. This development was possible through an understanding of the different kinetics of bacterial destruction and chemical reactions. The most recent development has been the introduction of ESL milk to satisfy the demand for a longer lasting market milk with a taste similar to that of pasteurised milk. Further developments in this technology can be expected, possibly in conjunction with some forms of non-thermal processing.

7.13 Appendix

7.13.1 *Some heat transfer properties*

Density (ρ, kg m^{-3}): mass/volume; for solids, must distinguish between bulk density and solid density; for foams, overrun is used. Natural convection is due to density differences.

Specific gravity: mass/mass; dimensionless quantity; the term original gravity is often used in brewing examples.

Viscosity (μ, Pa s or Nsm^{-2}): a measure of the internal friction within the fluid; the ratio of shear stress/shear rate; distinguish between Newtonian and non-Newtonian fluids.

Specific heat (c, J kg^{-1} K^{-1}): the amount of energy required to raise unit mass by unit temperature rise; water has a high value compared to other food components.

Latent heat (h_{fg}, kJ kg^{-1}): the amount of energy required to convert unit mass of material from solid to liquid or liquid to gas at a constant temperature; water has a high value.

Specific enthalpy (H, kJ kg^{-1}): enthalpy $= U + PV$, and specific enthalpy is the enthalpy/mass; at constant pressure, enthalpy changes are equivalent to heat changes.

Thermal conductivity (k, W m^{-1} K^{-1}): this is a measure of the rate of heat transfer through a solid; foods are poor conductors of heat.

Thermal diffusivity (α, m^2 s^{-1}): this measures how quickly a material changes in temperature, when energy is added or removed; it is evaluated from thermal conductivity/density $\times$ specific heat – ($k/\rho c$)

Heat film coefficient (h, W m^{-2} K^{-1}): a measure of heat transfer by convection, within a fluid; compare – gases (low), liquids (intermediate) and condensing vapours (high).

Overall heat transfer coefficient (OHTC) (U, W m^{-2} K^{-1}): a measure of the overall heat transfer performance of a heat exchanger, which accounts for all the individual resistances to heat transfer; high OHTC values are desirable, and OHTC is reduced by fouling.

7.13.2 *Definitions and equations*

Temperature conversion

$$^\circ C = (^\circ F - 32) \times 0.56$$

$$K = {}^\circ C + 273.15$$

Temperature difference

$$^\circ C = {}^\circ F \times 0.56$$

$$1^\circ C = 1K$$

Energy units

$$4.18\ J = 1\ calorie$$

$$1\ BTU = 1.055\ kJ$$

$$1\ therm = 10^5\ BTU$$

$$1\ kWh\ (unit) = 3.6 \times 10^6\ J\ or\ 3.6\ MJ$$

Heat exchanger design: rate of heat transfer $Q = UA\Delta\theta_m$

Q = duty (Js^{-1}); U = overall heat transfer coefficient (Wm^{-2} K^{-1}); A = surface area (m^2), $\Delta\theta_m$ = log mean temperature difference (K).

Batch pasteurisation time $(t) = \dfrac{Mc}{AU}\ln\left(\dfrac{\theta_h - \theta_I}{\theta_h - \theta_f}\right)$

M = mass (kg): c = specific heat (J kg^{-1} K^{-1}); A = surface area (m^2); U = OHTC (Wm^{-2} K^{-1}); h = heating medium, I = initial, f = final temperatures.

Regeneration efficiency $= \dfrac{\text{Energy supplied by regeneration}}{\text{Total energy required if there was no regeneration}} \times 100$

Values may be as high as 95%; high values reduce energy costs, but increase capital costs and reduce heating rates.

References

Andreini, R., Choidi, J., De Noni, I., Resmini, P., Battelli, G., Cecchi, L., Todesco, R., Cattaneo, T.M.P., Rampilli, M. & Foschino, R. (1990) [Evaluation of heat damage in UHT and in-bottle sterilized milk samples treated in Italy]. *Scienza E Tecnica Lattier-Caseria*, **41**, 472–492.

Andrews, A.T., Anderson, M. & Goodenough, P.W. (1987) A study of the heat stabilities of a number of indigenous milk enzymes. *Journal of Dairy Research*, **54**, 237–246.

Anonymous (1991) *Food and Drink – Good Manufacturing Practice: A Guide to its Responsible Management*, 3rd edn, Institute of Food Science and Technology, London.

Anonymous (2003a) *Dairy Facts and Figures – 2004*, Milk development Council, London.

Anonymous (2003b) *Proposed Draft Code of Hygienic Practice for Milk and Milk Products at Step 5* (ALINORM 03/13 A, Appendix III).

Ansari, I.A., Sahoo, P.K. & Datta, A.K. (2006) Fouling of heat exchangers in the dairy industry: an overview. *Indian Dairyman*, **56**, 41–47.

Ashton, T.R. & Romney, A.D.T. (1981) In-container sterilization. *Factors Affecting the Keeping Quality of Heat Treated Milk*, Document No. 130, pp. 55–70, International Dairy Federation, Brussels.

Barrett, N., Grandison, A.S. & Lewis, M.J. (1999) Contribution of lactoperoxidase to the keeping quality of pasteurized milk. *Journal of Dairy Research*, **66**, 73–80.

Biziak, R.B., Swartzel, K.R. & Jones, V.A. (1985) Energy use for continuous thermal processing of milk. *Journal of Food Science*, **50**, 1607–1610, 1614.

Blake, M.R., Weimer, B.C., McMahon, D.J. & Savello, P.A. (1995) Sensory and microbial quality of milk processed for extended shelf life by direct steam injection. *Journal of Food Protection*, **58**, 1007–1013.

Boumpa, T., Tsioulpas, A., Grandison, A. & Lewis, M. (2008) Sediment formation in UHT goat's milk. *Journal of Dairy Research*, **75**, 160–166.

Bramley, A.J. & McKinnon, C.H. (1990) The microbiology of raw milk. In: *Dairy Microbiology, Volume 1, The Microbiology of Milk* (ed. R.K. Robinson), pp. 163–208, Elsevier Applied Science, New York.

Brandes, T. (2000) ESL milk – solution with indirect heating applications. In: *Proceedings of the Conference on Extended Shelf Life Milk 2000*, International Dairy Federation, Brussels.

Brown, J.V., Ranjith, H.M.P. & Prentice, G.A. (1984) Comparative shelf lives of skimmed, semi-skimmed and whole milks. *Journal of the Society of Dairy Technology*, **37**, 132–134.

Brown, J.V., Wiles, R. & Prentice, G.A. (1979) The effect of a modified tyndallization process upon the sporeforming bacteria of milk and cream. *Journal of Society of Dairy Technology*, **32**, 109–112.

Browning, E., Lewis, M.J. & MacDougall, D. (2001) Predicting safety and quality parameters for UHT-processed milks. *International Journal of Dairy Technology*, **54**, 111–120.

Burton, H. (1988) *UHT Processing of Milk and Milk Products*, Elsevier Applied Science, London.

Busse, M. (1981) Pasteurized milk, factors of a bacteriological nature. *Factors affecting the Keeping Quality of Heat Treated Milk*, Document No. 130, pp. 38–41, International Dairy Federation, Brussels.

Carlson, B. (1998) Food processing equipment-Historical and modern designs. In: *Aseptic Processing and Packaging of Food* (eds J.R.D. David, R.H. Graves & V.R. Carlson), pp. 95–143, CRC Press, Boca Raton, FL.

Chavez, M., Negri, L., Taverna, M. & Cuatrin, A. (2004) Bovine milk composition affecting the ethanol stability. *Journal of Dairy Research*, **71**, 201–206.

Cronshaw, H.B. (1947) *Dairy Information*, Dairy Industries Ltd., London.

Datta, N. & Deeth, H.C. (2001) Age gelation of UHT milk – a review. *Food and Bioproducts Processing*, **79**, 197–210.

Datta, N. & Deeth, H.C. (2003). Diagnosing the cause of proteolysis in UHT milk. *Lebensmittel-Wissenschaft und Technology*, **36**, 173–182.

Datta, N., Elliott, A.J., Perkins, M.L. & Deeth, H.C. (2002) Ultra-High Temperature (UHT) treatment of milk: comparison of direct and indirect modes of heating. *Australian Journal of Dairy Technology*, **57**, 211–227.

Davis, J.G. (1955) *A Dictionary of Dairying*, 2nd edn, Leonard Hill, London.

Deeth, H.C. (2002) Lipolysis in milk and dairy products – a research journey. *Food Australia*, **54**, 433–436.

Deeth, H.C. & Fitz-Gerald, C.H. (2006) Lipolytic enzymes and hydrolytic rancidity. In: *Advanced Dairy Chemistry – 2 Lipids*, 3rd edn (eds P.F. Fox and P. McSweeney), pp. 481–556, Springer, New York.

Deeth, H.C., Khusniati, T., Datta, N. & Wallace, R.B. (2002) Spoilage patterns of skim and whole milks. *Journal of Dairy Research*, **68**, 227–241.

Dentener, A. (1984). Influence of milk protein on quality aspects of UHT milk products. *Food Technology in New Zealand*, **19**, 33–35.

Elliott, A.J., Datta, N. & Deeth, H.C. (2005) Heat-induced and other chemical changes in commercial UHT milks. *Journal of Dairy Research*, **72**, 442–446.

EU (1992) Council Directive 92/46/EEC of 16 June 1992 laying down the health rules for the production and placing on the market of raw milk, heat-treated milk and milk-based products. *Official Journal of the European Communities*, **35**, 1–32.

Fredsted, L.B., Rysstad, G. & Eie, T. (1996) Pure-Lac™: the new milk with protected freshness and extended shelflife. In: *Heat Treatments and Alternative Methods*, Special Issue 9602, pp. 104–125, International Dairy Federation, Brussels.

Gallmann, P.U. (2000) Possible quality improvement/benefits of ESL products. In: *Proceedings of the Conference on Extended Shelf Life Milk 2000*, International Dairy Federation, Brussels.

Geerts, J., Bekhof, J. & Scherjon, J. (1983) Determination of calcium ion activities in milk with an ion selective electrode. *Netherlands Milk and Dairy Journal*, **37**, 197–211

Gomez Barroso, B. (1997) *Effect of Raw Milk Quality on Keeping Quality of Pasteurized Milk*, MSc Dissertation, Department of Food Science and Technology, The University of Reading, Reading.

Gould, G.W. (1989) Heat induced injury and inactivation. In: *Mechanisms of Action of Food Preservation Procedures* (ed. G.W. Gould), pp. 11–42, Elsevier Applied Science, London.

Grant, I.R. (2006) *Mycobacterium avium* ssp. *paratuberculosis* in foods: current evidence and potential consequences. *International Journal of Dairy Technology*, **59**, 112–117.

Grant, I.R., Rowe, M.T., Dundee, L. & Hitchings, E. (2001) *Mycobacterium avium* ssp. *paratuberculosis*: its incidence, heat resistance and detection in raw milk and diary products. *International Journal of Dairy Technology*, **54**, 2–13.

Griffiths, M.W. (1986) Use of milk enzymes and indices of heat treatments. *Journal of Food Protection*, **49**, 696–705.

Hammer, P., Knappstein, K. & Hahn, G. (1998) *Significance of* Mycobacterium paratuberculosis *in Milk*, Document No. 330, pp. 12–16, International Dairy Federation, Brussels.

Hammer, P., Lembke, F., Suhren, G. & Heeschen, W. (1996) Characterisation of heat resistant mesophilic *Bacillus* species affecting the quality of UHT milk. *Heat Treatments and Alternative Methods*, Special Issue 9602, pp. 9–16, International Dairy Federation, Brussels.

Hoffmann, W., Kiesner, C., Clawin-Rädecker, I., Martin, D., Einhoff, K., Lorenzen, P.C., Meisel, H., Hammer, P., Suhren, G. & Teufel, P. (2006) Processing of extended shelf life milk using microfiltration. *International Journal of Dairy Technology*, **59**, 229–235.

Holdsworth, S.D. (1997) *Thermal Processing of Packaged Foods*, Blackie Academic and Professional, London, UK.

Horne, D. (2003) Ethanol stability. In: *Advanced Dairy Chemistry – Proteins*, Vol. 1, 3rd edn (eds P.F. Fox & P.L.H. McSweeney), pp. 975–1000, Kluwer Academic, New York.

Huijs, G., van Asselt, A., Verdurmen, R. & de Jong, P. (2004) High speed milk. *Dairy Industries International*, **69**(11), 30–32.

IDF (1984) *The Thermization of Milk*, Document No. 182, International Dairy Federation, Brussels.

IDF (1986) *Monograph on Pasteurised Milk*, Document No. 200, International Dairy Federation, Brussels.

IDF (1988) *Code of Practice for the Preservation of Milk by the Lactoperoxidase System*, Document No. 234, International Dairy Federation, Brussels.

IDF (1993) *Catalogue of Tests for the Detection of PPC of Milk,* Document No. 281, International Dairy Federation, Brussels.

IDF (1994) *Indigenous Antimicrobial Agents of Milk – Recent Developments*, Special Issue 9404, International Dairy Federation, Brussels.

IDF (1998) *Significance of* Mycobacterium paratuberculosis *in Milk*, Document No. 330, International Dairy Federation, Brussels.

IDF (2000) Bacillus sporothermodurans – *A* Bacillus *Forming Highly Heat-Resistant Spores*, Document No. 357, pp. 1–27, International Dairy Federation, Brussels.

IDF (2004) *Heat Resistance of Pathogenic Organisms*, Document No. 392, pp. 12–126, International Dairy Federation, Brussels.

Janzen, J.J., Bishop, J.R. & Bodine, A.B. (1982) Relationship of protease activity to shelf life of skim and whole milk. *Journal of Dairy Science*, **65**, 2237–2240.

Juffs, H. & Deeth, H.C. (2007) *Scientific Evaluation of Pasteurisation for Pathogen Reduction in Milk and Milk Products*. Food Standards Australia and New Zealand, Canberra.

Kelly, P.M. (2006) Innovation in milk powder technology. *International Journal of Dairy Technology*, **59**, 70–75.

Kessler, H.G. (1981) *Food Engineering and Dairy Technology*, Verlag A Kessler, Freising.

Kessler, H.G. (1989) Effect of thermal processing on milk. In: *Developments of Food Preservation – 5* (ed. S. Thorne), pp. 91–130, Elsevier Applied Science, New York.

Kessler, H.G. (1994) Heat resistance of microorganisms under seals. In: *Proceedings of the 24th International Dairy Congress (Melbourne)*, International Dairy Federation, Brussels.

Kessler, H.G. & Horak, F.P. (1984) Effect of heat treatment and storage conditions on keeping quality of pasteurized milk. *Milchwissenschaft*, **39**, 451–454.

Kjærulff, G. (2000) Infusion heating and application of the Pure-Lac™concept. In: *Proceedings of the Conference on Extended Shelf Life Milk 2000*, International Dairy Federation, Brussels.

Kocak, H.R. & Zadow, J.G. (1985) Age gelation of UHT whole milk as influenced by storage temperature. *Australian Journal of Dairy Technology*, **40**, 14–21.

Komorowski, E.S. (2006) New dairy hygiene legislation. *International Journal of Dairy Technology*, **59**, 97–101.

Larsen, P.H. (1996) Microfiltration for pasteurized milk. *Heat Treatments and Alternative Methods*, Special Issue 9602, pp. 232–239, International Dairy Federation, Brussels.

Lehmann, F.L. (1996) Recontamination of industrial ultrafiltration units and pasteurisers by thermoduric – thermophilic bacteria. *Heat Treatments and Alternative Methods*, Special Issue 9602, pp. 126–128, International Dairy Federation, Brussels.

Lehmann, F.L., Russell, P.S., Solomon, L.S. & Murphy, K.D. (1992) Bacterial growth during continuous pasteurization. *Australian Journal of Dairy Technology*, **47**, 28–32.

Lewis, M.J. (1990) *Physical Properties of Foods and Food Processing Systems*, Ellis Horwood Publishers (now Woodhead Publishing), Cambridge.

Lewis, M.J. & Heppell, N. (2000) *Continuous Thermal Processing of Foods: Pasteurization and UHT Sterilization*, Aspen Publishers, Gaithersburg.

Lin, M.-J., Lewis, M.J. & Grandison, A.S. (2006) Measurement of ionic calcium in milk. *International Journal of Dairy Technology*, 59, 192–199.

Marks, N.E., Grandison, A.S. & Lewis, M.J. (2001) Challenge testing of the lactoperoxidase system in pasteurized milk. *Journal of Applied Microbiology*, **91**, 735–741.

McCance, R.A. (2002) *McCance and Widdowson's The Composition of Foods*, 6th edn, pp. 88–123, The Royal Society of Chemistry, Cambridge, UK.

McMahon, D.J. (1996) Age-gelation of UHT milk: changes that occur during storage on shelf-life and the mechanism by which age-gelation occurs. *Heat Treatments and Alternative Methods*, Special Issue 9602, pp. 315–325, International Dairy Federation, Brussels.

Mehta, R.S. (1980) Milk processed at ultra-high-temperatures – a review. *Journal of Food Protection*, **43**, 212–225.

Montilla, A. & Calvo, M.M. (1997) Goat's milk stability during heat treatment: effect of pH and phosphates. *Journal of Agriculture and Food Chemistry*, **45**, 931–934.

Muir, D.D. (1996a) The shelf life of dairy products: 1 Factors influencing raw milk and fresh products. *Journal of the Society of Dairy Technology*, **49**, 24–32.

Muir, D.D. (1996b) The shelf life of dairy products: 2 Raw milk and fresh products. *Journal of the Society of Dairy Technology*, **49**, 44–48.

Peleg, M. (2006) It's time to revise thermal processing theories. *Food Technology*, **60**(7), 92.

Perkins, M.L., Elliott, A.J., D'Arcy, B.R. & Deeth, H.C. (2005) Stale flavour volatiles in Australian commercial UHT milk during storage. *Australian Journal of Dairy Technology*, **60**, 231–237.

Perkins, M.L., Zerdin, K., Rooney, M.L., D'Arcy, B.R. & Deeth H.C. (2006) Active packaging of UHT milk to prevent the development of stale flavour during storage. *Packaging Technology and Science*, **20**, 137–146.

Prakash, S., Datta, N., Lewis, M.J. & Deeth, H.C. (2006) Reducing fouling during UHT treatment of goat's milk. *Milchwissenschaft*, **62**, 16–19.

Ramsey, J.A. & Swartzel, K.R. (1984) Effect of UHT processing and storage conditions on rates of sedimentation and fat separation of aseptically packaged milk. *Journal of Food Science*, **49**, 257–262.

Ranjith, H.M.P. (2000) High temperature pasteurization. In: *Proceedings of the Conference on Extended Shelf Life Milk 2000*, International Dairy Federation, Brussels.

Ravanis, S. & Lewis, M.J. (1995) Observations on the effects of raw milk quality on the keeping quality of pasteurised milk. *Letters in Applied Microbiology*, **20**, 164–167.

Reiter, B. & Harnulv, B.G. (1982) The preservation of refrigerated and uncooled milk by its natural lactoperoxidase system. *Dairy Industries International*, **47**(5), 13–19.

Robertson, G.L. (2006) *Food Packaging: Principles and Practice*, 2nd edn, pp. 255–270, Taylor & Francis, Boca Raton, FL.

Rysstad, G. & Kolstad, J. (2006) Extended shelf life milk – advances in technology. *International Dairy Journal*, **59**, 85–96.

Rysstad, G. & Spikkestad, N. (2005) [Extended shelf life milk – technology and market chances], in German, *Deutsche Milchwirtschaft*, **56**, 584–587.5
Schaafsma, G. (1989) *Effects of Heat Treatment on the Nutrient Value of Milk*, Document No. 238, pp. 68–70, International Dairy Federation, Brussels.
Scheldeman, P., Goossens, K., Rodriguez-Diaz, M., Pil, A., Goris, J., Herman, L., de Vos, P., Logan, N.A. & Heyndrickx, M. (2004) *Paenibacillus lactis* sp. nov., isolated from raw and heat-treated milk. *International Journal of Systematic and Evolutionary Microbiology*, **54**, 885–891.
Schmidt, D., Cromie, S.J. & Dommett, T.W. (1989) Effects of pasteurisation and storage conditions on the shelf life and sensory quality of aseptically packaged milk. *Australian Journal of Dairy Technology*, **44**, 19–24.
Schroder, M.A. & Bland, M.A. (1984) Effect of pasteurisation temperature on keeping quality of whole milk. *Journal of Dairy Research*, **51**, 569–578.
Shew, D.I. (1981) *New Monograph on UHT Milk*, Document No. 133, pp. 115–121, International Dairy Federation, Brussels.
SI (Statutory Instruments) (1995) *Food Milk and Dairies, The Dairy Products (Hygiene) Regulations*, Statutory Instruments No. 1086, HMSO, London.
Tran, H., Datta, N., Lewis M.J. & Deeth H.C. (2008) Processing parameters and predicted product properties of industrial UHT milk processing plants in Australia. *International Dairy Journal*, **18**, 939–944.
Tsioulpas, A., Grandison, A. & Lewis, M.J. (2007a) Effect of minerals on casein micelle stability of cows' milk. *Journal of Dairy Research*, **74**, 167–173.
Tsioulpas, A., Grandison, A. & Lewis, M.J. (2007b) A study of the pH of individual milk samples. *International Journal of Dairy Teachnology*, **60**, 96–97.
von Bockelmann, B. & von Bockelmann, I. (1998) Aseptic packaging. *Long-Life Products: Heat-treated, Aseptically Packed: A Guide to Quality*, pp. 9–13, Bernhard von Bockelmann Publisher, Åkarp, Sweden.
Walstra, P. & Jenness, R. (1984) *Dairy Chemistry and Physics*, John Wiley & Sons, New York.
White, J. & Davies, D. (1958a) The relation between the chemical composition of milk and the stability of the caseinate complex. *Journal of Dairy Research*, **25**, 236–255.
White, J. & Davies, D. (1958b) The relation between the chemical composition of milk and the stability of the caseinate complex, II coagulation by ethanol. *Journal of Dairy Research*, **25**, 255–266.
Wilbey, R.A. (2002) Microbiology of cream and butter. In: *Dairy Microbiology Handbook – The Microbiology of Milk and Milk Products*, 3rd edn (ed. R.K. Robinson), pp. 123–174, John Wiley & Sons, New York.
Wirjantoro, T.I. & Lewis, M.J. (1996) Effect of nisin and high temperature pasteurisation on the shelf-life of whole milk. *Journal of the Society Dairy Technology*, **4**, 99–102.
Wirjantoro, T.I., Lewis, M.J., Grandison, A.S., Williams, G.C. & Delves-Broughton, J. (2001) The effect of nisin on the keeping quality of reduced heat-treated (RHT) milks. *Journal of Food Protection*, **64**, 213–219.
Zadow, J.G. (1978) The stability of milk of low pH towards UHT processing. In: *Proceedings of the 20th International Dairy Congress*, Vol. 3, pp. 711, International Dairy Federation, Brussels.
Zadow, J. & Hardham, J. (1981) Studies on the stability of concentrated and reconstituted concentrated skim milks towards UHT processing. *Australian Journal of Dairy Technology*, **36**, 30–33
Zadow, J., Hardham, J., Kocak, H. & Mayes, J. (1983) The stability of goat's milk to UHT processing. *Australian Journal of Dairy Technology*, **38**, 20–23.

8 Novel Methods of Milk Processing

M. Villamiel, M.A.I. Schutyser and P. de Jong

8.1 Introduction

This chapter evaluates a number of milk-processing technologies that can be considered novel or emergent. Traditionally, thermal treatments are applied in the dairy industry because of their efficacy in inactivating microorganisms and enzymes. However, during some of these thermal treatments important chemical changes take place giving rise to modifications not only in the sensorial characteristics, but more likely in the nutritional value of the food. These limitations are the drive for the dairy industry to search for new technologies to improve its products and develop new ones with high-quality and stable properties. Most of these novel technologies are not completely new and they were tried out several decades ago in the food industry, but without success. More recently, the technical-scientific progress together with consumer demands for minimally processed foods has led to their renaissance.

In this chapter, the authors have attempted to collect the most investigated technologies alternative to the conventional thermal treatments. General topics, effects on microorganisms and enzymes, as well as chemical and sensorial modifications are covered. In spite of the great number of developed studies, more research is needed to validate the effectiveness of these technologies and, therefore, to identify serious alternatives for the traditional treatments. Table 8.1 gives an overview of the development status of the different treatment methods reviewed, their efficacy in inactivation of bacteria, spores and enzymes and their influence on sensorial product properties as discussed in this chapter. In the following paragraphs, each novel method is described with respect to its operating principle, its effect on dairy products, the state of the art and its future opportunities (see also Tewari & Juneja, 2007).

8.2 Microwaves

Microwave energy is used for domestic applications and for several processing operations such as cooking, baking, drying and thawing (Young & Jolly, 1990; Datta & Davidson, 2000; Wäppling-Raaholt & Ohlsson, 2005). As compared to conventional heat treatment, during microwave heating, microwaves penetrate the food directly and rapidly without contact with hot surfaces, and therefore, less time is needed to reach the desired process temperature (Young & Jolly, 1990). In spite of the known advantages of microwave heating, the food industry has been reluctant to adopt this technology, probably due to the difficulty of predicting the heating uniformity (Ohlsson, 1983; Mudgett, 1986). This, together with the fact that microwave processing involves substantial capital costs and needs to be designed precisely

Table 8.1 Overview of different novel milk-processing methods and their efficacy for inactivation of bacteria, spores and enzymes, and their influence on the retention of sensorial characteristics of the treated product.

Milk-processing methods	Main process parameters	Development status for dairy products	Bacterial inactivation	Spores inactivation	Enzymes inactivation	Retention of sensorial characteristics
Microwave heating	Frequency, t, T	Laboratory scale	+	+/−	+/−	Improvement compared to conventional heating
High pressure (HP)	P, t, T	Pilot phase	+	+/−	−	Significant if operation is executed at room temperature
Pulsed electric fields (PEF)	Field strength, pulse number, frequency, and width	Pilot phase	+	−	−	Significant if operation is executed at room temperature
Ultrasound	Frequency, t	Laboratory scale	+/−	+/−	−	Ultrasound is effective in combination with increased temperature and pressure, but ultrasound itself has no significant effect
Microfiltration (MF)	Membrane type, flow conditions, P, cleaning frequency	Commercialised	+	+	−	Significant
Innovative steam injection (ISI)	t, T	Pilot phase	++	++	+	Improvement compared to ultra-high-temperature (UHT) treatment, similar to pasteurisation

Note: T = temperature; t = time; P = pressure; −negligible effect; +/− minor effect; + effect; ++ significant effect.

around an application, has slowed the development of industrial equipment (Akiyama, 2000; Tops, 2000; Orsat & Raghavan, 2005; Wäppling-Raaholt & Ohlsson, 2005).

8.2.1 *Operating principle*

Microwaves are produced in a magnetron and are part of the electromagnetic spectrum within the frequency range from 300 MHz to 300 GHz, which falls between the radio frequency and infrared regions (Decareau, 1985). 2450 MHz is commonly used for domestic ovens and commercial food-processing equipment, although 915 MHz is also permitted in USA and other countries (Young & Jolly, 1990). Microwaves are non-ionising radiation; they do not break chemical bonds or cause molecular changes in compounds by removal of electrons (Cross & Fung, 1982).

Since the beginning of microwave processing, there has been controversy on the possible 'athermal' effects of microwaves which might participate in the inactivation of microorganisms. However, it was demonstrated that this inactivation is exclusively attributable to the heat originated within the food (Heddleson & Doores, 1994), the shapes of the inactivation curves being similar to those for conventional heating (Datta & Davidson, 2000).

The two main mechanisms by which microwaves cause heat generation in foods are dipolar rotation and ionic polarization (Vasavada, 1990). Due to their dipolar nature, water molecules follow the electric field of microwaves and oscillate at very high frequencies, producing heat. The other mechanism is related to oscillatory migration of ions. In milk, it has been demonstrated that the proteins are the main contributors to the heating (Kudra *et al.*, 1991).

The main drawback of microwave treatment is the lack of uniform heating, which can lead to cold and/or hot spots within the product; these might cause, respectively, risk of microbial survival (Ohlsson, 1990) and acceleration of chemical reactions between components (Villamiel *et al.*, 1996a). Several features of the equipment (oven-type, frequency, power, batch or continuous system) and the food (dielectric properties, size and shape) affect microwave heating (Schiffmann, 1986; Datta & Hu, 1992; Sieber *et al.*, 1996a; Villamiel & Corzo, 1998). Many attempts have been made to improve the uniformity during microwave treatments (Datta & Davidson, 2000; Wäppling-Raaholt & Ohlsson, 2005).

8.2.2 *Effects on product properties*

Since the first reported application on milk pasteurisation (Hamid *et al.*, 1969), many studies have investigated the suitability of microwaves to inactivate microorganisms and alkaline phosphatase in milk. High-temperature and short-time (HTST) (Jaynes, 1975) and low-temperature and long-time (LTLT) (Merin & Rosenthal, 1984) treatments of milk have been successfully assayed. Studies on the effect of microwaves on *Salmonella typhimurium*, *Escherichia coli*, *Pseudomonas fluorescens*, *Staphylococcus aureus*, *Listeria monocytogenes*, *Yersinia enterocolitica* and *Campylobacter jejuni* in milk were not conclusive, since the inactivation was affected by the volume and composition of the sample, among other experimental factors. It should be noted that these treatments were performed in batch systems, where considerable uneven heat distribution can be produced (Stearns & Vasavada, 1986; Knutson *et al.*, 1988; Galuska *et al.*, 1989; Choi et *al.*, 1993a,b; Heddleson *et al.*, 1996).

Although most of the studies reported in the literature on pasteurisation and sterilisation have been conducted on cow's milk, other authors have suggested the usefulness of domestic microwave ovens for pasteurisation of buffalo (El-Shibiny *et al.*, 1982) and goat (Thompson & Thompson, 1990) milks, and infant formula (Kindle *et al.*, 1996).

As the experimental conditions and design of the equipment were demonstrated to be of paramount importance for successful inactivation of microorganisms, Villamiel *et al.* (1996b) developed a continuous-flow system, which provided a uniform heat distribution. According to the phosphatase, lactoperoxidase and total bacteria inactivation, the system developed was effective for pasteurisation of cow's and goat's milk. Microwave treatment provided milk not only with satisfactory microbial quality, but also with longer shelf life than that of milk heated in a plate heat exchanger under the same heating conditions (Villamiel *et al.*, 1996c). Recently, Clare *et al.* (2005) confirmed that microwave technology may be a useful alternative for the processing of milk.

One of the chemical modifications that have caused more polemic is the amino acid change during batch microwave heating of infant formula. Lubec *et al.* (1989) indicated that isomerisation of L-amino acids, supposedly originated after this treatment, could be neurotoxic for children. On the contrary, Segal (1990) and Bögl (1990) stated that D-amino acids are neurotoxic when they are injected in the brain of experimental animals; moreover, the human metabolism has enzymatic systems to modify these D-amino acids. Finally, Fay *et al.* (1991), Marchelli *et al.* (1992), Sieber *et al.* (1996a) and Vasson *et al.* (1998) clarified this controversy, and they pointed out that, under the usual conditions of microwave treatment of milk and infant formula, no significant changes are produced in amino acids.

The effect of microwaves on proteins and carbohydrates during batch heating of milk has been investigated. Comparative studies between microwave and conventional treatments have shown that more denaturation of whey proteins, extent of Maillard reaction and lactose isomerisation are produced in the former than in the latter (Merin & Rosenthal, 1984; Fernández-Márquez *et al.*, 1992; Villamiel *et al.*, 1996a), probably due to an acceleration of the reactions. However, Meissner & Erbersdobler (1996) did not observe significant differences in Maillard reaction extent and lactose isomerisation between the two types of treatment. These different results could be attributed to the temperature/time conditions used in the different assays.

Whey proteins, lysine, non-casein nitrogen and vitamins B_1 and C have been used as parameters to assess the lack of uniformity during batch microwave treatment of milk. By using these compounds, Sieber *et al.* (1993) and Eberhard *et al.* (1990) demonstrated the presence of important temperature gradients, with the upper part of the milk achieving the highest temperature.

To clarify the effects of microwaves on milk components under more practical conditions and to minimise the uneven heat distribution which characterises batch microwave systems, Villamiel *et al.* (1996b, 1998) and López-Fandiño *et al.* (1996b) developed a continuous-flow microwave equipment which allowed identical heating, holding and cooling phases as compared to a conventional process. No studies had previously addressed these aspects. At 85–120°C, the observed thermal whey protein denaturation (Table 8.2) and Maillard reaction extent were lower in the microwave system, probably due to a better heat distribution and the lack of hot surfaces contacting the milk. In addition, volatiles, such as aldehydes, ketones, alcohols, esters and aromatic hydrocarbons, were similar to those found in raw milk, indicating that microwave heating has no adverse effects on milk flavour (Valero *et al.*, 2000).

Table 8.2 Denaturation (%) of β-lactoglobulin in milk treated by microwave and conventional equipment at temperatures from 72 to 85°C after 0, 15 and 25 s of holding time.

Experimental conditions		β-Lactoglobulin denaturation (%)	
Temperature (°C)	Holding time (s)	Microwave	Conventional
72	0	0.9	3.2
72	15	1.6	3.6
72	25	2.8	4.2
80	0	9.2	9.6
80	15	13.8	13.3
80	25	16.9	17.3
85	0	18.8	23.3
85	15	26.4	27.8
85	25	28.0	33.0

During microwave and conventional treatments, the coagulation properties of milk have been studied. Vasavada *et al.* (1995) reported that cow's milk heated in a batch microwave oven had better coagulation properties than milk heated conventionally. However, other authors did not find significant differences in the coagulation parameters between conventional and microwave heating in a continuous-flow (Villamiel *et al.*, 1996d) or in a batch (de la Fuente *et al.*, 2002) system. de la Fuente *et al.* (2002) also observed that the mineral balance brought about by microwave heating did not differ from the changes produced by conventional heating.

The behaviour of vitamins (B_1, B_2, B_6, A, E and folic acid) during microwave treatment of milk has been studied by several researchers (Vidal-Valverde & Redondo, 1993; Medrano *et al.*, 1994; Ovesen *et al.*, 1996; Sieber *et al.*, 1996b; Sierra *et al.*, 1999; Sierra & Vidal-Valverde, 2000a,b, 2001). Although some conflicting results were obtained, in general, the reported findings indicated that the retention of vitamins during domestic and continuous-flow microwave treatments is comparable to that with conventional methods of heating.

8.2.3 *State of the art and opportunities*

It can be summarised that, according to the present knowledge, microwaves do not have specific effect on the chemistry of milk. The effect of microwaves on milk microorganisms and components is attributed to the heat generation within the food. The main drawback of microwave treatment is related to the uneven heat distribution, which can be improved by continuous-flow equipment; moreover, the scale-up of the technology is more feasible in continuous systems. To date, microwave equipment for the thermal treatment of milk on industrial scale is not available in the industry.

8.3 High Pressure

High pressure (HP) is a non-thermal treatment of foods that constitutes a clear alternative to traditional heat treatments. In fact, HP is regarded as having great potential not only for food preservation, but also for production of innovative foods with modified functional properties or structures. Some pressurised foods, such as fruit jams, jellies, sauces, juices and yoghurts, avocado pulp, guacamole and cooked ham, are already commercially available (Cheftel, 1995; Tauscher, 1995; Gould, 2001; Rademacher & Hinrichs, 2002; Hjelmqwist, 2005). Most of the treatments are carried out in batch or in semi-continuous equipment. The retention of the food's original freshness, colour, flavour, taste and nutritional quality during microorganism inactivation is, perhaps, the main advantage of HP processing (Tauscher, 1995; Datta & Deeth, 1999; Farkas & Hoover, 2000; Sierra *et al.*, 2000).

8.3.1 *Operating principle*

HP treatment of food is usually performed in the range 100–1000 MPa at room temperature or higher (up to 60–80°C to inactivate spores) during times up to 30 min. The pressure increases the temperature of foods approximately 2–3°C 100 MPa^{-1}. This slight increase of temperature is followed by a corresponding adiabatic cooling effect on depressurisation (Balci & Wilbey, 1999). Foods, in flexible and sealed packages, are placed into the pressure vessel filled with a pressure-transmitting medium, which is usually water. HP penetrates uniformly through the food, and there is no dependence between time and mass. Critical process factors of HP treatment include pressure, time of pressure, time to achieve treatment pressure, decompression duration, treatment temperature, product initial temperature, vessel temperature distribution, product pH and water activity, and packaging material integrity, among others (Farkas & Hoover, 2000).

The effects of pressure on biomolecules follow the Le Chatelier–Braun principle, which implies that the volume change that is imposed by high pressure evokes changes in the three-dimensional configuration of a biomolecule. Pressures used in the HP treatment of foods appear to have little effect on covalent and hydrophobic bonds, whereas the breaking of ionic and hydrogen bonds is favoured. Small molecules such as amino acids, vitamins and flavour components remain unaffected due to their relatively simple structures. However, proteins, enzymes, polysaccharides and nucleic acids may change under the influence of pressure (Balci & Wilbey, 1999). Such effects may result in the inactivation of microorganisms, denaturation of enzymes and textural changes in foods.

8.3.2 *Effect on product properties*

The effect of HP on milk bacteria was initially studied by Hite (1899), who found that microbial spoilage was delayed after HP treatment at 680 MPa for 10 min. Hite *et al.* (1914) indicated that complete sterilisation of milk by HP needs continuous pressurisation at 680 MPa for 7 days. Since these first investigations, several publications have focused not only on the microorganisms naturally present in milk, but also on microorganisms introduced to milk by contamination; this topic has been extensively reviewed. As examples of the reviews of Balci & Wilbey (1999), Datta & Deeth (1999) and Patterson (2005) can be mentioned. In general, *E. coli* and *L. monocytogenes* seem to be the most pressure-resistant species at room temperature.

Most of the vegetative forms of microorganisms can be destroyed at 600 MPa for 15 min and at 20–30°C (Thimson & Short, 1965); bacteria in the stationary phase being more resistant than those in the growth phase. Several mechanisms can be involved in the inactivation of microorganisms by HP: (a) permeabilisation of membrane, (b) concomitant leakage of

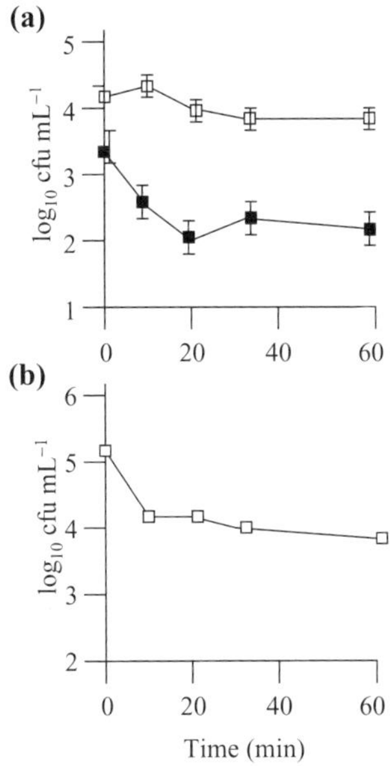

Fig. 8.1 Effect of 200 MPa (a) and 400 MPa (b) of pressure at different exposure times on total bacteria (□) and psychrotrophic (■) counts of milk. Note: The error bars ($n = 3$) are included. Taken from López-Fandiño *et al.* (1996a), and reproduced with permission of *Journal of Dairy Science*.

the contents of the cells and organelles and (c) changes in proteins, enzymes and nucleic acids (Hayashi, 1988; Hoover *et al.*, 1989; Isaacs *et al.*, 1995; Balci & Wilbey, 1999). The main factors that can influence the effectiveness of HP for microorganism inactivation are the type and state of growth, the pressure and time of application, and the composition and pH of the food. In general, spores of bacteria have been found to be more resistant to HP than vegetative cells.

Due to the different experimental conditions used in various studies, certain differences on the effect of HP on microorganisms can be found. However, it seems to be clear that the inactivation of bacteria by HP does not follow first-order kinetics, and some of the microorganisms remain viable after HP treatment. Some hypotheses to explain this are related to heterogeneity in the age of bacteria, genetic variability and formation of aggregates. However, the most probable explanation is the recovery of cells after each cycle of pressurisation, since HP might sub-lethally injure a fraction of the microorganisms (López-Fandiño *et al.*, 1996a; García-Risco *et al.*, 1998). The ability of injured bacteria to recover could depend on the substrates, pressure and time (Earnshaw *et al.*, 1995). As an example, Figure 8.1 shows the effects of exposure time on aerobic and psychotropic counts of milk treated under 200 and 400 MPa.

Since most authors have demonstrated the difficulty of inactivating microorganisms by HP, some investigations have addressed the possible combination of HP with additives (sorbic and benzoic acid, lysozyme, chitosan) (Popper and Knorr, 1990; Papineau *et al.*, 1991) or other treatments, such as moderate heating. In the case of spores, a rise in temperature can induce their germination, and the resulting cells can, subsequently, be inactivated by HP (Datta & Deeth, 1999).

In general, enzymes are much less affected by HP than by heat. However, some are inactivated either partially or completely. The effects of HP on milk native enzymes, such as alkaline phosphatase, lactoperoxidase, γ-glutamyltransferase and phosphohexoseisomerase, have been studied by several researchers (Jonhston, 1995; López-Fandiño *et al.*, 1996b; Mussa and Ramaswamy, 1997; Sionneau *et al.*, 1997; Kussendrager & van Hooijdonk, 2000; Ludikhuyze *et al.*, 2000, 2001; Claeys *et al.*, 2003). Taking into account the kinetic data, γ-glutamyltransferase seems to be the most reliable pressure-sensitive indicator, since its inactivation by HP at 500 MPa and 20°C is sufficiently close to that of *L. monocytogenes* and *E. coli*, i.e. the most pressure-resistant species (Rademacher & Hinrichs, 2006). Plasmin inactivation by HP requires values of pressure within the range of 400–600 MPa and temperatures higher than 20°C (García-Risco *et al.*, 1998; García-Risco *et al.*, 2000; Borda *et al.*, 2004a, b; Huppertz *et al.*, 2004).

The most important changes in milk during HP treatment take place in the protein fraction. Substantial changes in the mineral balance of milk can also be produced. The effect of HP on milk proteins and its influence on their functional properties have been extensively covered in several recent reviews (López-Fandiño, 2006a,b). Both caseins and whey proteins (especially β-lactoglobulin) are modified under HP depending on the pressure, time and temperature of the treatment. During HP treatment of milk, micelles of caseins can associate or dissociate, the dissociation being responsible for the changes of its visual aspect (Needs *et al.*, 2000). HP treatments at 200 MPa give rise to micellar disaggregation, whereas at 250 MPa aggregation occurs prior to dissociation. The predominant effect of higher pressures is the decrease in size of casein micelles (Regnault *et al.*, 2004). HP treatments may also induce the solubilisation of colloidal calcium phosphate during micellar fragmentation (Schader *et al.*, 1997), and dissociate the heat-induced crystalline calcium phosphate of heat-treated milk (de la Fuente *et al.*, 1999). The former effect could be reversible during storage of milk (Schader *et al.*, 1997), and the latter effect during decompression (Hubbard *et al.*, 2002). β-Lactoglobulin is denatured at pressures higher than 100 MPa at 25°C, whereas α-lactalbumin and cow's serum albumin can withstand pressures up to 400 MPa without denaturation (López-Fandiño *et al.*, 1996a). The resistance of α-lactalbumin to pressure can be attributed to the lack of free sulphydryl groups and to its rigid molecular structure. All these changes affect the coagulation properties of milk and enhance the cheese yield. The treatment of milk by HP favours acid coagulation, giving rise to acid gels whose structure depends on the extent of whey protein denaturation and the micellar sizes. Milk gels can also be originated from concentrated milk under HP, the structures obtained being different from those obtained conventionally (López-Fandiño, 2006a, b).

8.3.3 *State of the art and opportunities*

HP technology itself has greatly progressed over the last 20 years. The development of HP treatment has been compromised by engineering problems, and most of the subsequent progress has been achieved in the production of specialist engineering materials. The food industry has recognized the potential commercial applications of HP for food processing and for the production of new products.

Although the effect of HP on milk has been broadly studied, to date, no commercial pressurised milk is available. This may be due to the difficulty of developing a continuous system avoiding cross-contamination and corrosion (Itoh *et al.*, 1996; Mermelstein, 1998). Another weak point for the industrial application of HP to the processing of milk is the high capital cost of the installations. In addition, more research is needed to determine the effect of HP on the nutritional quality of milk.

8.4 Pulsed electric fields

The use of high-intensity pulsed electric fields (PEF) is seen as an interesting alternative to conventional thermal processing methods for liquid or semi-liquid foods. Application of PEF technology has been successfully demonstrated for the pasteurisation of foods such as juices, milk, yoghurt, soups and liquid eggs. The high inactivation levels of various spoilage and pathogenic microorganisms with minimal temperature increase, and thus minimal impact on quality and nutritional factors make PEF a promising technology. An additional advantage of PEF is the reduced fouling compared to conventional pasteurisation (de Jong & van Heesch, 1998). The disadvantage of PEF is the limited bacterial spore and enzyme inactivation. To increase the efficacy of PEF for inactivation of bacterial spores and enzymes, recent investigations have focused on the use of combined technologies, e.g. PEF and heat or PEF and the use of bacteriocins or acids. Application of PEF to milk and other liquid dairy products has been extensively reviewed in various recent publications (Sampedro *et al.*, 2005; Sepulveda *et al.*, 2005).

8.4.1 *Operating principle*

PEF involves the application of short duration (typically 2–300 ms), high electric field pulses. In the early 1960s, this technology was introduced for microbial inactivation. It has been observed that PEF treatment causes electroporation of the cell membrane. The cell membrane contains protein channels and pores responsible for maintaining the trans-membrane voltage difference. When a pulsed electric field is applied, the opening of many voltage-sensitive protein channels can prevent the breakdown of the lipid bilayer. Electric reversible breakdown of the lipid bilayer occurs when a trans-membrane voltage difference of 150–500 mV is exceeded, which is achieved when exposing the membrane to an electric field of 1–10 kV cm^{-1}. If the electric field strength is increased up to 25 kV cm^{-1}, irreversible breakdown of the cell membrane occurs (van Heesch *et al.*, 2000).

Critical process parameters for microbial inactivation with PEF are electric field strength, pulse length, pulse shape, number of pulses and initial temperature (Bendicho *et al.*, 2004). Various models have been proposed to describe the kinetics of microbial inactivation by PEF. Equation 8.1 assumes a linear relationship between the logarithm of the survival fraction and the electric field strength, as well as a linear relationship between the logarithm of the survival fraction and the logarithm of treatment time (Hulsheger *et al.*, 1981).

$$S = \left(\frac{t}{t_c}\right)^{-\frac{E-E_c}{k}} \qquad (8.1)$$

In this equation, S is the fraction of survival, E_c is related to the trans-membrane voltage difference at which the lipid bilayer is destroyed, t_c is the threshold value of the treatment time and k is an empirical factor. In correlation with theory, it is observed that microbial cells of larger diameter are killed at lower field strength than cells of small diameter (de Jong & van Heesch, 1998). Other authors have proposed Equation 8.2 to calculate the percentage of survivors as a function of the electric field and number of pulses (a) (Peleg, 1995).

$$S = \frac{1}{1 + \exp\left(\frac{E - E_c}{a}\right)} \tag{8.2}$$

In general, it is observed that short pulses, high electric field strength and high pulse frequencies (1–2 kHz) are most effective in microbial inactivation (de Jong & van Heesch, 1998). Another observation is that square-waved pulses are more effective than exponential decay pulses.

The configuration of the PEF treatment chamber is optimised to avoid local fluctuations in the electric field strength. Different chamber configurations have been applied for both batch and continuous operation (Barbosa-Cánovas *et al.*, 1998). PEF can potentially give rise to undesired dielectric breakdown of the treated product. This phenomenon is observed when the electrical field strength equals the dielectric strength of the product. The presence of suspended particles or gas bubbles can be troublesome as they affect local electrical field strength (de Jong & van Heesch, 1998). The risk of dielectric breakdown limits the application of PEF to primarily liquid products.

8.4.2 *Effect on product properties*

Both microbial and enzyme inactivation by PEF are known to be highly dependent on different product parameters, such as composition, conductivity, ionic strength and pH (Wouters *et al.*, 2001). High protein and/or fat concentrations appear to have a protective effect for bacteria during PEF treatment of milk, although some studies have not observed any influence (Bendicho *et al.*, 2004). PEF is found to be most effective for liquid foods with low conductivity, i.e. low salt concentration or ionic strength (Wouters *et al.*, 2001). In addition to reduced ionic strength, it is observed that low pH also favours microbial inactivation. For example, addition of acids prior to PEF treatment of skim milk is found to significantly ($\log_{10}$ reduction of 2.5) affect the inactivation of *Listeria innocua* (Fernández-Molina *et al.*, 2005a). It is hypothesised that a highly acidic environment may prevent sensitised microorganisms from recovery (Wouters *et al.*, 2001). The adjustment of physical and/or chemical parameters, such as acidification, to prevent microbial growth is called hurdle technology. Depending on the food application, combinations of hurdle technology and PEF could result in successful preservation methods.

Inactivation of vegetative cells by PEF is more effective than inactivation of microbial spores. Furthermore, vegetative cells that are in the logarithmic growth phase are more sensitive to PEF than cells in the stationary phase (Wouters *et al.*, 2001). The degree of inactivation is also dependent on the bacterial species; the following have been subject of tests: *E. coli*, *Pseudomonas* spp., *Bacillus* spp., *S. aureus*, *Lactobacillus* spp., *Salmonella* spp., *Listeria* spp. Sampedro *et al.* (2005) have recently summarised inactivation data in milk and similar systems. In general, the inactivation reached for vegetative microorganisms is high, with a maximum of 9 $\log_{10}$ reduction for *E. coli* in simulated milk ultrafiltrated (SMUF) (Qin *et al.*, 1995).

Enzyme inactivation with PEF is found less effective than microbial inactivation (Loey van *et al.*, 2001). With respect to milk, this can be of interest as PEF can be specifically applied to inactivate microorganisms while protein degradation is minimised. Inactivation kinetics by PEF of various enzymes naturally present in milk have been investigated, e.g. alkaline phosphatase, plasmin, lipase and peroxidase (Bendicho *et al.*, 2004). Varying effects of PEF on enzyme activity have been observed. Plasmin present in SMUF was found to be significantly inactivated, i.e. 90% decrease in activity (Vega-Mercado *et al.*, 1995), while cow's immunoglobulin G present in milk protein concentrated soymilk was not inactivated (Li *et al.*, 2005). In other studies, the effect of PEF on enzymes of bacterial origin has been investigated. Only 13% inactivation of lipase from *P. fluorescens* was found (Bendicho *et al.*, 2002). In contrast, high inactivation of protease from *Bacillus subtilis* by PEF was reported by Bendicho *et al.* (2003). The inactivation of the protease appeared to be less in whole milk than in skimmed milk; in the latter, a maximum inactivation of 81% was observed.

Most studies did not observe a significant influence of PEF on the physical, chemical or sensorial properties of milk (Sampedro *et al.*, 2005). The taste of PEF-treated milk was similar to that of pasteurised milk (Sepulveda *et al.*, 2005). Small changes in the vitamin content of milk before and after treatment were found only after intense PEF treatment with a large number (>100) of pulses (Bendicho *et al.*, 2004). Recently, however, it was discovered that PEF affects the consistency of protein components of milk, such as casein micelles (Floury *et al.*, 2006). This may explain the observed decrease in viscosity and enhanced coagulation properties of PEF-treated milk.

The shelf life of PEF-treated and thermally pasteurised milk stored under refrigerated conditions are comparable, i.e. 2 weeks (Qin *et al.*, 1995). Others showed that pasteurisation (72°C for 15 s) followed by PEF extended the shelf life even up to 60 days (Sepulveda *et al.*, 2005).

8.4.3 *State of the art and opportunities*

It can be concluded that PEF treatment is very effective in the inactivation of vegetative microorganisms. However, high inactivation of spores and enzymes requires different or combined treatment methods. Hitherto many PEF studies have been executed on laboratory scale, and involved treatment of only a few millilitres of liquid product. Only recently have pilots proven feasibility of the technology on a large scale (Min *et al.*, 2003). Large capital costs involved with PEF are limiting industrial application. Despite this, the first large-scale PEF facilities are available now with a capacity of 2000 L h^{-1} (http://www.foodtech-international.com/papers/PulsedElectricField.htm, accessed 18 April 2006). Applications of PEF for fruit juices and other acid liquid foods are expected in first instance. Industrial application of PEF requires further study to clarify the level to which product safety can be guaranteed. Potential risks are the formation of electrolytic products in the food product and release of electrode material in the stream.

8.5 Ultrasound

Ultrasound can be defined as waves with a frequency higher than 20 kHz that are able to travel through gas, liquid and solid materials. A distinction has to be made between low- and high-intensity ultrasound (Povey & McClements, 1988; McClements, 1995). Typical values for low-intensity ultrasound are low power levels ($\leq$1 W cm^{-2}) and high frequencies (0.1–20 MHz), which leave the system unchanged. This makes low-intensity ultrasound an excellent tool for measuring food properties (McClements, 1995). High-intensity ultrasound is characterised by high power levels (10–1000 W cm^{-2}) and relatively low frequencies ($\leq$0.1 MHz) that can give rise to physical and chemical alterations within the material to which it is applied, the applications being mainly focused on the processing of products (McClements, 1995). High-energy ultrasound has been used for degassing of liquid foods, induction of oxidation/reduction reactions, extraction of enzymes and proteins, enzyme inactivation and induction of nucleation for crystallization. As preservation technology, no commercial food products are currently available; however, ultrasound-assisted technologies for product modifications or process improvements exist (Welti-Chanes *et al.*, 2002; Knorr *et al.*, 2004). The main applications in milk and dairy products are inactivation of bacteria and enzymes, homogenisation of milk, extraction of enzymes and lactose hydrolysis (Villamiel *et al.*, 1999). The studies related to preservation of food have been mainly carried out on laboratory scale. This could be in part because the benefits of this technique in terms of nutritional safety are not fully known. Additionally, some authors have stated that the energy consumption of ultrasound necessary to kill microorganisms is higher than for conventional methods.

8.5.1 *Operating principle*

During ultrasonic treatment, the main active force is mechanical in nature (Shukla, 1992). When ultrasound waves travel through a liquid, alternating compression and expansion cycles are produced. During the expansion cycle, high-intensity ultrasound causes the growth of existing bubbles. When they attain a volume at which they do not absorb more energy, they implode violently. This phenomenon is known as cavitation. During implosion, locally very high temperatures (up to 5500°C) and pressures (50 MPa) are reached inside the bubbles. Most authors think that this is the main effect of ultrasound on microorganisms (Raso *et al.*, 1998). Other hypotheses consider the formation of free radicals as a contributor to microorganism inactivation (Alliger, 1975; Earnshaw *et al.*, 1995). Similar mechanisms can be expected for enzyme inactivation. Cavitation is also a heat-generating process, since input of mechanical energy causes molecular motion, raising the temperature (Berliner, 1984) and, moreover, part of the energy is absorbed for the heating of the sample (Floros & Liang, 1994).

8.5.2 *Effects on product properties*

It has been demonstrated that ultrasound on its own is not very effective for inactivating microorganisms and enzymes in milk, and most studies have focused on the combination of ultrasound with heat (thermosonication, TS), pressure (manosonication, MS) or both (manothermosonication, MTS) (Ordóñez *et al.*, 1987; García *et al.*, 1989; Wrigley & Llorca, 1992; Vercet *et al.*, 1997, 2002a; Pagán *et al.*, 1999) and a synergistic or additive effect has been found. These combinations have been demonstrated to be mainly useful for those microorganisms and enzymes liable to develop heat resistance.

Various factors should be considered with regard to the success of ultrasound treatment (with and without heat/pressure combination) for microorganism and enzyme inactivation. Key variables are assumed to be the amplitude of the ultrasonic waves, the exposure/contact time, the type of microorganism and enzyme, the volume of food to be processed, the composition and pH of the food, and the temperature and pressure of treatment (McClements, 1995; Sala *et al.*, 1995; Vercet *et al.*, 1998; Hoover, 2000).

The majority of studies on ultrasound and ultrasound-combined processes have been done in batch-type reactors, and as mentioned elsewhere, it is evident that continuous processes are easier to scale-up than batch treatments. Villamiel & de Jong (2000a) set up a continuous-flow apparatus and evaluated the effects of ultrasound with and without heat generation on important bacteria for the dairy industry (*P. fluorescens* and *Streptococcus thermophilus*). A comparative study was also performed using a conventional heating system under similar process conditions, and for both treatments, computer simulations were developed. During ultrasound treatment, as the heating is within the sample, the equipment wall temperature was lower than the liquid temperature. This could offer some advantages over conventional heating systems in terms of better temperature distribution and less fouling formation and, thus, the product quality might be improved (de Jong *et al.*, 1992; de Jong, 1996; Villamiel & de Jong, 2000a). Effects on fat, whey proteins (Figure 8.2), caseins, alkaline phosphatase, lactoperoxidase and γ-glutamyltransferase were also evaluated. The ultrasound continuous-flow system was proved to be an adequate method for preservation and homogenisation of milk (Villamiel & de Jong, 2000b). A continuously working pilot-plant-scale prototype was also assayed for bacterial decontamination of milk. At identical degrees of bacterial inactivation, the ultrasound-assisted thermal treatments required a lower processing temperature than a treatment with conventional thermal processing (Zenker *et al.*, 2003).

The effect of MTS on vitamins and non-enzymatic browning compounds was explored in a continuous system by Vercet *et al.* (2001). This process did not affect significantly the nutrient content of milk, whereas higher non-enzymatic browning was observed under MTS than under heat treatment. MTS has been also used to treat milk used for the production of yoghurt (Vercet *et al.*, 2002b). MTS yoghurts had stronger structures, probably due to the role of denatured whey protein in the increased consistency and viscosity of yoghurts.

8.5.3 *State of the art and opportunities*

Due to the slight effect on microorganisms and enzymes, it will be difficult for ultrasonic treatment to become a commercial process on its own, but in combination with other treatments (heat and/or pressure), it has more potential as a minimal processing method in the dairy industry. More research is needed to improve the processing equipment and to gain more insights into the effect of ultrasound on the main milk components. Studies on processing of milk by ultrasound in continuous-flow seem to be promising for the

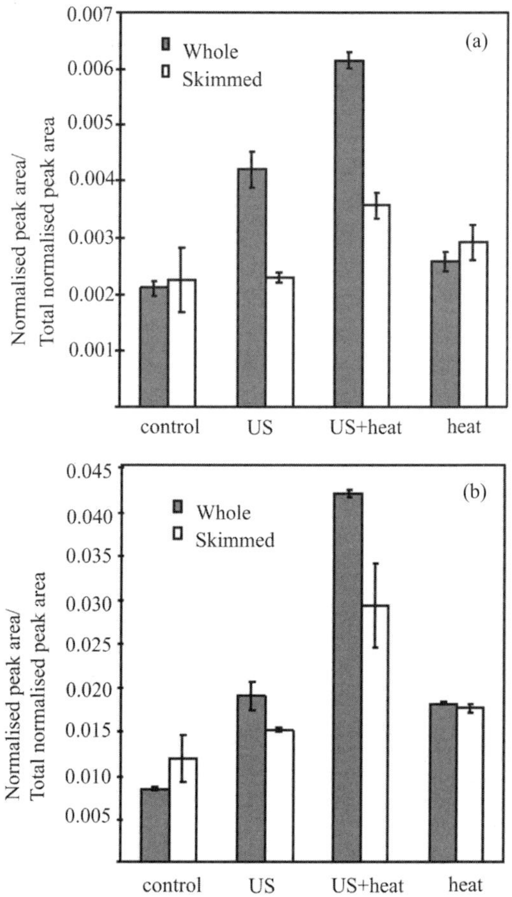

Fig. 8.2 Denatured whey proteins (α-lactalbumin (a) and β-lactoglobulin (b)) in the casein fraction precipitated at pH 4.6 of whole and skimmed milk subjected to continuous flow ultrasonic (US), ultrasonic and heat (75.5°C) and conventional heating (75.5°C) for 102.3 s.

scale-up of this technique as a preservation method alternative to the traditional thermal treatments.

8.6 Microfiltration

During membrane filtration particles suspended in a liquid are retained by a semi-permeable membrane. Typical microfiltration (MS) membranes contain pores with a diameter between 0.1 and 10 μm. The major industrial applications for MF in the dairy industry are removal of bacteria, whey defatting and micellar casein enrichment (Maubois, 2002). Besides being a non-thermal treatment method, MF has the specific advantage that it is very effective in the removal of bacterial spores in comparison with conventional pasteurisation. A major drawback of MF is fouling at the membrane surface, which adversely affects selectivity and throughput. Fouling also requires frequent rinsing and cleaning procedures, which can have a detrimental effect on the cost-effectiveness of this technology.

The breakthrough of MF in dairy industry came in the 1980s together with the development of ceramic MF membranes and improved membrane module design. The ceramic membrane modules made the uniform trans-membrane pressure (UTP) concept industrially feasible. In the UTP concept a constant trans-membrane pressure over the length of the membrane module is maintained by applying high cross-flow velocities at both sides of the membrane (Holm *et al.*, 1984). The UTP process concept resulted in an increase of permeation fluxes and improved selectivity. However, despite the improved module designs and improved processing concepts, fouling remains a major issue in the cost-effectiveness of this technology and, therefore, limits its application.

8.6.1 *Operating principle*

MF is a pressure-driven membrane process in which the feed is separated into a retentate (concentrate) and a permeate (filtrate). Parameters that determine the effectiveness of a MF process are selectivity or retention and the permeate flux. The way in which milk particles are affected by MF processing is determined by the average pore size of the membrane. It should be emphasised here that uniformity in pore size is an important factor in achieving a good combination of retention and permeation of components in the feed.

MF allows simultaneous removal of bacteria, bacterial spores and somatic cells from milk (te Giffel & van der Horst, 2004). Typical membranes for this application have an average pore diameter of 1.4 μm. In Table 8.3, the sizes of different components present in milk are shown. Both bacteria and somatic cells, which are naturally present in milk, are almost fully retained by MF. Note that the presence of somatic cells has an adverse effect on the shelf life of milk. From the size of casein micelles, it can be concluded that these are hardly retained by MF. Obviously, retention of single proteins is also very small (<1 g 100 g^{-1}). Fat globules (0.2–6 μm) are partially retained because of their similar size to bacteria. MF as a method to remove bacterial and somatic cells is therefore specifically used for treatment of defatted or skimmed milk. Typical membrane fluxes encountered are in the order of 500 L m^{-2} h^{-1}.

8.6.2 *Effect on product properties*

MF offers the dairy industry opportunities to produce milk and milk products that have sensorial characteristics similar to fresh milk together with improved shelf life. Besides

Table 8.3 Overview of sizes of different components in milk.

Component	Size (μm)
Casein micelles	0.03–0.3
Fat globules	0.2–6
Bacteria	0.2–7
Somatic cells	6–15

production of consumption milk with extended shelf life, this method can be used as pretreatment of skimmed milk in the production of raw milk cheeses and acid cheese milk (Zoon & Hup, 1991). Studies on MF on skimmed milk show $\log_{10}$ reductions for *L. monocytogenes*, *Brucella abortii*, *S. typhimurium* and *Mycobacterium tuberculosis* of 3.4, 4.0, 3.5 and 3.7, respectively. It is evident that the removal of somatic cells is nearly complete. The removal of somatic cells from milk is especially important in the production of cheese, since the presence of a large number of these cells affect the composition of milk, coagulation time and the yield, quality, flavour and texture of cheese.

Several varieties of MF liquid milks are available on the market. Perceived freshness and improved shelf life are the reason for their commercial success. Shelf life of these MF milks stored under refrigerated conditions is reported to be in the order of 15 days (Saboya & Maubois, 2000). The shelf life can be extended up to 35 days if MF is combined with pasteurisation. The raw cheeses that are produced with MF-treated skim milk can be considered at least as safe from a hygienic point of view as cheeses made from pasteurised milk. Furthermore, the MF step significantly reduces presence of the spore-forming *Clostridium tyrobutyricum*, which can cause late blowing during the maturation of the cheese. On the other hand, it remains questionable whether the cheeses produced with MF milk can be still called raw milk cheese, since, for example, the natural flora present in raw milk plays an important role in the development of flavour and texture of these cheeses.

8.6.3 *State of the art and opportunities*

With MF, the dairy industry has a versatile tool to drastically improve the hygienic safety of milk products while applying minimal heat treatment. Tetra Pak and APV offer complete process systems comprising MF for removal of bacteria from milk, i.e. the Bactocatch and APV Invensys process, respectively. Both systems are applied in the production of market milk and cheese. The Bactocatch process involves MF of skimmed milk (te Giffel & van der Horst, 2004). Prior to filtration, whole milk is preheated to ~50°C and the cream is separated. Subsequently, the cream is mixed together with the retentate from the MF step, and sterilised before being remixed with the skimmed milk permeate. In the APV process the retentate of the filtrated skimmed milk is remixed with the whole milk that is fed to the preheating step. In this process, only the cream is sterilised.

Besides the application of membrane filtration for the removal of bacteria, spores and somatic cells, further opportunities are identified in the complete fractionation of milk. The possibility to separate milk into well-defined fractions by MF and other membrane filtration steps could lead to optimal use of the different milk components and their functional properties (Brans *et al.*, 2004).

8.7 Innovative steam injection – a novel heating method

Heat treatment is the most conventional processing method to guarantee microbial safety of dairy products. Throughout the twentieth century, various heat treatment methods were developed. Each heat treatment is designed with a specific temperature–time trajectory to achieve a certain inactivation of microorganisms, spores and enzymes, while preserving

Table 8.4 Typical temperature–time specification for the different thermal treatments.

Treatment	Time–temperature specification	Inactivation
Thermisation	20 s, 65°C	Psychrotrophic microorganisms
Low pasteurisation	20 s, 74°C	Pathogenic microorganisms
High pasteurisation	20 s, 85°C	All organisms except spores
Sterilisation	30 min, 110°C	All organisms plus spores
Ultra-high temperature (UHT)	5–10 s, 140°C	All organisms plus spores
Innovative steam injection (ISI)	0.1 s, 150–200°C	All organisms plus spores

sensorial and nutritional characteristics as much as possible (Table 8.4). The sensorial and nutritional characteristics of heat-treated dairy products are affected because of a large number of chemical, physical and biochemical reactions that take place during heating. Different reactions can be distinguished, such as destruction of microorganisms, inactivation of enzymes, denaturation of proteins, loss of nutrients and formation of new components. The nutritional value is, for example, affected due to inactivation of thiamin (vitamin B_1). Complex Maillard reactions lead, for example, to the formation of pigments responsible for the brown colour of heated milk.

Consumers demand long shelf life and safe products, but they also want foods as fresh as possible. For conventional heat treatment processes, it is, therefore, of major importance to minimise heat load during processing, while still guaranteeing a certain microbial inactivation. In this respect, the innovative steam injection (ISI) approach is being developed as a novel heating technology to be applied for liquid foods (Verdurmen *et al.*, 2002; Huijs *et al.*, 2004). ISI is intended to serve as an alternative for conventional pasteurisation, while achieving a drastically improved shelf life similar to UHT-treated products. The advantage of ISI is that sufficient microbial inactivation is achieved within very short processing time (Table 8.4), while proteins and vitamins remain largely intact.

8.7.1 *Operating principle*

The sophisticated design of the ISI heater allows fast injection of high pressure steam combined with a subsequent flash cooling step in a vacuum vessel (Figure 8.3). This design makes it possible to achieve very short residence times at a high temperature (Huijs *et al.*, 2004).

8.7.2 *Effect on product properties*

From pilot plant experiments at a scale of 300 L h^{-1}, it is observed that ISI achieves significant inactivation of heat-resistant spores such as *Bacillus cereus*, while functionality of important ingredients is preserved (Huijs *et al.*, 2004). The combination of short residence time and high temperature leads to sufficient inactivation of spores (6 log reduction) while denaturation of whey proteins is very limited, 20–25% compared to 50–60% with conventional UHT. Although the taste of ISI milk does differ from pasteurised milk in a few taste attributes such as caramel, in two out of three cases a tasting panel could not indicate a significant preference between pasteurised and ISI milk (Huijs *et al.*, 2004). The

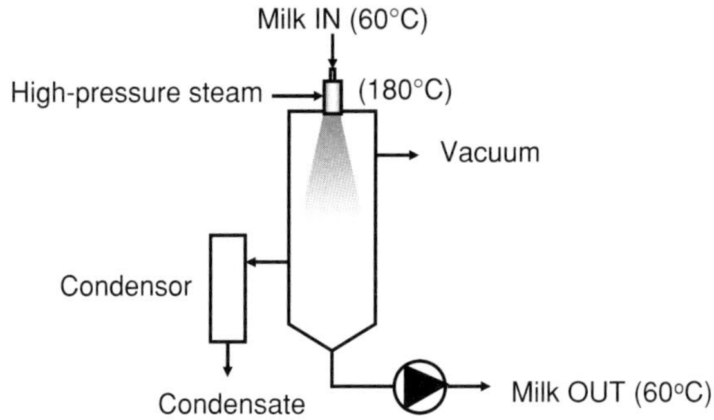

Fig. 8.3 Schematic overview of the innovative steam injection (ISI) heater.

shelf life of ISI-treated milk stored under refrigerated conditions appears to be 60 days. At room temperature, the ISI-treated milk is not stable, which can be explained by the incomplete inactivation of the enzyme plasmin, resulting in a bitter tasting product after some time.

8.7.3 *State of the art and opportunities*

The process costs of the ISI heating technology are expected to be 10% higher than those of the conventional UHT process, which makes the ISI technology substantially less expensive than alternative non-thermal technologies. The development of ISI technology is progressing, with the current focus on the scale-up of this technology. If scale-up of ISI technology is successful, it is believed that this technology may lead to a new generation of fresh and safe dairy products.

8.8 Combined technologies

Most non-thermal preservation techniques appear to be highly effective in inactivation of vegetative cells of bacteria, yeasts and fungi. However, inactivation of bacterial spores and enzymes remains difficult and therefore application of these novel technologies remains primarily limited to foods where enzymatic reactions do not affect product properties or where spore germination is unlikely. Combinations of non-thermal preservation processes with conventional or other emerging processes could extend their application. In some cases, combination of different technologies could allow a milder use of single treatments. The total preservation effect can be merely the sum of the effects of the individual treatments, but in terms of food quality and safety, a synergistic effect is preferable. For design of combined processes, it is of primary importance to distinguish factors that sensitise microorganisms or enzymes and factors that provoke complete inactivation of sensitised microorganisms or enzymes. In Table 8.5, an overview is given of different combinations of non-thermal technologies with heat, pressure, pH, use of antimicrobials and storage at low temperature. Mathematical models could be used as a tool to design highly efficient combined treatments and determine processing conditions (Manas & Pagan, 2005).

Table 8.5 Combined treatments that have been investigated for their efficacy in inactivation of bacteria, spores and enzymes.

Combined treatments	Mild heat treatment[a]			Elevated pressure			Reduced pH			Antimicrobials			Storage at low temperature		
effect on	V	S	E	V	S	E	V	S	E	V	S	E	V	S	E
High pressure	++	+	+	+	++	++	++	+	−	+/++	+	−	+/++	+	+
Ultrasound[b]				+	+	+									
Microwave	+	+	+/−												
Pulsed electric fields	++	+	+/−	+	++	++	++	+	−	+/++	+/−	−	+/++	+	+
Microfiltration	+	+	+/−												

[a] V = vegetative cells; S = spores; E = enzymes; − negligible effect; +/− minor effect; + positive effect; ++ significant effect.

[b] Manothermosonication (MTS) is considered here, consisting of a combination of ultrasound, pressure and heat.

8.8.1 *Effect on product properties*

High-pressure treatment (>400 MPa) results in significant damage to cell membranes of vegetative cells. Inactivation of bacterial spores and enzymes by HP is unfortunately very limited. Combined treatment of HP with heat has been widely investigated to enhance the efficacy of HP (Raso & Barbosa-Cánovas, 2003). During HP treatment, bacterial spores are sensitised due to their germination. During the heating step, further inactivation of the spores is achieved. Subsequent heat treatment enhances the efficacy of the HP treatment. Combined pressure and heat treatment is especially effective at temperatures causing direct inactivation of germinated spores (>60°C). An example is the inactivation of *L. monocytogenes* in UHT milk, for which a $\log_{10}$ reduction of 2 was observed at $T = 25$°C, while a $\log_{10}$ reduction of more than 7 was observed at $T = 45$°C. For *S. aureus* in UHT milk, $\log_{10}$ reductions of 1 and 6 were observed at 20 and 50°C, respectively. The effect of combined heat and pressure treatment on enzyme activity is only moderate; products, therefore, still require storage under refrigerated conditions to maintain quality and sufficient shelf life.

It has been demonstrated that efficacy of microbial inactivation is enhanced when ultrasound treatment is applied at elevated pressure (manosonication, MS). Moreover, the combination of ultrasound, pressure and heat, i.e. MTS, is found to enhance efficacy even more. The effect for most vegetative cells MTS has been found to be additive, while for spores (e.g. *Bacillus subtilis*) the effect is synergistic (Raso & Barbosa-Cánovas, 2003).

PEF treatment is more effective in combination with high temperature. Enhanced microbial inactivation is related to the temperature effect on the cell membrane properties. The high temperature reduces the thickness of the lipid bilayer and makes it more susceptible to damage. This method resulted in the milk with a shelf life up to 30 days (Fernández-Molina *et al.*, 2005b).

MF is applied as a mild preservation method is combined with heat treatment in the production of pasteurised milk with extended shelf life (ESL). It should be emphasised here that in the production of ESL milk not only the combined treatment is responsible for the extended shelf life, but also the optimisation as regards hygiene of the entire processing chain, comprising processing, packaging systems and distribution. The typical shelf life of liquid ESL milk is 15–25 days stored at 7°C.

8.8.2 *Preservation in the presence of antimicrobials*

HP, PEF and ultrasound treatments can sensitize bacterial cells causing sub-lethal damage. Generally, sensitised cells are more susceptible to antimicrobials, and therefore, combined physical treatment and use of antimicrobials may be attractive for certain foods, although studies in this field are scarce. Application of antimicrobials is mainly hampered due to these compounds interacting with food ingredients and causing changes in the organoleptic properties of the food product (Devlieghere *et al.*, 2004). The dislike of chemical additives by consumers has stimulated the search for natural additives by the food industry. Typical natural additives that have been investigated in food products are lactoperoxidase, lysozyme, different types of bacteriocins and organic acids.

Lactoperoxidase (LPS) is a naturally present enzyme in cow's milk. LPS has no antimicrobial properties, but it can facilitate formation of antimicrobial compounds in the presence of an oxidisable substrate and hydrogen peroxide (H_2O_2). H_2O_2 is naturally present at small concentrations in milk, but may also be formed by action of, for example, xanthine oxidase, glucose oxidase and ascorbic acid in the presence of free oxygen. The oxidisable substrate is thiocyanate, which is also naturally present in milk, and conversion results in the formation of hypothiocyanate. This compound attacks free sulphydryl groups of proteins in the cell membrane of bacteria and, as a consequence, causes cell disruption. LPS is primarily active against H_2O_2-producing bacteria, such as lactococci and lactobacilli (Barbosa-Cánovas *et al.*, 1998). The LPS system can be used for the preservation of milk, infant formula and liquid whole eggs. The shelf life of pasteurised milk could be extended by 20 days by the use of LPS. There are also some drawbacks for the use of LPS, one of which is its affinity for glass surfaces.

Lysozyme is an enzyme that is naturally present in milk, but also in plant tissues and eggs, and causes degradation of peptidoglycan present in bacterial cell walls. Lysozyme can be used as a food additive and is, for example, used in the production of some hard cheeses to prevent gassing and blowing due to *Clostridium*.

Bacteriocins are protein-containing macromolecules that have antimicrobial properties against certain bacteria. A well-known bacteriocin is nisin, which is produced during fermentation of *Lactococcus lactis* subsp. *lactis* in milk media. Nisin exhibits specific antimicrobial properties that cause damage to gram-positive bacteria. Desirable properties of nisin are its non-toxicity and heat stability. Several studies have reported synergistic effects of nisin in combination with other treatments (Calderon-Miranda *et al.*, 1999a,b). Lysozyme and nisin have been found effective in combination with pressure treatment for *E. coli* (Hauben *et al.*, 1996). Combination of PEF, heat, nisin and lysozyme has been found very successful for treatment of raw skimmed milk (Smith *et al.*, 2001). Although some successful studies have been reported, also several authors have emphasised the disadvantages of bacteriocins, which have a narrow activity spectrum, meaning that the compounds are only active against specific bacteria, and inactivation of bacteriocins may occur through interaction with other food ingredients or proteolytic activity (Devlieghere *et al.*, 2004).

The effect of the presence of organic acids, such as acetic acid or propionic acid, has been widely studied in combination with other processing methods. A synergistic effect of PEF and addition of acid is frequently encountered, although in some cases only an additive effect was observed, e.g. for *P. fluorescens* in skim milk (Fernández-Molina *et al.*, 2005a).

8.8.3 *State of the art and opportunities*

The application of non-thermal processing technologies has been widely investigated during the last 20 years. These technologies have been found successful in microbial inactivation while preserving many sensorial and nutritional properties. However, their industrial application appears limited due to the high resistance of bacterial spores and enzymes to inactivation. In several studies, a promising synergistic effect is found by combining different treatments (Raso & Barbosa-Cánovas, 2003). More in-depth studies are required to understand the mechanisms behind different treatments and their combinations. Application of antimicrobials is mainly hampered due to their interaction with food ingredients and the changes in organoleptic properties of the food product. Future studies will need to find out which specific products could benefit more from the use of antimicrobials.

8.9 Computer models for design and implementation of novel milk-processing methods

The dairy industry is continuously searching for novel technologies to improve its products and develop new ones. The key to the design of novel milk-processing methods is to achieve optimal product quality while ensuring safety and long shelf life (de Jong *et al*., 2005). The design of novel processing technologies requires the availability of kinetic models that describe, for example, the inactivation of microorganisms and enzymes as a function of the processing parameters. In addition, models that can predict the effect of combined treatments by considering lethal events or events leading to sub-lethal damage would greatly facilitate the further development and application of novel milk-processing methods (Manas & Pagan, 2005).

8.9.1 *Optimisation of milk-processing methods with models*

Mathematical models have been extensively applied in the dairy industry for optimisation of processes. Various models are available that can, for example, predict microbial growth and fouling in process lines. The amount of microorganisms remaining in the treated product can be simulated from initial contamination, process design and parameters. Subsequently, the process line can be optimised to ensure microbial inactivation with minimal energy costs. Moreover, models can also be applied to simultaneously optimise microbial and enzymatic inactivation, and nutrient retention, while, at the same time, production costs can be minimised by avoiding protein fouling. Taking into account many different aspects of dairy processing methods, it is argued that mathematical models are very useful for their integral optimisation. Instead of using models for process optimisation also another approach can be followed. In the alternative approach, it is possible to take the desired product properties and functionalities as starting point for the design of novel processing concepts (Verschueren & de Jong, 2006). The quality and functionality of a product determines its success in the market. Conventional processing concepts are normally the basis for development of new products. This assumption leads to the development of suboptimal products since concessions are required to cope with limitations of existing processing equipment. Product- or even better consumer-oriented design of novel processing methods could lead to new successful products. A product-oriented approach has led to the development of the ISI technology, described in Section 8.7 (Verschueren & de Jong, 2006). The basis for the ISI technology is the question how short and high should be the heating step to achieve a high-quality product? The hypothesis behind this question is that short- and high-heating results in an improvement of quality and functionality, while safety is still guaranteed. In Figure 8.4, calculated time and temperature dependencies of protein denaturation and inactivation of two bacteria species are shown together with the operational window for the ISI technology. With ISI significant decimal reduction of bacteria is achieved, while protein denaturation is minimal.

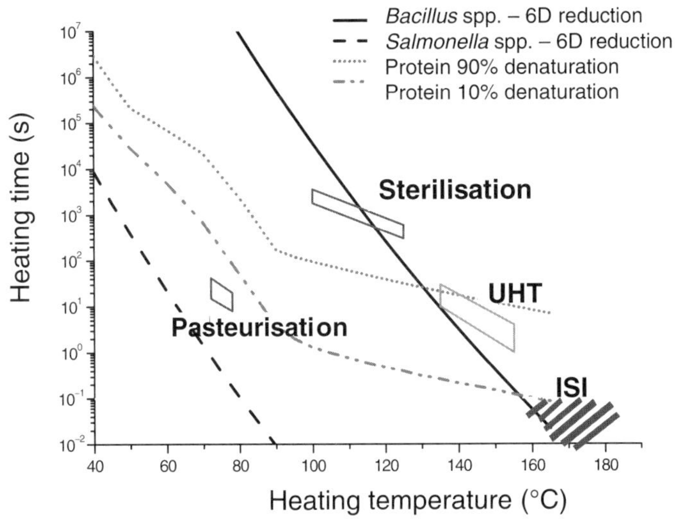

Fig. 8.4 Temperature–time dependency of several heat-induced reactions based on model calculations and operational windows of heating processes (D = decimal reduction).

8.9.2 *Availability of models*

Three different types of models can be distinguished that are applied for optimisation of dairy processing methods:

- Process models including energy and mass balances result in a prediction of the treatment intensity–time history of the product.
- Product models that predict transformation or inactivation of food components and contaminants related to the food properties perceived by the consumer. For example, the denaturation and aggregation of proteins, the inactivation of enzymes, bacteria and spore inactivation, contamination and the formation of reaction products (pigments, (off-) flavours).
- Cost models that predict operating costs related to processing conditions. For example, cleaning costs due to microbial and/or physical fouling.

Many different process, product and cost models can be implemented in one software application (e.g. Nizo Premia). This software allows optimisation of dairy processes with various raw materials, processing equipment, temperature, processing times, etc. In Figure 8.5, a schematic illustration of the structure of the Premia software package is shown. Numerous optimisation studies with Premia have been executed and have resulted in significant reduction of operational costs and/or in improved product quality for many different dairy processes (de Jong *et al.*, 2005). It is expected that the industrial application of novel non-thermal processes could be facilitated with the increased availability of kinetic models that describe the effect of treatment on, for example, inactivation of bacteria or enzymes. Availability of these models could allow integral optimisation of combinations of novel and

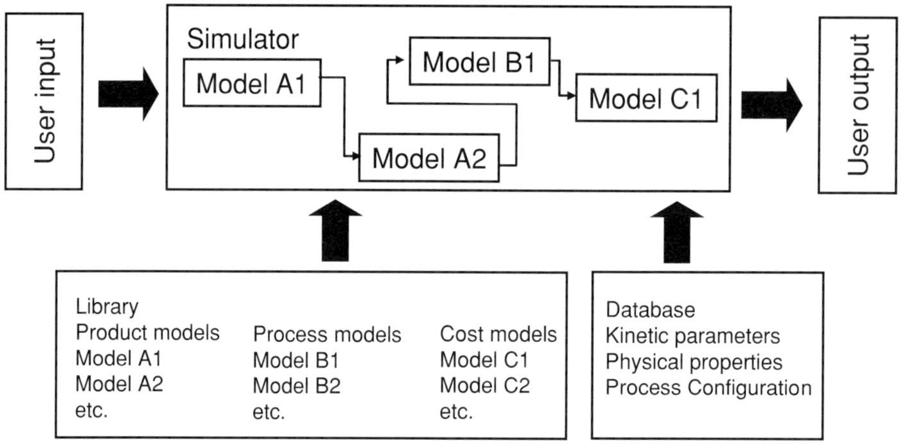

Fig. 8.5 Schematic representation of the structure of the Premia software package, including product, process and cost models and a database with kinetic parameters, physical properties and process configurations; the simulator represents a series of connected models that together describe the different aspects of a process.

conventional processing methods and thus also stimulate the industrial application of these processing methods.

References

Akiyama, H. (2000) Otsuka Chemical Co. Ltd., 463 Kagasuno Kawauchi, CHO, Tokushima 771–0193 Japan.

Alliger, H. (1975) Ultrasonic disruption. *American Laboratory*, **10**(10), 75–85.

Balci, A.T. & Wilbey, R.A. (1999) High-pressure processing of milk-the first 100 years in the development of a new technology. *International Journal of Dairy Technology*, **52**, 149–155.

Barbosa-Cánovas, G.V., Pothakamury, U.R., Palou, E. & Swanson, B.G. (1998) *Non-thermal Preservation of Foods*, Marcel Dekker, New York.

Bendicho, S., Barbosa-Cánovas, G.V. & Martín, O. (2003) Reduction of protease activity in milk by high-intensity pulsed electric field treatments. *Journal of Dairy Science*, **86**, 697–703.

Bendicho, S., Barbosa-Cánovas, G.V. & Martín, O. (2004) Milk processing by high intensity pulsed electric fields. *Trends in Food Science & Technology*, **13**, 195–204.

Bendicho, S., Estela, C., Giner, J., Barbosa-Cánovas, G.V. & Martín, O. (2002) Effects of high intensity pulsed electric field and thermal treatments on a lipase from *Pseudomonas fluorescens*. *Journal of Dairy Science*, **85**, 19–27.

Berliner, S. (1984) Applications of ultrasonic processors. *International Biotechnology Laboratory*, **2**(March/April), 42–49.

Bögl, K.W. (1990) Possible health milk. *View of the Federal Public Heath Office – Bundesgesundheitsblatt*, **33**, 342–343.

Borda, D., Indrawati, Smout, C., van Loey, A. & Hendrickx, M. (2004a) High pressure thermal inactivation of a plasmin system. *Journal of Dairy Science*, **87**, 2351–2358.

Borda, D., van Loey, A., Smout, C. & Hendrickx, M. (2004b) Mathematical models for combined high pressure and thermal plasmin inactivation kinetics in two model systems. *Journal of Dairy Science*, **87**, 4042–4049.

Brans, G., Schroën, C.G.P.H., Van Der Sman, R.G.M. & Boom, R.M. (2004) Membrane fractionation of milk: state of the art and challenges. *Journal of Membrane Sciences*, **243**, 263–272.

Calderon-Miranda, M.L., Barbosa-Cánovas, G.V. & Swanson, B.G. (1999a) Inactivation of *Listeria innocua* in skim milk by pulsed electric fields and nisin. *International Journal of Food Microbiology*, **51**, 19–30.

Calderon-Miranda, M.L., Barbosa-Cánovas, G.V. & Swanson, B.G. (1999b) Transmission electron microscopy of *Listeria innocua* treated by pulsed electric fields and nisin in skimmed milk. *International Journal of Food Microbiology*, **51**, 31–38.

Cheftel, J.C. (1995) Review: high-pressure, microbial inactivation and food preservation. *Food Science and Technology International*, **1**, 75–90.

Choi, K., Marth, E.H. & Vasavada, P.C. (1993a) Use of microwave energy to inactivate *Yersinia enterocolitica* and *Campilobacter yeyuni* in milk. *Milchwissenchaft*, **48**, 134–136.

Choi, K., Marth, E.H. & Vasavada, P.C. (1993b) Use of microwave energy to inactivate *Listeria monocytogenes* in milk. *Milchwissenchaft*, **48**, 200–203.

Claeys, W.L., Indrawati, van Loey, A.M. & Hendrickx, M.E. (2003) Review: are intrinsic TTIs for thermically processed milk applicable for high-pressure processing assessment? *Innovative Food Science and Emerging Technologies*, **4**, 1–14.

Clare, D.A., Bang, W.S., Cartwright, G., Drake, M.A., Coronel, P. & Simunovic, J. (2005) Comparison of sensory, microbiological, and biochemical parameters of microwave versus indirect UHT fluid skim milk during storage. *Journal of Dairy Science*, **88**, 4172–4182.

Cross, G.A. & Fung, D.Y.C. (1982) The effect of microwaves on nutrient values of food. *CRC Critical Reviews in Food Science and Nutrition*, **16**, 355–381.

Datta, A.K. & Davidson, P.M. (2000) Microwave and radio frequency processing. *Journal of Food Science* (Supplement kinetics of microbial inactivation for alternative food processing technologies), **65**, 32–41.

Datta, N. & Deeth, H.C. (1999) High pressure processing of milk and dairy products. *Australian Journal of Dairy Technology*, **54**, 41–48.

Datta, A.K. & Hu, W. (1992) Quality optimisation of dielectric heating processes. *Food Technology*, **46**(12), 53–56.

Decareau, R.V. (1985) Microwaves in the food processing industry. In: *Dielectric Properties of Foods – 19* (ed. B.S. Schweigert), pp. 15–37, Academic Press, New York.

de Jong, P. (1996) *Modelling and Optimisation of Thermal Treatments in the Dairy Industry*, PhD thesis ISBN 90–9009034-7, University of Technology, Delft, The Netherlands.

de Jong, P., Bouman, S. & van der Linden, H.J.L.J. (1992) Fouling of heat treatment equipment in relation to the denaturation of β-lactoglobulin. *Journal of Society of Dairy Technology*, **45**, 3–9.

de Jong, P., te Giffel, M.C., de Jong, C., Vissers, M. & Smid, E. (2005) Houdbaarheid verdriedubbelen. *Voedingsmiddelentechnologie*, **38**(11), 14–16.

de Jong, P. & van Heesch, E.J.M. (1998) Review: effect of pulsed electric fields on the quality of food products. *Milchwissenshaft*, **53**, 4–8.

de la Fuente, M.A., Olano, A., Casal, V. & Juárez, M. (1999) Effects of high-pressure and heat treatment on the mineral balance of goat's milk. *Journal of Dairy Research*, **66**, 65–72.

de la Fuente, M.A, Olano, A. & Juárez, M. (2002) Mineral balance in milk heated using microwave energy. *Journal of Agricultural and Food Chemistry*, **50**, 2274–2277.

Devlieghere, F., Vermeiren, L. & Debevere, J. (2004) New preservation technologies: possibilities and limitations. *International Dairy Journal*, **14**, 273–285.

Earnshaw, R.G., Appleyard, J. & Hurst, R.M. (1995) Understanding physical inactivation processes: combined preservation opportunities using heat, ultrasound and pressure. *International Journal of Food Microbiology*, **28**, 197–219.

Eberhard, P., Strahm, W. & Sieber, R. (1990) Pasteurisation von milch und rahm in mikrowellen-haushaltsgeräten. *Milchwissenchaft*, **45**, 768–771.

El-Shibiny, S, Sabour, M.M., El-Alamy, H.A. & Alam, S. (1982) Effect of microwaves on buffalo's milk. *Egyptian Journal of Dairy Science*, **10**, 29–34.
Farkas, D.F. & Hoover, D.G. (2000) High pressure processing. *Journal of Food Science* (Supplement Kinetics of microbial inactivation for alternative food processing technologies), **65**, 47–64.
Fay, L., Richli, U. & Liardon, R. (1991) Evidence of the absence of amino acid isomerization in microwave-heated milk and infant formulas. *Journal of Agricultural and Food Chemistry*, **39**, 1857–1859.
Fernández-Márquez, M, Ruiz-López, M.D. & García-Villanova, B. (1992) Determination of total solids in milk by microwave drying and its effects on hydroxymethylfurfural formation. *Australian Journal of Dairy Technology*, **47**, 56–57.
Fernández-Molina, J.J., Altunakar, B., Bermúdez-Aguirre, D., Swanson, B.G. & Barbosa-Cánovas, G.V. (2005a) Inactivation of *Pseudomonas fluorescens* in skim milk by combinations of pulsed electric fields and organic acids. *Journal of Food Protection*, **68**, 1232–1235.
Fernández-Molina, J.J., Fernández-Gutiérrez, S.A., Altunakar, B., Bermúdez-Aguirre, D., Swanson, B.G. & Barbosa-Cánovas, G.V. (2005b) The combined effect of pulsed electric fields and conventional heating on the microbial quality and shelf life of skim milk. *Journal of Food Processing and Preservation*, **29**, 390–406.
Floros, J.D. & Liang, H. (1994) Accoustically assisted diffusion through membranes and biomaterials. *Food Technology*, **48**(12), 79–84.
Floury, J., Grosset, N., Leconte, N., Pasco, M., Madec, M. & Jeantet, R. (2006) Continuous raw skim milk processing by pulsed electric field at non-lethal temperature: effect on microbial inactivation and functional properties. *Lait*, **86**, 43–57.
Galuska, P.J., Kolarik, R.W., Vasavada, P.C. & Marth, E.H. (1989) Inactivation of *Listeria monocytogenes* by microwave treatment. *Journal of Dairy Science*, **72**, 139–149.
García, M.L., Burgos, J., Sanz, B. & Ordóñez, J.A. (1989) Effect of heat and ultrasonic waves on the survival of two strains of *Bacillus subtilis*. *Journal of Applied Bacteriology*, **67**, 619–628.
García-Risco, M.R., Cortés, E., Carrascosa, A.V. & López-Fandiño, R. (1998) Microbiological and chemical changes in high-pressure-treated milk during refrigerated storage. *Journal of Food Protection*, **61**, 735–737.
García-Risco, M.R., Olano, A., Ramos, M. & López-Fandiño, R. (2000) Micellar changes induced by high-pressure. Influence in the proteolytic activity and organoleptic properties of milk. *Journal of Dairy Science*, **83**, 2184–2189.
Gould, G.W. (2001) The evolution of high-pressure processing of foods. In: *Ultra High Pressure Treatment of Food* (eds M.E.G. Hendrickx & D. Knorr), pp. 3–21, Kluwer Academic, New York.
Hamid, M.A.K., Boulanger, R.J., Tong, S.C., Gallop, R.A. & Pereira, R.R. (1969) Microwave pasteurization of raw milk. *Journal of Microwave Power*, **4**, 272–275.
Hauben, K.J.A., Wuytack, E.Y., Soontjens, C.C.F. & Michiels, C.W. (1996) High-pressure transient sensitization of *Escherichia coli* to lysozyme and nisin by disruption of outer-membrane permeability. *Journal of Food Protection*, **59**, 350–355.
Hayashi, R. (1988) Utilization of the effect of high pressures on organisms, organic compounds and biopolymers (as applied to food processing, e.g. preservation, maturation, enzyme regulation, pest control and microbial inhibition/destruction). *Bioscience and Industry*, **46**, 3931–3933.
Heddleson, R.A. & Doores, S. (1994) Factors affecting microwave heating of foods and microwave induced destruction of foodborne pathogens – a review. *Journal of Food Protection*, **57**, 1025–1037.
Heddleson, R.A., Doores, S., Anantheswaran, R.C. & Kuhn, G.D. (1996) Viability loss of *Salmonella* species, *Staphylococcus aureus*, and *Listeria monocytogenes* in complex foods heated by microwave energy. *Journal of Food Protection*, **59**, 813–818.
Hite, B.H. (1899) The effect of pressure in the preservation of milk. *West Virginia Agricultural Experimental Station Bulletin*, **58**, 15–35.

Hite, B.H., Giddings, N.J. & Weakley, C.E. (1914) The effect of pressure on certain microorganisms encountered in the preservation of fruits and vegetables. *Bulletin of the West Virginia Agricultural Experimental Station*, **146**, 1–67.

Hjelmqwist, J. (2005) Commercial high-pressure equipment. In: *Novel Food Processing Technologies* (eds G.V. Barbosa-Cánovas, M.S. Tapia & M.P. Cano), pp. 361–373, CRC Press, Boca Raton, FL.

Holm, S., Malmberg, R. & Svensson, K. (1984) Method and plant for producing milk with a low bacterial content. Alfa-Laval Food & Dairy Engineering AB, European Patent 0194286.

Hoover, D.G. (2000) Ultrasound. *Journal of Food Science* (Supplement: Kinetics of microbial inactivation for alternative food processing technologies), **65**, 93–95.

Hoover, D.G., Metrick, C., Papineau, A.M., Farkas, D.F. Knorr, D. (1989) Biological effects on high hydrostatic pressure on food microorganisms. *Food Technology*, **43**(3), 99–107.

Hubbard, C.D., Caswell, D., Lüdermann, H.-D. & Arnold, M. (2002) Characterization of pressure-treated skimmed milk powder dispersions: application of NMR spectroscopy. *Journal of the Science of Food and Agriculture*, **82**, 1107–1114.

Huijs, G., van Asselt, A., Verdurmen, R.E.M. & de Jong, P. (2004) High speed milk. *Dairy Industries International*, **69**(11), 30–31.

Hulsheger, H., Potel, J. & Niemann, E.G. (1981) Killing of bacteria with electric pulses of high field strength. *Radiation Environmental Biophysics*, 20, 53–65.

Huppertz, T., Fox, P.F. & Kelly, A.L. (2004) Plasmin activity and proteolysis in high-pressure treated bovine milk. *Lait*, 84, 297–304.

Isaacs, N.S., Chilton, P. & Mackey, B. (1995) Studies on the inactivation by high pressure of microorganisms. In: *High Pressure Processing of Foods* (eds D.A. Ledward, D.E. Johnston, R.G. Earnshaw & A.P.M. Hasting), pp. 65–79, Nottingham University Press, Loughborough, UK.

Itoh, S., Yoshioka, K., Terakawa, M. & Nagano, I. (1996) Continuous high pressure system for liquid food. In: *High Pressure Bioscience and Biotechnology* (eds R. Hayashi & C. Balny), pp. 455–462, Elsevier Science B.V., Amsterdam.

Jaynes, H.O. (1975) Microwave pasteurisation of milk. *Journal Milk and Food Technology*, **38**, 386–387.

Jonhston, D.E. (1995) High pressure effects on milk and meat. In: High *Pressure Processing of Foods* (eds D.A. Ledward, R.G. Earnshaw & A.P.M. Hasting), pp. 99–121, Nottingham University Press, Loughborough, UK.

Kindle, G., Busse, A., Kampa, D., MeyerKonig, U. & Daschner, F.D. (1996) Killing activity of microwaves in milk. *Journal of Hospital Infection*, **33**, 273–278.

Knorr, D., Zenker, M., Heinz, V. & Lee, D.-U. (2004) Applications and potential of ultrasonics in food processing. *Trends in Food Science & Technology*, **15**, 261–266.

Knutson, K.M., Marth, E.H. & Wagner M.K. (1988) Use of microwave oven to pasteurize milk. *Journal of Food Protection*, **51**, 715–719.

Kudra, T., van de Voort, F.R., Raghavan, G.S.V. & Ramaswamy, H.S. (1991) Heating characteristics of milk constituents in a microwave pasteurization system. *Journal of Food Science*, **56**, 931–934, 937.

Kussendrager, K.D. & van Hooijdonk, A.C.M. (2000) Lactoperoxidase: physico-chemical properties, occurrence, mechanism of action and application. *British Journal of Nutrition*, **84** (Suppl. 1), 19–25.

Li, S., Bomser, J.A. & Zhang, Q.H. (2005) Effects of pulsed electric fields and heat treatment on stability and secondary structure of bovine immunoglobulin G. *Journal of Agricultural and Food Chemistry*, **53**, 663–670.

Loey van, A., Verachtert, B. & Hendrickx, M. (2001) Effects of high electric field pulses on enzymes. *Trends in Food Science & Technology*, **12**, 94–102.

López-Fandiño, R. (2006a) High-pressure-induced changes in milk proteins and possible applications in dairy technology. *International Dairy Journal*. **16**(10), 1119–1131.

López-Fandiño, R. (2006b) Functional improvement of milk whey proteins induced by high-hydrostatic pressure. *Critical Reviews in Food Science and Nutrition*, **46**, 1–13.

López-Fandiño, R., Carrascosa, A.V. & Olano, A. (1996a) The effect of high pressure on whey protein denaturation and cheese making properties of raw milk. *Journal of Dairy Science*, **79**, 929–936.

López-Fandiño, R., Villamiel, M., Corzo, N. & Olano, A. (1996b) Assessment of the thermal treatment of milk during continuous microwave and conventional heating. *Journal of Food Protection*, **59**, 889–892.

Lubec, G., Wolf, C. & Bartosch, B. (1989) Aminoacids isomerization and microwave exposure. *Lancet*, **334**, 1392–1393.

Ludikhuyze, L., Claeys, W. & Hendrickx, M. (2000) Combined pressure–temperature inactivation of alkaline phosphatase in bovine milk: a kinetic study. *Journal of Food Science*, **65**, 155–160.

Ludikhuyze, L., Claeys, W. & Hendrickx, M. (2001) Effect of temperature and/or pressure on lactoperoxidase activity in bovine milk and acid whey. *Journal of Dairy Research*, **68**, 625–637.

Manas, P. & Pagan, R. (2005) Microbial inactivation by new technologies of food preservation. *Journal of Applied Microbiology*, **98**, 1387–1399.

Marchelli, R., Dossena, A., Palla, G., Audhuy-Peaudecerf, M., Lefeuvre, S., Carnevali, P. & Freddi, M. (1992) D-Amino acids in reconstituted infant formula: a comparison between conventional and microwave heating. *Journal of the Science of Food and Agriculture*, **59**, 217–226.

Maubois, J.-L. (2002) Membrane microfiltration: a tool for a new approach in dairy technology. *Australian Journal of Dairy Technology*, **57**, 92–96.

McClements, D.J. (1995) Advances in the application of ultrasound in food analysis and processing. *Trends in Food Science & Technology*, **6**, 293–299.

Medrano, A., Hernández, A., Prodanov, M. & Vidal-Valverde, C. (1994) Riboflavin, a-tocopherol and retinol retention in milk after microwave heating. *Lait*, **74**, 153–159.

Meissner, K. & Erbersdobler, H.F. (1996) Maillard reaction in microwave cooking: comparison of early Maillard products in conventionally and microwave heated milk. *Journal of the Science of Food and Agriculture*, **70**, 307–310.

Merin, U. & Rosenthal, I. (1984) Pasteurization of milk by microwave irradiation. *Milchwissenchaft*, **39**, 643–644.

Mermelstein, N.H. (1998) High-pressure processing begins. *Food Technology*, **52**(6), 104, 105, 108.

Min, S., Jin, Z.T. & Zhang, Q.H. (2003) Commercial scale pulsed electric field processing of tomato juice. *Journal Agricultural and Food Chemistry*, **51**, 3338–3344.

Mudgett, R.E. (1986) Microwave properties and heating characteristics of foods. *Food Technology*, **40**(6), 84–93.

Mussa, D.M. & Ramaswamy, H.S. (1997) Ultra high pressure pasteurisation of milk: kinetics of microbial destruction and changes in physico-chemical characteristics. *Lebensmittel Wissenschaft und Technologie*, **30**, 551–557.

Needs, E.C., Stenning, R.A., Gill, A.L., Ferragut, V. & Rich, G.T. (2000) High-pressure treatment of milk: effects on casein micelle structure and on enzymic coagulation. *Journal of Dairy Research*, **67**, 31–42.

Ohlsson, T. (1983) Fundamentals of microwave cooking. *Microwave World*, **4**(2), 4–9.

Ohlsson, T. (1990) Controlling heating uniformity-the key to successful microwave products. *European Food and Drink Review*, **Summer**, 7–11.

Ordóñez, J.A., Aguilera, M.A., Garcia, M.L. & Sanz, B. (1987) Effect of combined ultrasonic and heat treatment (thermoultrasonication) on the survival of a strain of *Staphylococcus aureus*. *Journal of Dairy Research*, **54**, 61–67.

Orsat, V. & Raghavan, V. (2005) Microwave technology for food processing: an overview. In: *The Microwave Processing of Foods* (eds H. Schubert & M. Regier), pp. 105–118, CRC Press, Boca Raton, FL.

Ovesen, L., Jakobsen, J., Leth, T. & Reinholdt, J. (1996) The effect of microwave heating on vitamins B_1 and E, and linoleic and linolenic acids, and immunoglobulins in human milk. *International Journal of Food Sciences and Nutrition*, **47**, 427–436.

Pagán, R., Mañas, P., Alvarez, I. & Condón, S. (1999) Resistance of *Listeria monocytogenes* to ultrasonic waves under pressure at sublethal (manosonication) and lethal (manothermosonication) temperatures. *Food Microbiology*, **16**, 139–148.

Papineau, A.M., Hoover, D.G., Knorr, D. & Farkas D.F. (1991) Antimicrobial effect of water soluble chitosans with high hydrostatic pressure. *Food Biotechnology*, **5**, 45–47.

Patterson, M.F. (2005) Microbiology of pressure-treated foods. *Journal of Applied Microbiology*, **98**, 1400–1409.

Peleg, M. (1995) A model of microbial survival after exposure to pulsed electric fields. *Journal of the Science of Food and Agriculture*, **67**, 93–99.

Popper, L. & Knorr, D. (1990) Applications of high-pressure homogenisation for food preservation. *Food Technology*, **44**(7), 84–89.

Povey, M.J.W. & McClements, D.J. (1988) Ultrasonic in food engineering. Part I: introduction and experimental methods. *Journal of Food Engineering*, **8**, 217–245.

Qin, B.L., Pothakamury, U.R., Vega-Mercado, H., Martin-Belloso, O.M., Barbosa-Cánovas, G.V. & Swanson, B.G. (1995) Food pasteurization using high-intensity pulsed electric fields. *Food Technology*, **49**(12), 55–60.

Rademacher, B. & Hinrichs, J. (2002) *Ultra High Pressure Technology for Dairy Products*. The Emerging Technologies, Bulletin No. 374, pp. 12–18, International Dairy Federation, Brussels.

Rademacher, B. & Hinrichs, J. (2006) Effects of high pressure treatment on indigenous enzymes in bovine milk: reaction kinetics, inactivation and potential application. *International Dairy Journal*, **16**(6) 655–661.

Raso, J. & Barbosa-Cánovas, G.V. (2003) Nonthermal preservation of foods using combined processing techniques. *Critical Reviews in Food Science and Nutrition*, **43**, 265–285.

Raso, J., Pagán, R., Condón, S. & Sala, F.J. (1998) Influence of temperature and pressure on the lethality of ultrasound. *Applied and Environmental Microbiology*, **64**, 465–471.

Regnault, S., Thiebaud, M., Dumay, E. & Cheftel, J.C. (2004) Pressurization of raw skim milk and of a dispersion of phosphocaseinate at 9°C or 20°C: effects on casein micelle size distribution. *International Dairy Journal*, **14**, 55–68.

Saboya, L.V. & Maubois, J. (2000) Current developments of microfiltration technology in the dairy industry. *Lait*, **80**, 541–553.

Sala, F.J., Burgos, J., Condón, S., López, P. & Raso, J. (1995) Effect of heat and ultrasound on microorganisms and encimes. In: *New Methods of Food Preservation* (ed. G.W. Gould), pp. 176–204, Blackie Academic & Professional, London.

Sampedro, F., Rodrigo, M., Martínez, A., Rodrigo, D. & Barbosa-Cánovas, G.V. (2005) Quality and safety aspects of PEF application in milk and milk products. *Critical Reviews in Food Science and Nutrition*, **45**, 25–47.

Schader, K., Buchheim, W. & Morr, C.V. (1997) High pressure effects on the colloidal calcium phosphate and the structural integrity of micellar casein in milk. Part 1. High pressure dissolution of colloidal calcium phosphate in heated milk systems. *Nahrung*, **41**, S133–S138.

Schiffmann, R.F. (1986) Food product development for microwave processing. *Food Technology*, **40**(6), 94–98.

Segal, W. (1990) Microwave heating of milk. *Lancet*, **335**, 470.

Sepulveda, D.R., Góngora-Nieto, M.M., Guerrero, J.A. & Barbosa-Cánovas, G.V. (2005) Production of extended-shelf life milk by processing pasteurized milk with pulsed electric fields. *Journal of Food Engineering*, **67**, 81–86.

Shukla, T.P. (1992) Microwave ultrasonic in food processing. *Cereal Food World*, **37**, 332–333.

Sieber, R., Eberhard, P., Fuchs, D., Gallmann, P.U. & Strahm, W. (1996b) Effect of microwave heating on vitamins A, E, B_1, B_2 and B_6 in milk. *Journal of Dairy Research*, **63**, 169–172.

Sieber, R., Eberhard, P. & Gallmann, P.U. (1996a) Heat treatment of milk in domestic microwave ovens. *International Dairy Journal*, **6**, 231–246.
Sieber, R., Eberhard, W. & Strahm, P. (1993) Mikrowellenerhitzung von milch: zonale unterschiede und einflub des rührens. *Ernährung/Nutrition*, **17**, 83–86.
Sierra, I. & Vidal-Valverde, C. (2001) Vitamin B_1 and B_6 retention in milk after continuous-flow microwave and conventional heating at high temperatures. *Journal of Food Protection*, **64**, 890–894.
Sierra, I. & Vidal-Valverde, C. (2000a) Effect of storage on the vitamin B_1 content of microwave and conventionally treated milk. *Milchwissenchaft*, **55**, 253–255.
Sierra, I. & Vidal-Valverde, C. (2000b) Influence of heating conditions in continuous-flow microwave or tubular heat exchange systems on the vitamin B_1 and B_2 content of milk. *Lait*, **80**, 601–608.
Sierra, I., Vidal-Valverde, C. & López-Fandiño, R. (2000) Effect of high-pressure on the vitamin B1 and B6 content of milk. *Milchwissenschaft*, **55**, 365–367.
Sierra, I., Vidal-Valverde, C. & Olano, A. (1999) The effects of continuous flow microwave treatment and conventional heating on the nutritional value of milk as shown by influence on vitamin B-1 retention. *European Food Research and Technology*, **209**, 352–354.
Sionneau, M., Vasseur, J., Bouix, M. & Pontvianne, P.-M. (1997) Continuous high pressure of milk. In: *High Pressure Research in the Biosciences and Biotechnology* (ed. K. Heremans), pp. 451–453, Leuven University Press, Leuven, Belgium.
Smith, K., Mittal, G.S. & Griffiths, M.W. (2001) Pasteurization of raw skim milk by pulsed electric field and antimicrobials, *Nizo Dairy Conference on Food Microbes*, 13–15 June 2001, Ede, The Netherlands.
Stearns, G. & Vasavada, P.C. (1986) Effect of microwave processing on quality of milk. *Journal of Food Protection*, **49**, 853.
Tauscher, B. (1995) Review: pasteurization of food by hydrostatic pressure: chemical aspects. *Zeitschrift für Lebensmittel-Untersuchung und-Forschung*, **200**, 3–13.
te Giffel, M. & van der Horst, C. (2004) Comparison between bactofugation and microfiltration regarding efficiency of somatic cell and bacteria removal. *Advances in Fractionation and Separation: Processes for Novel Dairy Applications*, Bulletin No. 389, pp. 49–53, International Dairy Federation, Brussels.
Tewari, G. & Juneja, V.K. (eds) (2007) *Advances in Thermal and Non-Thermal Food Preservation*, Blackwell Publishing, Oxford.
Thimson, W.J. & Short, A.J. (1965) Resistance of microorganisms to hydrostatic pressure. *Biotechnology and Bioengineering*, **7**, 139–159.
Thompson, J.S. & Thompson, A. (1990) In-home pasteurisation of raw goat's milk by microwave treatment. *Journal of Food Microbiology*, **10**, 59–64.
Tops, R. (2000) Managing Director, Tops Foods, Lammerides 26, Olen Belgium Phone: 00–32-(0)14 28 55 60, Fax: 00–32-(0)14 22 61 50.
Valero, E., Villamiel, M., Sanz, J. & Martínez-Castro, I. (2000) Chemical and sensorial changes in milk pasteurized by microwave and conventional systems during cold storage. *Food Chemistry*, **70**, 77–81.
van Heesch, E.J.M., Pemen, A.J.M., Huijbrechts, P., van der Laan, P.C.T., Ptasinski, K.J., Zanstra, G.J. & de Jong, P. (2000) A fast pulsed power source applied to treatment of conducting liquids and air. *IEEE Transactions on Plasma Science*, **28**, 137–142.
Vasavada, P.C. (1990) Microwave processing for the dairy industry. *Food Australia*, **42**, 562–564.
Vasavada, P.C., Bastian, E.D. & Marth, E.H. (1995) Chemical and functional changes in milk caused by heating in a waterbath or microwave-oven. *Milchwissenchaft*, **50**, 553–556.
Vasson, M.P., Farges, M.C., Sarret, A. & Cynober, L. (1998) Free amino acid concentrations in milk: effects of microwave versus conventional heating. *Amino Acids*, **15**, 385–388.
Vega-Mercado, H., Powers, J., Barbosa-Cánovas, G.V. & Swanson, B.G. (1995) Plasmin inactivation with pulsed electric fields. *Journal of Food Science*, **60**, 1143–1146.

Vercet, A., López, P. & Burgos, J. (1997) Inactivation of heat-resistant lipase and protease from *Pseudomonas fluorescens* by manonthermoultrasonication. *Journal of Dairy Science*, **80**, 29–36.

Vercet, A., López, P. & Burgos, J. (1998) Free radical production by manothermosonication. *Ultrasonics*, **36**, 615–618.

Vercet, A., Burgos, J. & López-Buesa, P. (2001) Manothermosonication of foods and food-resembling systems: effect on nutrient content and non-enzymatic browning. *Journal of Agricultural and Food Chemistry*, **49**, 483–489.

Vercet, A, Burgos, J. & López-Buesa, P. (2002a) Manothermosonication of heat-resistant lipase and protease from Pseudomonas fluorescens: effect of pH and sonication parameters. *Journal of Dairy Research*, **69**, 243–254.

Vercet, A., Oria, R., Marquina, P., Crelier, S. & López-Buesa, P. (2002b) Rheological properties of yoghurt made with milk submitted to manothermosonication. *Journal of Agricultural and Food Chemistry*, **50**, 6165–6171.

Verdurmen, R.E.M., te Giffel, M. & de Jong, P. (2002) Inactiveren van bacteriële sporen – nu ook met behoud van smaak? [Inactivation of microbial spores – now with flavour preservation?]. *Voedingsmiddelentechnologie*, **35**(25), 10–12.

Verschueren, M. & de Jong, P. (2006) Consumentgericht ontwerpen [Consumer-oriented design]. *Voedingsmiddelen technologie*, **39**(7), 25–27.

Vidal-Valverde, C. & Redondo, P. (1993) Effect of microwave heating on the thiamine content of cow's milk. *Journal of Dairy Research*, **60**, 259–262.

Villamiel, M. & Corzo, N. (1998) Lactulose formation during batch microwave treatment of different types of dairy products. *Milchwissenschaft*, **53**, 487–490.

Villamiel, M., Corzo, N., Martínez-Castro, I. & Olano, A. (1996a) Chemical changes during microwave treatment of milk. *Food Chemistry*, **56**, 385–388.

Villamiel, M. & de Jong, P. (2000a) Inactivation of *Pseudomonas fluorescens* and *Streptococcus thermophilus* in Trypticase® Soy Broth and total bacteria in milk by continuous-flow ultrasonic treatment and conventional heating. *Journal of Food Engineering*, **45**, 171–179.

Villamiel, M. & de Jong, P. (2000b) Influence of high-intensity ultrasound and heat treatment in continuous flow on fat, proteins, and native enzymes of milk. *Journal of Agricultural and Food Chemistry*, **48**, 472–478, 3068.

Villamiel, M., López-Fandiño, R., Corzo, N., Martínez-Castro, I. & Olano, A. (1996b) Effects of continuous-flow microwave treatment on chemical and microbiological characteristics of milk. *Z-Lebensmittel Unters Forch*, **202**, 15–18.

Villamiel, M., López-Fandiño, R. & Olano, A. (1996c) Microwave pasteurization in a continuous flow unit. Shelf life of cow's milk. *Milchwissenchaft*, **51**, 674–677.

Villamiel, M., López-Fandiño, R. & Olano, A. (1996d) Microwave pasteurization in a continuous flow unit. Effects on the cheese-making properties of goat's milk. *Milchwissenchaft*, **52**, 29–32.

Villamiel, M., Muñoz, M.M., Hernández, A. & Corzo, N. (1998) β-Lactoglobulin denaturation and furosine formation during continuous microwave treatment of milk at temperatures of 90–120°C. *Milchwissenchaft*, **53**, 434–436.

Villamiel, M., van Hamersveld, E.H. & de Jong, P. (1999) Review: effect of ultrasound processing on the quality of dairy products. *Milchwissenschaft*, **54**, 69–73.

Wäppling-Raaholt, B. & Ohlsson, T. (2005) Improving the heating uniformity in microwave processing. In: *The Microwave Processing of Foods* (eds H. Schubert & M. Regier), pp. 292–312, CRC Press, Boca Raton, FL.

Welti-Chanes, J., Barbosa-Canovas, G. & Aguilera, J.M. (2002) Engineering and food for the 21st century. In: *Food Preservation Technology Series* (eds J. Welti-Chanes, G. Barbosa-Canovas & J.M. Aguilera), pp. 927–948, CRC Press, Boca Raton, FL.

Wouters, P.C., Alvarez, I. & Raso, J. (2001) Critical factors determining inactivation kinetics by pulsed electric field food processing. *Trends in Food Science & Technology*, **12**, 112–121.

Wrigley, D.M. & Llorca, N.G. (1992) Decrease of *Salmonella typhimurium* in skim milk and egg by heat and ultrasonic wave treatment. *Journal of Food Protection*, **55**, 678–680.

Young, G.S. & Jolly, PG. (1990) Microwaves: the potential for use in dairy processing. *Australian Journal of Dairy Technology*, **45**, 34–37.

Zenker, M, Heinz, V. & Knorr, D. (2003) Application of ultrasound-assisted thermal processing for preservation and quality retention of liquid foods. *Journal of Food Protection*, **66**, 1642–1649.

Zoon, P. & Hup G. (1991) *Process for the Preparation of Raw Milk Cheese*, European Patent 0546641A1.

9 Hygiene Practices in Liquid Milk Dairies

A.J. van Asselt and M.C. te Giffel

9.1 Introduction

Milk and dairy products are highly nutritious media, in which microorganisms can multiply and cause spoilage. The levels and types of microorganisms in milk and dairy products depend on the microbial quality of the raw materials, the conditions under which the products are produced and the temperature and duration of storage. The most common spoilage microorganisms of milk and dairy products are gram-negative rod-shaped bacteria (e.g. *Pseudomonas* spp., coliforms), gram-positive spore-forming bacteria (e.g. *Bacillus* spp., *Clostridium* spp.), lactic-acid-producing bacteria (e.g. *Lactococcus* spp.) and yeasts and moulds.

Milk and milk products are also, to a limited extent, associated with foodborne illness. In the USA, milk was involved in 1.1–1.7% of the foodborne disease outbreaks with a known vehicle in the period of 1998–2002, i.e. 199–704 reported cases per year. More than 90% of all reported cases of dairy-related illness are of bacterial origin. Disease is mainly due to consumption of unpasteurised milk containing pathogenic microorganisms (e.g. *Salmonella* spp., *Listeria monocytogenes* or *Campylobacter* spp.) (te Giffel, 2003).

In this chapter, the principal microbial hazards concerning milk and milk processing, focused on heat-treated liquid milk products, are described. Subsequently, effective control measures, mainly aspects related to hygiene during processing, are discussed.

9.2 Principal hazards

9.2.1 *Raw milk*

In principle, milk produced by healthy cows contains very low concentrations of microorganisms. The teat canal acts as a barrier for the microorganisms present in the environment of the cow to enter the udder. This implies that most of the bacteria present in milk originate from the environment or the outside of the udder. Sources of contamination are bedding, milking equipment, feed and manure. Therefore, the initial concentration of microorganisms varies between less than 100 to several thousand per millilitre. The concentration is higher at the start of the milking process and decreases during the milking as a result of a washing effect.

Due to implementation of hygiene programs, the quality of the raw milk has increased significantly over time. In 1977 in Germany, it was reported that raw milk had an average count of 5×10^5 colony-forming units (cfu) mL^{-1}, and in 2002, an average count of 2×10^4 cfu mL^{-1} (Suhren & Reichmuth, 2003). The main reasons for the reduction in microbial counts are the introduction of closed milking systems, and the use of refrigerated bulk tanks to store and transport the raw milk. In addition, the improvement and reliability of cooling systems have contributed to the quality improvement of raw milk.

This has also resulted in a shift in microflora from mainly gram-positive-acid-producing bacteria to gram-negative psychrotrophic microorganisms, mainly species of *Pseudomonas, Aeromonas, Flavobacterium, Alcaligines and Acinetobacter* (Dogan & Boor, 2003). These microorganisms grow rapidly at refrigerating temperatures, produce heat-resistant proteolytic and lipolytic enzymes, and limit the duration of storage at the farm and length of transportation from farm to factory prior to heat treatment. Furthermore, a wide variety of pathogens all associated with animals, equipment, personnel and environment is found in raw milk including *Mycobacterium* spp., *Salmonella* spp., *Campylobacter* spp., *Escherichia coli, Staphylococcus aureus, Listeria monocytogenes* and *Yersinia enterocolitica.* Depending on region, climate and season, raw milk can contain one or more of these species. Surveys, e.g. in Belgium, showed that the percentages of raw milk containing pathogens and the contamination levels are low.

9.2.2 *Pasteurised milk*

The pasteurisation of milk is defined by international organisations, such as the International Dairy federation (IDF, 1986), World Health Organization of the United Nations (WHO) and Food and Agriculture Organization of the United Nations (FAO, 2004), and is carried out to reduce the levels of vegetative pathogens and ensure food safety. Simultaneously, spoilage bacteria and yeasts are inactivated. However, as from 1 January 2006, five new regulations are effective (EU, 2002, 2004a–d), and replace the former Milk Hygiene Directive 92/46/EEC. One of the main implications of the new directives is that the specific time–temperature combinations are no longer specified. In practice, the former settings are still applied. Therefore, pasteurisation can be carried out by heating the product either for 30 min at 62.7°C (low temperature long time – LTLT) or for 15 s at 71.7°C (high temperature short time – HTST). The concentration of the vegetative microorganisms after pasteurisation is normally reduced by a factor of 1000. The inactivation of the vegetative (pathogenic) microorganisms (*L. monocytogenes* and *Salmonella* spp.) and bacterial spores (*Bacillus sporothermodurans* and *Bacillus cereus*) is shown in Figure 9.1; this illustration shows that bacterial spores are hardly affected by the thermisation and pasteurisation. These processes will inactivate only the vegetative microorganisms present in raw milk.

However, spoilage of pasteurised milk is caused by:

- Growth of surviving spore-forming bacteria (*Bacillus* spp. and *Clostridium* spp.)
- Growth of thermoduric bacteria (e.g. *Micrococcus* (Hileman *et al.*, 1941; Chadwick Hayes & Boor, 2001), *Streptococcus* spp., *Enterococcus* spp., *Lactobacillus* spp. and sometimes gram-negative rods)
- Recontamination and growth of psychrotrophic (gram-negative) bacteria (e.g. *Pseudomonas* spp.)
- Activity of heat-stable enzymes produced prepasteurisation.

Comparable with the hygiene trend at the farm, the improved milk handling at milk-processing facilities (i.e. extended refrigerated storage, higher heat treatment temperatures,

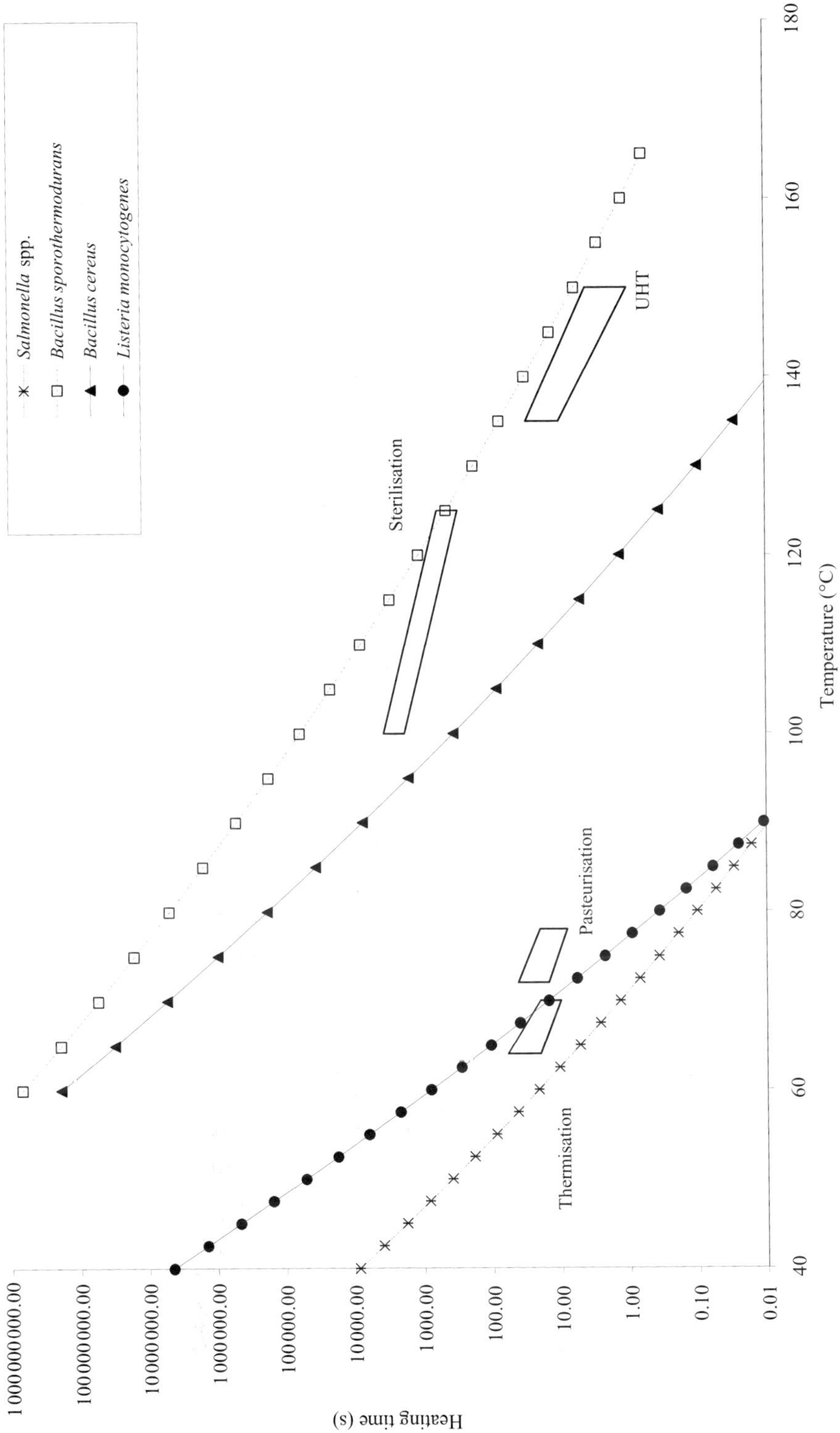

Fig. 9.1 Six decimal inactivation of different milk-specific microorganisms by various heat treatments applied in the processing of milk and dairy products.

control of re-contamination after heat treatment) enhanced the importance of heat-resistant spore formers with respect to shelf life of the product. *B. cereus* limits the shelf life of pasteurised milk, and is associated with sweet curdling, i.e. caused by proteolytic and lipolytic enzymes produced by the microorganism. As *B. cereus* survives the pasteurisation process, reduction can be achieved only by reducing the initial concentration in raw milk. Currently, the dairy industry aims at maximum spore levels of 10^3 spores L^{-1} raw milk to assure quality and safety of pasteurised milk and dairy products.

9.2.3 *Sterilised milk*

Sterilised milk can be stored at ambient temperatures (20–25°C) for at least 4–6 months. In order to achieve this, the aim is to reduce the viable counts *Bacillus stearothermophilus* (now *Geobacillus stearothermophilus*) and similar heat-resistant spore-forming bacteria by 9 decimals. This heat treatment is also sufficient to decrease *Clostridium botulinum* levels with 12 decimal reductions, thus ensuring food safety.

Sterilisation is carried out either by in-package sterilisation or by continuous sterilisation in heat exchangers combined with aseptic packaging. The latter is also known as ultra-high-temperature (UHT) treatment, which is normally in the range of 135–150°C in combination with appropriate holding times necessary to achieve commercial sterility (FAO, 2004). Microbial spoilage of UHT milk may occur by the outgrowth of heat-resistant spores that have survived the heat treatment. During the last 20 years, highly heat-resistant spores seem to occur more frequently. For example *Bacillus subtilis* is able to survive conventional sterilisation processes. *G. stearothermophilus* and *Bacillus sporothermodurans* are known for spoilage of commercial UHT milk (Huemer *et al.*, 1998). The latter organism is responsible for causing spoilage and un-sterility in UHT products, and can survive heat treatments at 142°C for 15 s. Outgrowth of the microorganism does not alter the product characteristics, although in some cases a pinkish colour may occur. *B. sporothermodurans* is not considered to be a threat to public health. However, it does imply problems for dairies due to legal requirements and to avoid trade restrictions. In addition to the microbial stability of UHT milk, the activity of heat-resistant endogenous milk enzymes like plasmin is relevant when it concerns physical stability of the milk. Insufficient inactivation causes bitter-tasting milk, resulting in product loss for producers and consumers of milk. To avoid bitterness, an inactivation to achieve a 99% reduction of the plasmin present is commonly applied. Irreversible inactivation starts at about 65°C (Metwalli *et al.*, 1998), but above 92°C the activation energy increases only slightly with temperature. This means that very high temperature, ultra-short-time treatments, like UHT treatments, can be sufficient to inactivate microorganisms and obtain a sterile product, but provide insufficient heat load to reduce plasmin activity and obtain a stable product. Therefore, preheating (e.g. 50 s at 90°C) is applied prior to the UHT treatment to reduce the amount of active plasmin.

9.3 Hygienic processing

9.3.1 *General*

Hygienic processing is often directly linked with hygienic design. Although the design of equipment plays an important role in hygienic processing, an integral approach of the total installation, varying from the design of the building, logistic routes, location of the production facility to maintenance and use of processing aids like lubricants and cleaning agents. In addition, when processes are designed, systems and components are frequently put together in such a way that new hazards arise. As a result, excessive product and economic losses may be experienced by users of the equipment. General requirements for designing food-processing equipment, including dairy equipment, are described in various guidelines and directives like the machine safety directive (EU, 1989), and the hygiene of foodstuffs. Furthermore, organisations like the European Hygienic Engineering Design Group (EHEDG), 3-A sanitary standards and NSF (the public Health and Safety Company) publish a range of guidelines concerning the safe production of various foods. Guidance on design, construction and installation of equipment, cleaning-in-place (CIP) systems and plant is given in various IDF documents (IDF, 1979, 1980, 1985, 1987, 1992, 1994, 1996, 1997a,b). Summaries of EHEDG guideline documents are published in '*Trends in Food Science and Technology*' (EHEDG, 1992, 1993a–i, 1994a,b, 1995a,b, 1997a,b, 2001a–c, 2007a–d; Mostert *et al.*, 1993; Benezech, 2001; Moens-Go Yanko, 2002, 2003, 2006a,b, 2007; Maller, 2007). An overview of all the titles available and a short description is given in the EHEDG yearbook 2007 (EHEDG, 2007e). 3-A sanitary standards are available for many types of equipment, from fittings to silo tanks. Relevant aspects when (re)designing a production process are discussed in this paragraph whilst referring to these documents.

9.3.2 *Building construction*

When a building is designed, the purpose and needs of the building have to be addressed already. It is important to realise that the building is the first barrier for insects, birds and dust. The best way to maintain the integrity is to avoid any unnecessary openings. This implies that all doors, windows and other transits through roofs or walls like ventilation shafts and exhaust pipes should be strictly functional. The location and construction of the pipes should be chosen for the most hygienic route (i.e. exhaust pipes not opposite the inlet for fresh air into the packaging room).

In addition to the construction of the outside of the building, the inner side of the construction is of equal importance. It is obvious that walls (including doors and windows) and ceilings need to be easy cleanable. When insulating, it is essential to avoid condensation within the insulating layer as product or processes may be (re-)contaminated. The best solution is to insulating on the outside of the roof or wall. Floors should be self-drainable (1° slope), and resistant against physical (e.g. lorry transport) and chemical (raw materials, cleaning agents) loads.

Most designing is logical thinking, but in practice, logic is overruled by a shortage of time and money, often resulting in a bad starting position when it concerns the hygienic design of the building.

9.3.3 *Zoning*

Originally, zoning was introduced for separating areas with different purposes like separation of raw materials and end products, and separation of wet and dry material; in other words, keep away anything that has nothing to do with the product. More recently, zoning is applied to separate areas with different levels and needs of hygiene. In general, four different levels can be distinguished: first zone is the factory site, second zone is the factory building, third zone is a high risk or high hygiene zone and fourth zone is a product enclosure zone. The core of the system is that the degree of control increases in such a way that fully processed products are handled in controlled areas where possible hazards are actively excluded (Holah, 2005). For example, a lorry truck transporting raw materials cannot trespass a high hygiene area where end product is filled and packaged, or the same lorry may not be used for even transporting the final product.

9.3.4 *Processing equipment*

At all stages of processing, good hygiene of the manufacturing plant is essential to ensure that the product stream is not (re-)contaminated after heat treatment of the raw milk (pasteurisation, sterilisation or UHT). Sources of post-pasteurisation contamination include equipment, packaging materials, air, aerosols, (condensed) water and lubricants. Pasteurisation equipment should be properly designed, installed, maintained and operated to ensure that the milk is heated to at least the specified temperature for the specified time. Biofilms present on the surface of milk-processing equipment threaten the quality and safety of dairy products. Dead ends, corners, cracks, crevices, gaskets, valves and joints are vulnerable points for microbial fouling. Fouling in a dairy-processing plant depends on the type of microorganism, the type of product being processed, the operational conditions of the plant (temperatures, length of production runs), and the type of surface. The hygienic design of processing equipment is of great importance in avoiding biofilm formation. Biofilm control also relies on well-defined cleaning and sanitising procedures and the effectiveness of these procedures. Bacteria within biofilms are more difficult to eliminate than free-living cells and once established they can act as a source of (re-)contamination. Contamination attributed to biofilm development has been reported in general milk processing (e.g. pasteurisation and milk transfer line) and the manufacture of cheese, whey and milk powder.

Pathogenic microorganisms, including *L. monocytogenes*, *Salmonella typhimurium* and *Y. enterocolitica*, will also attach to surfaces in dairy-processing environments, e.g. stainless steel. Subsequently, dairy products may be (re-)contaminated. However, attachment of microorganisms may be promoted or inhibited in dairy fluids depending on the composition of the product and the type of bacteria. In the presence of whole milk, attachment may be inhibited, but the presence of lactose and non-casein protein solutions can enhance microbial fouling. An association of the bacteria with milk fat globules or the effect of natural antibodies has been suggested as possible explanations. The pasteuriser can be a source of contamination with *Bacillus* spp., especially after non-production days like weekends. In addition, the growth of species of bacteria, e.g. *Streptococcus thermophilus*, on the surface in the regeneration section of plate heat exchangers can contaminate milk with counts ranging <100 to $>10^6$ cfu mL^{-1} and/or their metabolic products.

This affects the quality of products manufactured from this milk, but fouling also leads to an increased use of energy resulting in a decrease in production time, obstruction and corrosion causing considerable economic losses. The filling machine can also be a significant source of post-pasteurisation. The presence of spoilage psychrotrophs (*Acinetobacter* spp., *Pseudomonas* spp. and *Flavobacterium* spp.) in pasteurised milk may occur after pasteurisation and indicates inadequate cleaning (te Giffel, 2003).

9.3.5 *Cleaning*

Most of the equipment used for handling milk and milk products is cleaned and disinfected by CIP systems at least daily (see Tamime, 2008). CIP is an important unit-operation that is commonly applied throughout the whole food industry. In the dairy industry, CIP is applied daily for the major part of the equipment. This ensures a constant product quality, efficient heat transfer in heat exchangers and avoids possible growth of microorganisms. Start-up of closed processing lines in the dairy industry is usually done by circulating hot water in order to have an additional decontamination of the equipment. Monitoring CIP systems, i.e. concentrations of the cleaning agents, temperatures, flow, pressure and circulation time, is necessary to ensure the efficiency of cleaning.

The choice of a cleaning procedure strongly depends on the type of fouling. In the dairy industry, the fouling of milk-processing equipment is caused by the formation of a deposit on the surface, mainly consisting of proteins and minerals. Especially, heat exchangers are fouled by this deposit, but even at room temperature, a monolayer of proteins can adhere to the surface immediately. In general, proteins are the most difficult to remove especially when the protein is unfolded or denatured. Whey proteins in particular act as fouling intermediates as a result of the denaturation and aggregation of β-lactoglobulin. The activated molecule acts like a radical causing the deposition of proteins and other milk components on the surface of the equipment. The deposition of calcium phosphate salts is mainly dependent on the solubility of calcium salts. The higher the temperature, the less soluble the mineral becomes. In combination with the activated β-lactoglobulin, the insoluble calcium phosphate is responsible for the build-up of the fouling layer on surfaces of heat exchangers.

The removal of protein fouling is primarily achieved by cleaning with alkaline solutions. The cleaning solution penetrates into the spongy structure of the deposit layer and causes cracks in the layer. This facilitates the further penetration of the cleaning solution. The swollen layer is subsequently disrupted by the shear stress and/or pressure fluctuations in the fast-flowing cleaning liquid. The remaining mineral scaling of the surface (mainly calcium phosphate) can easily be removed by acid solutions. In milk-processing systems, the first cleaning phase is an alkaline phase followed by an acid phase since the deposit contains mainly proteins and fat. For whey-processing systems, the first cleaning phase is an acid phase since the components of the deposit are mainly minerals.

Factors that influence the effectiveness of cleaning are (a) time, (b) concentration of detergent and/or sterilising agent(s), (c) temperature and (d) flow. The efficiency of cleaning can be improved by the right combination of temperature, concentration of cleaning solution and flow. An optimum temperature for cleaning of pasteurisation equipment is between 65 and 70°C. Temperatures above 80°C do not always imply a better cleaning result, whereas energy use is increased (steam, hot water), corrosion may occur on the surface of the equipment, and it might lead to reactions of proteins in the deposit. Concerning alkaline solutions, a concentration of 0.5 g 100 g^{-1} is optimal to remove organic fouling from the surface. Higher concentrations (>1 g 100 g^{-1}) will lead to impermeable rubber-like layers that are difficult to remove. Concentrations lower than 0.3 g 100 g^{-1} are not sufficient to remove the fouling from the surface. Sufficient flow is required for taking away the loosened deposit. However, it is not necessary to have turbulent flow as there is no theoretical justification for this. The continuous monitoring of those factors will result in an objective analysis of the cleaning procedure. Although these findings are common knowledge in the dairy industry, still many plants do not use the optimal cleaning procedures. Even when cleaning processes are adjusted (e.g. in capacity) efficiency is not optimised automatically. Therefore, significant improvements in cleaning efficiency can be achieved (van Asselt *et al.*, 2002).

9.3.6 *Packaging materials*

In the food industry, packaging materials, carton-forming mandrels, filling heads and airborne microorganisms were identified as potential contamination sources. Food grade paper and board used in the dairy industry are usually of high hygienic quality and microbial counts are generally well below the limits set by the FDA (2003), ≤ 1 cfu cm^{-2} or 250 cfu g^{-1}. In a study by Pirttijarvi *et al.* (1996), it was demonstrated that the contamination of the inner surface of cartons intended for liquid foods rarely exceeded 10 cfu per package of 1 L. However, reusable glass milk bottles have been shown to be contaminated by spore-forming organisms, such as *B. cereus*, in concentrations of $<$10–250 cfu 100 mL^{-1} rinsing water.

9.3.7 *Processing aids*

Lubricants, grease and oil are essential components for the lubrication, heat transfer, pressure transmission and corrosion protection of processing equipment. For applications within the food industry, specific food grade lubricants are developed. Food grade means in this case that people will not become ill when consuming these products. However, even by using food grade lubricants, the product may become contaminated accidentally. Assuming that contact with product cannot fully be excluded, a serious entrance check of these materials is necessary to avoid mixing with non-food grade as contamination of the product with the latter encompasses serious health risks (Steenaard, 2001).

9.4 Monitoring and control

Control of production processes in the food industry has always focused on examination of end products. However, feedback of test results to the production process is generally not possible because it takes too long before the results of the analyses are known. Moreover, high numbers of samples have to be analysed to obtain statistically reliable results and inspection of end products enables defects only to be observed; their causes cannot be established.

Therefore, analysis at the end of the process has shifted to control of the process by the introduction of good manufacturing practice (GMP) and hazard appraisal (analysis) critical control points (HACCP) systems. The use of a continuous, preferably in-line, monitoring system is necessary to make sure that the critical points in the process are controlled. This enables rapid detection and correction of slight deviations of process parameters yielding increased productivity and profitability. In addition, large margins that are used, e.g. in heat treatments, to prevent safety issues can be minimised to improve quality aspects such as nutritional value and taste.

At present, sensors and instrumentation are commercially available and applied in the food industry for in-line and online measurement of physical and physical–chemical properties of products, e.g. temperature, pressure, flow and levels in tanks. However, sensors are not yet much applied to determine product composition and quality and safety aspects, such as complex textures, concentrations of chemical compounds and microorganisms (te Giffel, 2006).

9.4.1 *In-line detection of microorganisms*

Microbiological testing in the dairy plant is critical to ensure the quality of raw milk and final products. The traditional methods usually require several hours or days of culturing, but provide relatively low limits of detection (in the range of 1 cfu mL^{-1} or lower). Rapid methods can provide results within ~10 min. However, the limits of detection are in the range of 10^4–10^5 cfu mL^{-1}. Despite extensive efforts to develop reliable and rapid analysis technology, no commercial methods are available as yet.

NIZO food research investigated the potential of rapid detection methods for in-line control of pasteurisation. The running time of continuous-flow process equipment like pasteurisers is limited mainly by thermoresistent streptococci (TRS), which are not completely inactivated by pasteurisation, and grow at temperatures between 30 and 50°C, i.e. in the regenerative section of heat exchangers. During processing, the attached TRS are released into the product flow. Depending on the initial level in the raw milk, running times of ~4 to over 11 h can be realised before the critical level of 10^5 cfu TRS mL^{-1} is reached and the pasteuriser must be cleaned. Bactoscan (a technique combining staining of live and dead bacteria and microscopic detection) and ATP measurements (detecting bacterial activity of living cells) were tested for indication of reaching the critical level. The results showed that, owing to the detection limits (10^4–10^5 cfu mL^{-1}), both systems are only suited as an 'emergency break', but not as a timely indicator that cleaning is necessary. If processing equipment could be fitted with a suitable filtration–concentration unit, the technology can probably reach the target detection limits

9.4.2 *Optimisation of CIP procedures*

Cleaning and disinfection are essential to assure quality and safety in the food industry. Most dairy processes require at least daily cleaning. The applied procedures are often based on experience. Large margins are chosen for the intensity and length of the cleaning steps to ensure food safety. With production batches getting smaller and product diversity increasing, flexibility in CIP processes gains in relevance. Strategies based on in-line and online monitoring of cleaning steps can save energy and time, and decrease consumption of water and raw materials.

NIZO food research developed a monitoring system, called OPTI-CIP, based on in- and at-line measurements of removal of deposits and cleaning agents, and parameters like temperature, flow, conductivity and valve settings. Cases in the dairy industry demonstrated that the efficiency of cleaning can be improved by reducing the cleaning time by 50% (Figure 9.2) (van Asselt *et al.*, 2002)

However, this device measures removal of fouling at-line. In-line monitoring offers advantages, especially to production facilities with a large variety of products and short run times. Therefore, the application of a turbidity sensor for monitoring of organic fouling removal was tested during a two-step cleaning process in an evaporator.

Figure 9.3 shows that in-line and offline measurements provided comparable results: the peaks are observed at the same time. The in-line method measured higher levels of turbidity, i.e. organic fouling removal. This is due to higher background levels and foaming of the cleaning solution affecting the sensor measurements. When the in-line results were corrected for this, both methods were comparable. These data demonstrate that real-time monitoring of cleaning procedures is possible. For application in production plants, more research is needed with respect to robustness of the system.

A different way of optimising cleaning and disinfection is the combination of databases and predictive modelling. NIZO Premia™ is an example that combines research knowledge with predictive modelling. It is a software platform that is used for optimisation of product properties or process performance. For example, fouling is mainly caused by denaturation of proteins and precipitation of minerals. The denaturation process of β-lactoglobulin (an important whey protein) can be described as a consecutive set of reactions (de Jong, 1996). This knowledge can be used to predict the fouling behaviour in heat exchangers of different dairy products. By predicting the amount of fouling produced, the optimum running time for heat exchangers can be determined. In addition, the composition of the fouling layer is known; this enables the operator to choose the right cleaning procedures (i.e. cleaning agents, temperature, flow rate). Industrial trials showed that, after optimisation with NIZO Premia™, the amount of fouling could be reduced by 50–80%, resulting in longer running times and higher process efficiency (de Jong *et al.*, 2002b). Another possible use of predictive modelling is the design of new processing lines by estimating the extent of fouling during processing and final product properties. A typical example is the development of a new type of evaporator at a Dutch dairy company, where the use of NIZO Premia™ resulted in 70% less energy use compared to standard designed evaporators (Vissers *et al.*, 2002). A third option is the use of predictive modelling for assessing the opportunities and risks of the formation of biofilms. The process of biofilm formation (adherence, growth and release) can be quantified using predictive models (de Jong *et al.*, 2002a, b, den Aantrekker *et al.*, 2003). These models can be used either for preventing the formation of biofilms or for defining the right strategy for removal of the biofilm. Thus, predictive modelling is a powerful tool to analyse and optimise critical processes within the food industry.

9.5 Concluding remarks

Hygienic processing is often directly linked with hygienic design. An integrated approach to the total installation, varying from the design of the building, logistic routes, location

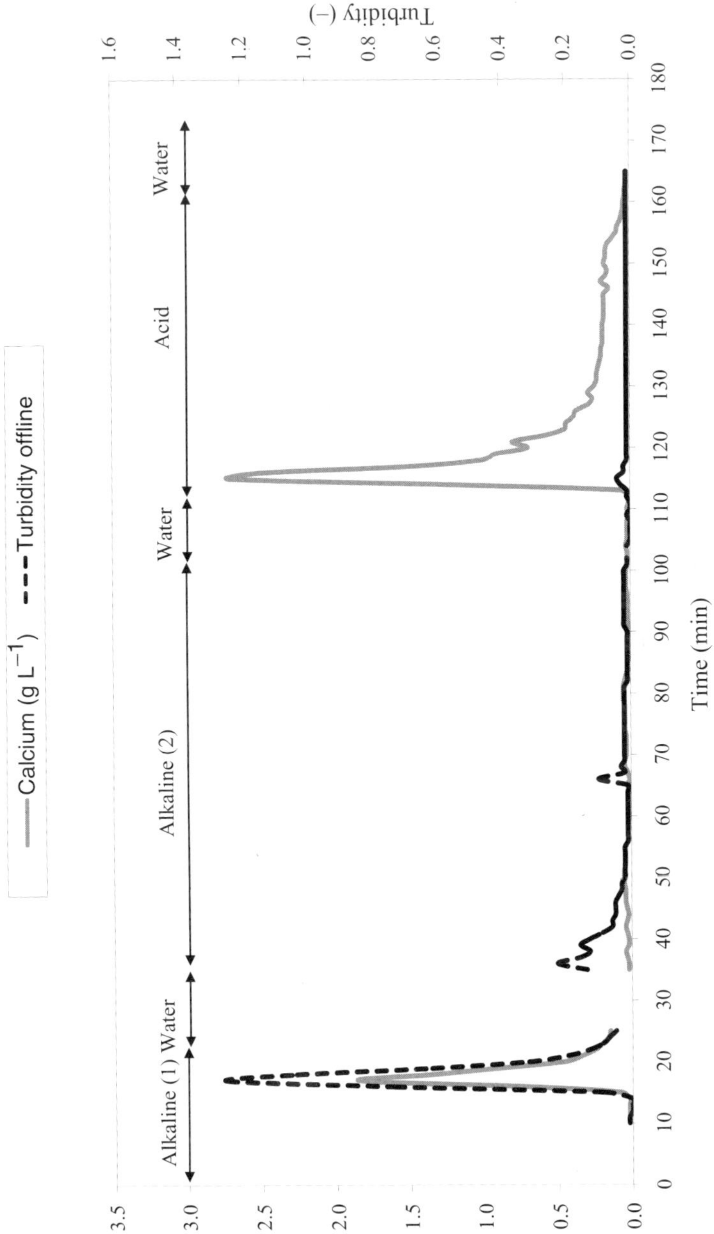

Fig. 9.2 Optimisation of cleaning-in-place (CIP) of an evaporator based on in-line and at-line measurement (2a = before optimisation; 2b = after optimisation). Note: Turbidity corresponds to the quantity of undissolved organic and inorganic material that is removed by alkaline cleaning; calcium concentration indicates the quantity of an inorganic contamination removed via acid cleaning.

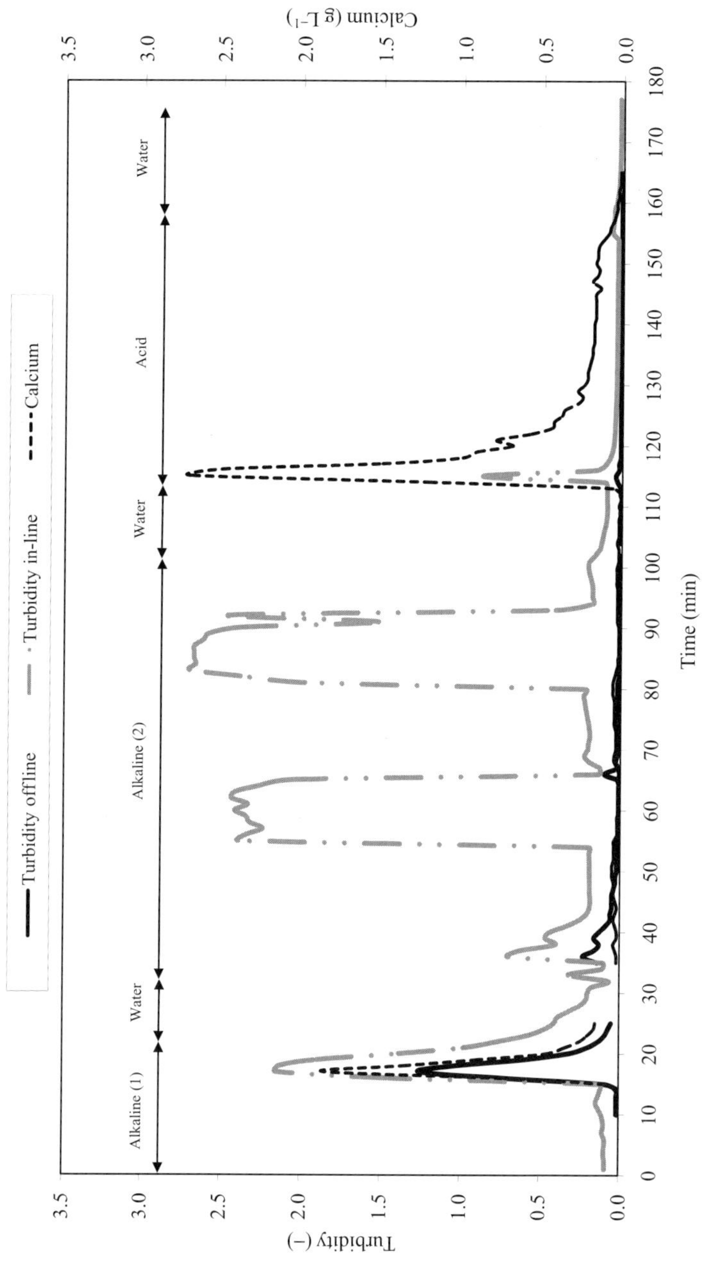

Fig. 9.3 Monitoring turbidity offline versus in-line during cleaning-in-place (CIP) of an evaporator.

of the production facility to maintenance and use of processing aids like lubricants and cleaning agents, contributes significantly to control of the microbial quality and safety of milk and dairy products.

Furthermore, optimisation of production processes is of great importance to the dairy industry as this may lead to quality improvements of products and processes, and to cost savings. Integrated process and product development by applying rapid detection methods for critical process parameters, in-line monitoring devices and predictive models, therefore, offers new opportunities for the dairy industry in the coming years. The reader is referred to the following websites for a more complete discussion: http://ec.europa.eu/food/food/biosafety/hygienelegislation/index_en.htm, http://eur-lex.europa.eu/nl/index.htm, http://www.ehedg.org/, http://www.3-a.org/, http://www.nsf.org/ and/or http://www.fil-idf.org/.

References

Benezech, T. (2001) A method for assessing the bacterial retention ability of hydrophobic membrane filters. *Trends in Food Science & Technology*, **12**, 36–38.

Chadwick Hayes, M. & Boor, K. (2001) Raw milk and fluid milk products. In: *Applied Dairy Microbiology* (eds E.H. Marth & J.L. Steele), pp. 59–76, Marcel Dekker, New York.

de Jong, P. (1996) *Modelling and Optimisation of Thermal Processes in the Dairy Industry*, PhD thesis, Delft University of Technology, Delft, The Netherlands.

de Jong, P., te Giffel, M.C. & Kiezebrink, E.H. (2002a) Prediction of the adherence, growth and release of thermoresistant streptococci in production chains. *International Journal of Microbiology*, **74**, 13–25.

de Jong, P., te Giffel, M.C., Straatsma, H. & Vissers, M.M.M. (2002b) Reduction of fouling and contamination by predictive kinetic models. *International Dairy Journal*, **12**, 285–292.

den Aantrekker, E.D., Vernooij, W.W., Reij, M.W., Zwietering, M.H., Beumer, R.R., van Schothorst, M. & Boom, R.M. (2003) A biofilm model for flowing systems in the food industry. *Journal of Food Protection*, **66**, 1432–1438.

Dogan, B. & Boor, K.J. (2003) Genetic diversity and spoilage potentials among *Pseudomonas* spp. isolated from fluid milk products and dairy processing plants. *Applied and Environmental Microbiology*, **69**, 130–138.

EHEDG (1992) Microbiologically safe continuous pasteurization of liquid foods. *Trends in Food Science & Technology*, **3**, 303–307.

EHEDG (1993a) The EC machinery directive and food-processing equipment. *Trends in Food Science & Technology*, **4**, 153–154.

EHEDG (1993b) Hygienic design of closed equipment for the processing of liquid food. *Trends in Food Science & Technology*, **4**, 375–379.

EHEDG (1993c) Hygienic equipment design criteria. *Trends in Food Science & Technology*, **4**, 225–229.

EHEDG (1993d) Hygienic packing of food products. *Trends in Food Science & Technology*, **4**, 406–411.

EHEDG (1993e) A method for the assessment of bacteria tightness of food-processing equipment. *Trends in Food Science & Technology*, **4**, 190–192.

EHEDG (1993f) A method for the assessment of in-line pasteurization of food-processing equipment. *Trends in Food Science & Technology*, **4**, 52–55.

EHEDG (1993g) A method for the assessment of in-line steam sterilizability of food-processing equipment. *Trends in Food Science & Technology*, **4**, 80–82.

EHEDG (1993h) The microbiologically safe continuous-flow thermal sterilization of liquid foods. *Trends in Food Science & Technology*, **4**, 115–121.

EHEDG (1993i) Welding stainless steel to meet hygienic requirements. *Trends in Food Science & Technology*, **4**, 306–310.

EHEDG (1994a) The continuous or semicontinuous flow thermal treatment of particulate foods. *Trends in Food Science & Technology*, **5**, 88–95.

EHEDG (1994b) Hygienic design of valves for food processing. *Trends in Food Science & Technology*, **5**, 169–171.

EHEDG (1995a) Experimental rigs are available for the EHEDG test methods. *Trends in Food Science & Technology*, **6**, 132–134.

EHEDG (1995b) Hygienic design of equipment for open processing. *Trends in Food Science & Technology*, **6**, 305–310.

EHEDG (1997a) Hygienic pipe couplings. *Trends in Food Science & Technology*, **8**, 88–92.

EHEDG (1997b) A method for the assessment of in-place cleanability of moderately-sized food processing equipment. *Trends in Food Science & Technology*, **8**, 54–57.

EHEDG (2001a) Challenge tests for the evaluation of hygienic characteristics of packing machines for liquid and semi-liquid products. *Trends in Food Science & Technology*, **12**, 244–248.

EHEDG (2001b) General hygienic design criteria for the safe processing of dry particulate materials. *Trends in Food Science & Technology*, **12**, 296–301.

EHEDG (2001c) Hygienic design and safe use of double-seat mixproof valves. *Trends in Food Science & Technology*, **12**, 203–206.

EHEDG (2007a) Materials of construction for equipment in contact with food. *Trends in Food Science & Technology*, **18**, S40–S50.

EHEDG (2007b) A method for assessing the in-place cleanability of food-processing equipment. *Trends in Food Science & Technology*, **18**, S54–S58.

EHEDG (2007c) Production and use of food-grade lubricants. *Trends in Food Science & Technology*, **18**, S79–S83.

EHEDG (2007d) Safe and hygienic water treatment in food factories. *Trends in Food Science & Technology*, **18**, S93–S98.

EHEDG (2007e) Titles available and description of EHEDG Guidelines. *Trends in Food Science & Technology*, **18**, S20–S25.

EU (1989) Council Directive 89/392/EEC of 14 June 1989 on *The Approximation of the Laws of the Member States Relating to Machinery*.

EU (2002) Regulation (EC) No 178/2002 of the European Parliament and of the Council of 28 January 2002 *Laying Down the General Principles and Requirements of Food Law, Establishing the European Food Safety Authority and Laying down Procedures in Matters of Food Safety*.

EU (2004a) Regulation (EC) No 852/2004 of the European Parliament and of the Council of 29 April 2004 on *The Hygiene of Foodstuffs*.

EU (2004b) Regulation (EC) No 853/2004 of the European Parliament and of the Council of 29 April 2004 *Laying Down Specific Hygiene Rules for Food of Animal Origin*.

EU (2004c) Regulation (EC) No 854/2004 of the European Parliament and of the Council of 29 April 2004 *Laying Down Specific Rules for the Organisation of Official Controls on Products of Animal Origin Intended for Human Consumption*.

EU (2004d) Regulation (EC) No 882/2004 of the European Parliament and of the Council of 29 April 2004 on *Official Controls Performed to Ensure the Verification of Compliance with Feed and Food Law, Animal Health and Animal Welfare Rules*.

FAO (2004) *Code of Hygienic Practice for Milk and Milk Products*, Food and Agriculture Organization of the United Nations, Rome.

FDA (2003) *Grade "A" Pasteurized Milk Ordinance Nutrition*, Centre for Food Safety and Applied Nutrition, Washington, DC.

Hileman, J.L., Leber, H. & Speck, M.L. (1941) Thermoduric bacteria in pasteurized milk. II. Studies on the bacteria surviving pasteurization, with special reference to high-temperature, short-time pasteurization. *Journal of Dairy Science*, **24**, 305–315.

Holah, J. (2005) Improving zoning within food processing plants. In: *Handbook of Hygiene Control in the Food Industry* (eds H.L. Lelieveld, M.A. Mostert & J. Holah), pp. 148–167, Woodhead Publishing, Cambridge.

Huemer, I.A., Klijn, N., Vogelsang, H.W.J. & Langeveld, L.P.M. (1998) Thermal death kinetics of spores of *Bacillus sporothermodurans* isolated from UHT milk. *International Dairy Journal*, **8**, 851–855.

IDF (1979) *Design and Use of CIP Systems in the Dairy Industry*, Document No. 117, International Dairy Federation, Brussels.

IDF (1980) *General Code of Hygiene Practice for the Dairy Industry*, Document No. 123, International Dairy Federation, Brussels.

IDF (1985) *Surface Finishes of Stainless Steels – New Stainless Steels*, Document No. 189, International Dairy Federation, Brussels.

IDF (1986) *Pasteurized milk*, Document No. 200, International Dairy Federation, Brussels.

IDF (1987) *Hygienic Design of Dairy Processing Equipment*, Document No. 218, International Dairy Federation, Brussels.

IDF (1992) *Hygienic Management in Dairy Plants*, Document No. 276, International Dairy Federation, Brussels.

IDF (1994) *Recommendations for the Hygienic Manufacture of Milk and Milk Based Products*, Document No. 292, International Dairy Federation, Brussels.

IDF (1996) *Codex Standards in the Context of World Trade Agreements/IDF General Recommendations for the Hygienic Design of Dairy Equipment*, Document No. 310, pp. 26–31, International Dairy Federation, Brussels.

IDF (1997a) *Hygienic Design and Mainenance of Dairy Buildings and Services*, Document No. 324, International Dairy Federation, Brussels.

IDF (1997b) *Fouling and Cleaning of Heat Treatment Equipment*, Document No. 328, International Dairy Federation, Brussels.

Maller, R.R. (2007) Passivation of stainless steel. *Trends in Food Science & Technology*, **18**, S112–S115.

Metwalli, A.A.M., de Jongh, H.H.J. & van Boekel, M.A.J.S. (1998) Heat inactivation of bovine plasmin. *International Dairy Journal*, **8**, 47–56.

Moens-Go Yanko, E. (2002) The prevention and control of Legionella spp. (including Legionnaires' disease) in food factories. *Trends in Food Science & Technology*, **13**, 380–384.

Moens-Go Yanko, E. (2003) Design of mechanical seals for hygienic and aseptic applications. *Trends in Food Science & Technology*, **14**, 478–481.

Moens-Go Yanko, E. (2006a) Guidelines on air handling in the food industry. *Trends in Food Science & Technology*, **17**, 331–336.

Moens-Go Yanko, E. (2006b) Hygienic engineering of fluid bed and spray dryer plants. *Trends in Food Science & Technology*, **17**, 621–625.

Moens-Go Yanko, E. (2007) Integration of hygienic and aseptic systems. *Trends in Food Science & Technology*, **18**, 48–58.

Mostert, M.A., Buteux, G., Harvey, P.C., Hugelshofer, W., Mellbin, P., Nassauer, J., Reinecke, G., Weber, W. & Wilke, B. (1993) Microbiologically safe aseptic packing of food products. *Trends in Food Science & Technology*, **4**, 21–25.

Pirttijarvi, T.S.M., Graeffe, T.H., Salkinoja-Salonen, M.S. & Pirttija, T.S.M. (1996) Bacterial contaminants in liquid packaging boards: assessment of potential for food spoilage. *Journal of Applied Bacteriology*, **81**, 445–458.

Steenaard, P. (2001) Food grade lubricants can be registered again. *Food in Europe, Building on Safety*, Valncia, Spain, EHEDG.

Suhren, G. & Reichmuth, J. (2003) Measurability and development of the hygienic value of the raw milk material. *Kieler Milchwirtschaftliche Forschungsberichte*, **55**, 5–36.

Tamime, A.Y. (2008) *Cleaning-In-Place (CIP)*, 3rd edn, Blackwell Publishing, Oxford.

te Giffel, M.C. (2003) Good Hygienic practice in milk processing. In: *Dairy Processing, Improving Quality* (ed. G. Smit), pp. 68–103, Woodhead Publishing, Cambridge.

te Giffel, M.C. (2006) Process control: the quicker, the better. *New Food*, **9**, 56–59.

van Asselt, A.J., van Houwelingen, G. & te Giffel, M.C. (2002) Monitoring system for improving cleaning efficiency of cleaning-in-place processes in dairy environments. *Food and Bioproducts Processing*, **80**, 276–280.

Vissers, M.M.M., de Jong, P. & de Wolff, J.J. (2002) Nieuwe indampertechniek resulteert in 70 procent energiebesparing. *Voedingsmiddelentechnologie*, **35**, 16.

10 Hazard Analysis (Appraisal) Critical Control Point (HACCP) in Milk Processing – A Practical Overview

A. Sayler

10.1 Introduction

The use of the Hazard analysis (appraisal) critical control point (HACCP) system has become a mature food safety system used widely in the international dairy-processing industry. HACCP is a logical, effective, scientifically based and highly structured *system* of food safety management designed to assist plant HACCP teams in producing a programme to minimise, manage or control hazards. One of the *key* advantages of the HACCP concept is to enable a dairy-food-manufacturing company to move away from a philosophy of control based on testing (i.e. testing for failure) to a preventive approach, whereby potential hazards are identified and controlled in the manufacturing environment (i.e. prevention of product failure).

Many countries including most members of the European Union (EU), Brazil, Canada and the USA have government-required or recommended dairy HACCP programmes for domestic dairy plants, dairy exports or dairy imports. HACCP has international recognition as the most effective means of controlling foodborne illness, with HACCP standards developed by the joint FAO/WHO Codex Alimentarius Commission, International Standards Organization (ISO) and the USA National Advisory Committee on Microbiological Criteria in Foods (NACMCF).

The concepts of HACCP were borrowed by the food industry as a 'spin-off' of the USA manned space programme during the 1960s. The National Aeronautics and Space Administration (NASA) used HACCP to provide assurance of the highest quality available for components for space vehicles. This programme, to develop assurance of product safety, was carried over into the development of foods for astronauts. NASA contracted with the Pillsbury Company to design and produce the first foods used in space. Pillsbury, biggest challenge came in trying to come as close as possible to 100% assurance that the foods they produced would be free of bacterial or viral pathogens. They realised that to be successful, they would have to implement a programme which established effective control over their process, raw materials, environment, and their people. In 1971, they introduced HACCP as a preventive system that enabled manufacturers to produce foods with a high degree of assurance that the foods were produced safely. *If the HACCP system is correctly implemented, there will be little requirement for the testing of final product other than for internal plant verification purposes.*

The success of an HACCP system mandates educating and training management and production personnel in the importance of their role in manufacturing safe dairy foods. This training should also include information for the control of foodborne hazards related to all stages of the food chain. It is important to recognise that production personnel must first

Fig. 10.1 Plant HACCP team meeting.

understand what HACCP is, and then learn skills necessary to make it function properly. Specific training activities should include working instructions and procedures that outline the tasks of production personnel monitoring each critical control point (CCP), and Figure 10.1 illustrates HACCP team members verifying a cheese-processing factory.

10.2 General aspects of the benefits of the HACCP system

Food safety is the primary concern in a dairy product HACCP system. However, the wholesomeness and quality of the product may also be enhanced by proper implementation of HACCP and associated prerequisite programmes. The overall and specific benefits of an HACCP system include:

- Focus on prevention.
- Utilises science-based food safety data and principles.
- Provides a high level of assurance of dairy product safety.
- Focuses appropriate technical resources and control on critical points in the production process.
- Lessens emphasis on end product testing.
- Places the primary responsibility for food safety on processors, where it belongs.
- Meets customer needs and expectations.
- Increased consumer confidence in dairy products.
- Assured brand integrity.
- Decreased numbers of consumer complaints.
- Reduced incidence of product holds and/or recalls.
- Increased sales opportunities.

What is evident from this list is that the successful implementation of many aspects of any proposed HACCP system is going to depend on the subjective conclusions reached by the HACCP team. The extent of end-product testing, for example, may vary from factory to factory even though the chemical composition of the products is identical, for the final decisions about the frequency of testing can only be made on-site. Indeed, comparisons between the guidelines suggested in textbooks often prove confusing rather than helpful, and for this reason, the provision of references to the literature of HACCP has been avoided. A student writing a thesis about HACCP may well need to discuss areas of possible confusion/conflict based on theoretical accounts of how HACCP should be employed, but employees in the dairy industry are mainly interested to know what systems have worked effectively in other factories. It is with such readers in mind that this chapter has been written.

10.3 HACCP definitions

HACCP has its own language, with terms either unique to HACCP or terms uniquely defined when used in the HACCP context. It is important that government officials overseeing dairy plants operating under HACCP, as well as dairy plant personnel and dairy customers, utilise HACCP terms properly in order to clearly communicate HACCP information. Below are a few key HACCP terms and their common definitions.

Allergen – A substance (usually the protein component) which, when consumed in a dairy product, causes an adverse reaction which involves the human immune system.

Audit – An evaluation against a dairy plant's written HACCP programme to determine conformance. An audit may also include evaluation of compliance with dairy customer or government HACCP requirements.

CCPs decision tree – A sequence of questions to determine whether a particular processing step or ingredient is a CCP.

Control – First, to manage the conditions of an operation to maintain compliance with established criteria, and second, the state where correct procedures are being followed and criteria are being met.

Control measure – Any action or activity that can be used to prevent, eliminate, or reduce a significant hazard.

Control point (CP) – Any step at which biological, chemical or physical factors can be controlled.

Corrective action – Procedures followed when a deviation occurs.

Critical – A deficiency or non-conformity that is likely to result in an adverse health consequence if left unmanaged.

CCP – Any point, step, or procedure at which control can be applied and a dairy food safety hazard can be prevented, eliminated or reduced to acceptable levels. A CCP is one that is concerned *only* with safety or health considerations.

Critical limit – A maximum and/or minimum value to which a biological, chemical or physical parameter must be controlled at a CCP to prevent, eliminate or reduce the potential of a food safety hazard to an acceptable level.

Deficiency – An element inadequate or missing from the requirements of the written HACCP programme.

Deviation – A failure to meet a critical limit for a CCP.
Documentation – Information/records available in a written, electronic or other form which can be utilised to detect trends assisting the verification and validation of the prerequisite programme and HACCP plan.
HACCP – A systematic approach to the identification, evaluation and control of significant food safety hazards.
HACCP plan – The written document specific to a product and process, which identifies CCP(s), establishes critical limits, control and documents and delineates procedures to be followed to assure control on the basis of the seven principles of HACCP.
HACCP programme – The preliminary steps, prerequisite programme, good manufacturing practices (GMPs) and written HACCP plan.
HACCP system – The implementation of the written HACCP programme.
HACCP team – The group of people who are responsible for developing, implementing and maintaining the HACCP system.
Hazard – A biological, chemical or physical agent that is of sufficient severity or is reasonably likely to cause illness or injury in the absence of its control.
Hazard analysis – The written and documented process of collecting and evaluating information on hazards associated with the food under consideration to decide which are severe or of sufficient likelihood to occur and must be addressed in the HACCP plan or prerequisite programme.
Monitor – To conduct a planned sequence of observations or measurements to assess whether a CCP is under control and to produce an accurate record for future use in verification. (Note: this activity is not 'verification'.)
Non-conformity – A failure to meet specified requirements of the HACCP prerequisite programme.
Performance standards – Specific regulatory requirements which are required to be incorporated into the HACCP programme by a customer, government or internal dairy company requirements, even if not justified by a plant's hazard analysis (i.e. finished product vitamin testing).
Prerequisite programmes – Procedures, including facility and equipment maintenance, basic sanitation, GMP, training, and/or recall programmes that address operational conditions providing the foundation for the HACCP programme. Prerequisite programmes are generally broad-based to manage related hazards at more than one location in a dairy plant.
Preventive measures – Physical, chemical or other factors that can be used to manage an identified hazard.
Potentially hazardous food – Food that requires time/temperature control for safety (TCS) to limit pathogenic microorganism growth or toxin formation.
Reasonably likely to occur – A hazard for which a prudent (operator) would establish controls because of experience, illness data, scientific reports or other information provide (*sic*) a basis to conclude that there is a reasonable possibility that, in the absence of these controls, the hazard will occur in the particular type of product being processed.
Risk – An estimate of the likely occurrence of a hazard, i.e. usually the responsibility of governments.
Sanitation standard operating procedure (SSOP) – The portion of the written prerequisite program which is mandated by the regulatory authority (e.g. water safety, condition and

cleanliness of food contact surfaces, protection from adulteration, prevention of cross-contamination, employee health and hygiene, pest control program, maintenance of hand washing and toilet facilities, and proper labelling, storage and use of toxic compounds). In some countries, SSOPs are used interchangeably with, and have the same meaning as, prerequisite programmes.

Sensitive ingredient – Any ingredient known to have been associated with a hazard and for which there is reason for concern; this includes ingredients that are known allergens.

Severity – The seriousness of a hazard.

Shall – Procedures that are mandatory.

Should – Procedures that are recommended or advisory.

Step – A point, procedure, operation or stage in the food-processing system from primary production to final consumption.

Target levels – Criteria which are more stringent than critical limits and which are used by an operator to make precise adjustments to reduce the risk of a deviation.

Toxic compounds – Those substances that, at their commercially supplied concentration, are a hazard to human health when inhaled, swallowed or absorbed through the skin. This determination is dependent on the substance's concentration during exposure.

Validation – The element of verification focused on collecting and evaluating scientific and technical information to determine whether the prerequisite programme and HACCP plan, when properly implemented, will effectively control the hazard(s). Validation is conducted initially when the HACCP programme is implemented, at a fixed frequency or based on availability of new scientific information, significant changes in the product, process, or hazards associated with the HACCP programme.

Verification – The use of methods, procedures or tests, in addition to those used in monitoring, to determine if the HACCP system has properly implemented the prerequisite programme and HACCP plan and/or whether there is a need for modification and revalidation.

10.4 Implementation and maintenance of the HACCP programme

The successful implementation of an HACCP programme in, for example, any dairy factory (see Figure 10.2) is facilitated by commitment from top management. This commitment is necessary in order to support the need for additional resources for start-up, initiation and implementation of the programme. The next step is to establish a plan that describes the individuals responsible for developing, implementing and maintaining the HACCP system. Initially, the HACCP coordinator and the cross-functional HACCP team are selected and trained. An important aspect in developing these teams is to assure there are representatives from all major plant operational areas. The team is then responsible for developing the initial programme and coordinating its implementation. Different product teams can be appointed to develop the HACCP programme for specific products. Upon completion of the HACCP programme, operator procedures, forms and procedures for monitoring and corrective action are developed. The production personnel, who will be responsible for monitoring and documenting, need to be trained. It is necessary that a timeline be established for the activities involved in the initial implementation of the HACCP programme. Implementation of

(a)

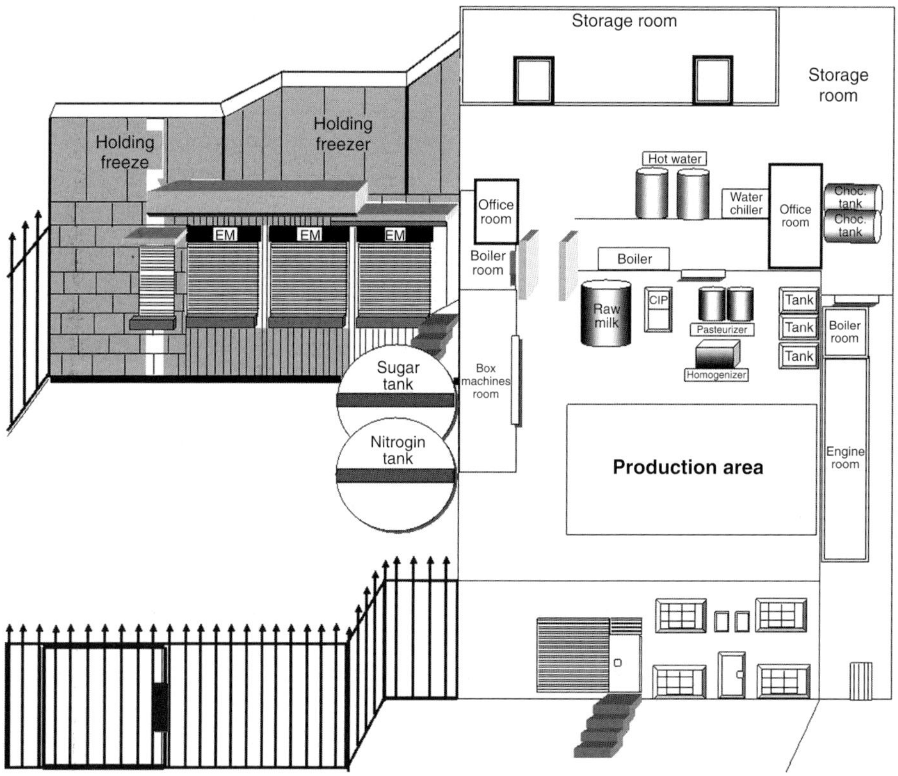

(b)

Fig. 10.2 (a) Typical dairy plant. (b) A schematic illustration of a typical dairy plant.

the HACCP system involves the continual application of the monitoring, record-keeping, corrective action procedures and other activities as described in the HACCP programme. An important aspect of maintaining the HACCP system is to assure that all individuals involved are properly trained so they understand their role and can effectively fulfil their responsibilities.

A separate HACCP programme must be developed for each dairy food. However, it may not be necessary to have a separate HACCP programme for each dairy food if multiple foods have similar product characteristics, share similar production processes and do not have significantly different hazards. If there is doubt, it is always recommended to develop a separate HACCP programme for each dairy food.

10.4.1 *Commitment by management*

Prior to proceeding to the HACCP team selection, it is extremely important to get full commitment to the HACCP initiative from all levels of management. Without a firm commitment of time, personnel and resources, the HACCP programme may be difficult, if not impossible, to implement effectively.

It is recommended that plant management agree to:

- Commit short-term resources to support development, training and implementation of an HACCP programme.
- Provide long-term commitment of support and resources.
- Create an environment to encourage a change in culture.
- Establish a tracking system to measure the progress and benefits of the HACCP programme.

An example of a management letter of endorsement is as follows:

Date:
To: Name
Title
Address

10.4.2 *Letter of endorsement of an HACCP programme*

As part of our continuing efforts to manufacture food under the safest possible conditions that meet or exceed customer, company and government standards, the ABC Milk Plant has adopted the HACCP principles for food safety.

Both corporate and plant management are fully committed to these principles, and will provide the necessary resources to implement a comprehensive HACCP food safety system. This system will include the development of a brief written prerequisite programme, hazard analysis, monitoring, records and verification programme as well as employee training. We recognise that changes in processing equipment, product formulation, scientific information and experience of the HACCP team may require modification of the written and implemented HACCP programme. It is also understood that periodic updating of this 'living' programme will be necessary to maintain its effectiveness. All affected company personnel are encouraged to support the development and implementation of the HACCP programme since its success is dependent on the commitment and contribution of our employees.

Approved by: ....
Plant Manager: .... Date: ....
Corporate Production Manager: .... Date: ....
President:. .... Date: ....

10.5 Steps to HACCP implementation

The preliminary tasks in the development of an HACCP plan include the following: (a) assemble the HACCP team, (b) describe the food and its distribution, (c) describe the intended use and consumers of the food, (d) develop a flow diagram, which describes the process and (e) verify the flow diagram.

10.5.1 *Assemble the HACCP team*

The first task in developing an HACCP programme is to assemble an HACCP team consisting of individuals who have specific knowledge and expertise appropriate to the dairy product and process (see Figure 10.3 and Table 10.1). It is the team's responsibility to develop the HACCP programme. The team should be cross-functional and include individuals from areas such as engineering, production, sanitation and quality assurance. The team should

Fig. 10.3 Dairy plant HACCP team verifying flow diagram.

Table 10.1 Plant HACCP team summary.

Name	Job title and role on HACCP team	Experience	HACCP programme development responsibilities	Task completion date
	An example is illustrated below			
Allen Smith	Quality control supervisor/HACCP team leader	BSc Microbiology, 12 years in various dairy plants	Oversee development and maintenance of plant HACCP programme – develop and finalise written verification and validation programme	June 2007
Name of drafter:................................ Date of document:..............................				

also include plant personnel, who are involved in the operation, as they are more familiar with the variability and limitations of the operation. In addition, this fosters a sense of ownership among those who must implement the plan. The HACCP team may need assistance from outside experts who are knowledgeable in the potential biological, chemical and/or physical hazards associated with the product and the process. However, a programme that is developed totally by outside sources, which may be erroneous, incomplete and lacking in support at the local level. Due to the technical nature of the information required for hazard analysis, it is recommended that experts who are knowledgeable in the dairy product process should either participate in or verify the completeness of the hazard analysis and the HACCP programme.

The HACCP team needs to be multi-disciplinary with knowledge and experience in the following areas: (a) quality assurance, (b) engineering, (c) production, (d) sanitation, (e) microbiology and (f) outside experts (if necessary). These individuals should have the knowledge and experience to correctly

- conduct a hazard analysis,
- identify potential hazards,
- evaluate potential hazards and how they will be managed,
- recommend controls, critical limits and procedures for monitoring and verification
- recommend appropriate corrective actions when a deviation occurs,
- recommend research related to the HACCP programme if important information is not known and
- verify and validate the HACCP programme.

Once the HACCP team has been selected, their next responsibilities are to

- select a trained HACCP coordinator,
- prepare a statement of intent,
- develop an 'action plan',
- identify resource needs and
- establish a timetable for completion of written documentation, implementation and oversight.

Table 10.2 An example of a product description form.

Plant name:	
Street address:	
State or county:	
Country:	
Zip or area code:	
Formal product name	Cheddar cheese
Food safety characteristic	pH 4.9–5.4
Ingredients	Raw milk, skimmed milk powder (SMP), bulk starter culture, culture, enzymes, colour, salt, natamycin
Packaging used	~18 kg block, vacuum sealed in poly bag and stored in a corrugated box
Labelling requirements	Keep refrigerated
Storage and distribution	Product is stored at $\leq$7.2°C; distributed for further processing in refrigerated trucks ($\leq$7.2°C)
Intended consumers	Consumers of all ages consume this product
Intended use	Ready to serve product; may also be used as an ingredient in preparing meals; may be processed further into small chunks or shredded
Shelf life	3–12 months under proper refrigeration
Approved by:	
Date:	

As the team conducts work, keep records (documentation) of points discussed and maintain an action plan. Documentation of decisions may become very useful later. Use Table 10.1 to capture important information about the HACCP team and identify whether there is a need for additional expertise, outside consultants or team members.

10.5.2 *Describe the food and its distribution*

First, the HACCP team describes each dairy product manufactured. This consists of a general description of the dairy product, ingredients and processing methods as well as other pertinent information.

The product description form (see Table 10.2) should contain the following, as a minimum:

- Name of dairy product
- Food safety characteristics
- Ingredients used
- Packaging used
- Labelling requirements
- Storage and distribution
- Intended consumer
- Intended use
- Shelf life

10.5.3 *Intended use and consumers of the food (see form above)*

The intended use of the dairy food should be based upon intended use, the end-user(s), i.e. consumers, consumer target groups (see Table 10.2). The intended consumers may be the general public or a particular segment of the population (i.e. infants or elderly). The use of the food may be for direct consumption, as an ingredient in another food or for non-food uses.

10.5.4 *Develop a flow diagram that describes the process*

The purpose of a flow diagram is to provide a clear, simple outline of the steps involved in the manufacturing process. The scope of the flow diagram must cover all the steps in the process, which are directly under the control of the establishment. In addition, the flow diagram can include steps in the process, which are before and after the processing that occurs in the HACCP plan. The flow diagram need not be as complex as engineering drawings. A block type flow diagram is sufficiently descriptive. Also, a simple schematic of the facility is often useful in understanding and evaluating product and process flow. Points to consider in establishing your diagram may include:

- All process steps where raw materials/ingredients and packaging are used.
- All raw materials/ingredients.
- All process steps in production (generally does not include valves and pumps).
- Product recycle/rework loops.
- Storage and distribution.
- Gases and water used in contact with product.

10.5.5 *Verify the flow diagram*

The HACCP team should perform an on-site review of the operation to verify the accuracy and completeness of the flow diagram. The flow diagram should be reviewed periodically, modified, updated and documented, as necessary. *It is recommended to take the diagram out to the production floor and walk through the steps to ensure the diagram's accuracy.* An example of an effective flow diagram of pasteurised liquid milk is shown in Figure 10.4.

10.6 Hazard components

Hazards to be considered in the hazard analysis are those reasonably likely to occur in a dairy-processing facility that is developing a written HACCP programme. Careful consideration must be given to all ingredients, every step in the process and the finished product packaging and storage. The hazard analysis for each processing facility, and for each product type within that facility, will be unique to that facility. When conducting the hazard analysis, the HACCP team should review adverse related events, utilise past experience, and consult food safety experts to establish which hazards are 'reasonably likely to occur' (see Section 10.3).

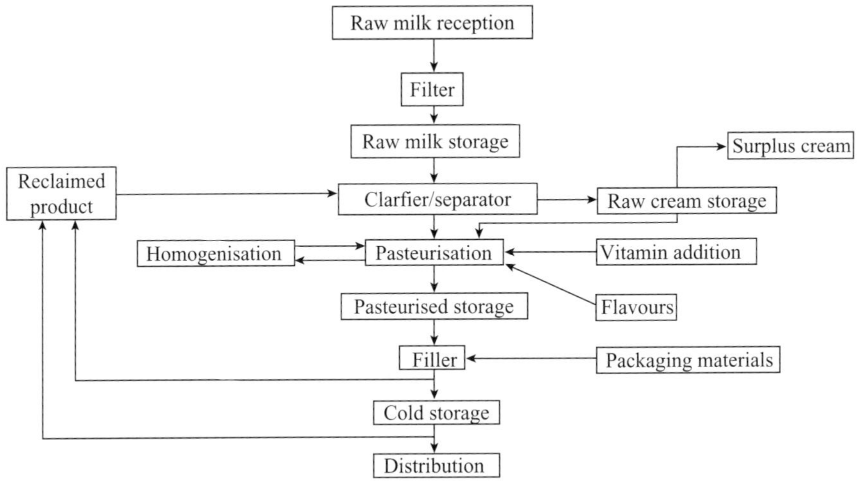

Fig. 10.4 Typical HACCP of pasteurised milk flow diagram.

Remember, *hazards*, as defined within HACCP relate to *product safety*. Also, hazards that are included in the hazard analysis should be reasonably likely to occur (see Section 10.3) in the process being evaluated, associated with the product being processed and evaluated for each product and process within the facility. In brief, the different hazards within an HACCP programme are as follows.

10.6.1 *Biological hazards*

Biological hazards for dairy processors might include pathogenic bacteria, viruses or parasites/protozoa. The mere presence of microorganisms may not result in a hazard. Specific pathogenic bacteria that have been linked to foodborne illness outbreaks associated with dairy products include *Escherichia coli*, *Listeria monocytogenes*, *Salmonella* spp. *Campylobacter* spp. and *Staphylococcus aureus*. Tables 10.3 and 10.4 illustrate the potential biological hazards found in foods; however, it might be more useful to group organisms by

Table 10.3 Some examples of biological hazards.

Severe	Moderate with potentially extensive spread	Moderate with limited spread
Brucella spp.	*Salmonella* spp.	*Bacillus cereus*
Clostridium botulinum	Enterotoxigenic *Escherichia coli*	*Clostridium perfringens*
Listeria monocytogenese	Enteroinvasive *Escherichia coli*	*Staphylococcus aureus*
Salmonella typhi, *paratyphi*, and *dublin*	*Shigella* spp.	*Campylobacter jejuni* and other species
Shigella dysenteriae	Viruses	*Aeromonas* spp.
Hepatitis A and E	Protozoa (*Cryptosporidium* spp.)	*Yersinia enterocolitica*
Escherichia coli 0157:H7	Protozoa (*Giardia* spp.)	Parasites

Table 10.4 Limiting conditions for the growth of certain pathogenic microorganisms.

Microorganisms	Minimum water activity (a_w)	Minimum pH	Maximum pH	Maximum salt (g 100 g^{-1})	Minimum temperature (°C)	Maximum temperature (°C)	Oxygen requirement
Bacillus cereus	0.92	4.3	9.3	18	4	55	Aerobe
Campylobacter jejuni	0.987	4.9	9.5	1.5	30	45	Micro-aerophilic[a]
Clostridium botulinum (type A, and proteolytic B and F)	0.935	4.6	9.0	10	10	48	Anaerobe[b]
Clostridium botulinum (type E, and nonproteolytic B and F)	0.97	5.0	9.0	5	3.3	45	Anaerobe
Clostridium perfringens	0.93	5.0	9.0	7	10	52	Anaerobe
Escherichia coli (pathogenic strains)	0.95	4.0	9.0	6.5	7.0	49.4	Facultative anaerobe[c]
Listeria monocytogenes	0.92	4.4	9.4	10	−0.4	45	Facultative anaerobe
Salmonella spp.	0.94	3.7	9.5	8	5.2	46.2	Facultative anaerobe
Shigella spp.	0.96	4.8	9.3	5.2	6.1	47.1	Facultative anaerobe
Staphylococcus aureus							Facultative anaerobe
growth	0.83	4.0	10.0	25	7	50	
toxin production	0.85	4.0	9.8	10	10	48	
Vibrio cholerae	0.97	5.0	10.0	6	10	43	Facultative anaerobe
Vibrio parahaemolyticus	0.94	4.8	11.0	10	5	44	Facultative anaerobe
Vibrio vulnificus	0.96	5.0	10.0	5	8	43	Facultative anaerobe
Yersinia enterocolitica	0.945	4.2	10.0	7	−1.3	42	Facultative anaerobe

[a] Requires limited level of oxygen.
[b] Requires the absence of oxygen.
[c] Grows either with or without oxygen.
Data adapted from J. Rushing (NC State University for International Dairy Foods Association, personal communication).

Table 10.5 Main examples of potential chemical hazards.

Chemical compound	Chemical hazard
Natural toxins	Mycotoxins Acute (e.g. ochratoxin, trichothecene, zearalenone or aflatoxin) Chronic (e.g. aflatoxin, sterigmatocystin or patulin) Other natural toxins (e.g. thyro-toxicosis)
Metals	Copper Cadmium Mercury Lead
Drug residues	β-lactams Sulphonamides Tetracyclines Macrolides Others
Cleaner/sanitiser residues	Nitrates Phosphates Chlorinated organics Iodophors Others
Pesticide residues	Organo-phosphates Fumigants Others
Allergens and sensitivities	Egg and egg products Milk and milk products Peanuts and peanut products Seafood/shellfish Seeds Soy and soy products Tree nuts Wheat and wheat products Sulphites
Food additives	Vitamins Colours Aspartame
Inadvertent or toxic chemicals	Equipment cleaning chemicals Lubricants Boiler additives Water treatment additives Others

characteristics necessary for growth and destruction. For example, most microorganisms in the vegetative state are easily destroyed by pasteurisation temperatures (e.g. 72°C for 15 s); however, higher heat treatment(s) may trigger outgrowth of certain spore-forming organisms. The outgrowth of spores from spore-forming microorganisms is generally inhibited by lower pH. Toxin-producing organisms usually require mesophilic growth conditions to achieve large enough populations to produce toxins.

Table 10.6 Some examples of potential physical hazards.

Glass fragments
Wood fragments
Plastic fragments, such as pieces, shavings
Metal fragments, such as bolts, nuts, bag clips/locks, shavings
Personal effects, such as jewellery, earrings, buttons, pens, bandages
Other extraneous materials (e.g. nut shells, fruit pits (cherry, peach), fruit material (stems, caps, seeds), gasket, o-rings, insect parts, others)

10.6.2 *Chemical hazards*

Chemical hazards that might be considered in a dairy plant hazard analysis are listed in Table 10.5 because they have the potential to cause illness in susceptible individuals, if not properly addressed.

10.6.3 *Physical hazards*

Physical hazards are those materials that are likely to cause injury or choking, and must also be evaluated within each dairy plant. Examples of physical hazards to consider would include glass, plastic or metal fragments – particularly from packaging materials and processing equipment (see Table 10.6). Employee practices might also influence the types of physical hazards to be considered in the processing facility.

Hard or sharp foreign objects in food may cause traumatic injury including laceration and perforation of tissues of the mouth, tongue, throat, stomach and intestine as well as damage to the teeth and gums. Hard or sharp natural components of a food (e.g. bones in seafood, shell in nut products) are unlikely to cause injury because of awareness on the part of the consumer that the component is a natural and intrinsic component of a particular product. The exception occurs when the food label represents that the hard or sharp component has been removed from the food, for example, pitted olives. The presence of the naturally occurring hard or sharp object in those situations (e.g. pit fragments in pitted olives) is unexpected and may cause injury.

Guidance from the US Food & Drug Administration regarding the size of hard or foreign objects which, if present in a food, render the food adulterated include:

- If the product contains a hard or sharp foreign object that measures 7–25 mm in length, and the product is ready-to-eat, or according to instructions or other guidance or requirements, it requires only minimal preparation steps (e.g. heating) that would not eliminate, invalidate or neutralise the physical hazard prior to consumption.
- If the product contains a hard or sharp foreign object less than 7 mm in length and if a special-risk group is among the intended consumers of the product.
- The product contains a hard or sharp foreign object over 25 mm (~1 in.) in length.

Table 10.7 Some examples of potential hazards identified by source.

	Potential hazard		
Product	Biological	Chemical	Physical
Cheese	*Listeria monocytogenes* *Salmonella* spp. *Brucella* spp. *Escherichia coli* (pathogenic type) *Staphylococcus aureus* Staphylococcal enterotoxin *Campylobacter* spp. *Shigella* spp. *Clostridium botulinum* (shelf-stable processed cheese types)	Nitrates Nitrites Aflatoxin Pesticides	
Chocolate and chocolate syrup	*Salmonella* spp.		
Raw cream	*Salmonella* spp. *Listeria monocytogenes* *Staphylococcus aureus* Staphylococcal enterotoxin *Clostridium perfringens* *Escherichia coli* (pathogenic type) *Yersinia* spp. *Campylobacter* spp. *Bacillus cereus* *Brucella* spp.	Therapeutic drug residues Pesticides Sulphonamides	Extraneous material
Dried milk products	*Salmonella* spp. *Listeria monocytogenes* *Staphylococcus aureus* Staphylococcal enterotoxin *Clostridium perfringens* *Clostridium botulinum* *Escherichia coli*	Sulphonamides Therapeutic drug residues Pesticides	
Evaporated/ sweetened and condensed milk products	*Salmonella* spp. *Listeria monocytogenes* *Staphylococcus aureus* Staphylococcal enterotoxin *Clostridium perfringens* *Escherichia coli* (pathogenic type) *Yersinia* spp. *Campylobacter* spp. *Bacillus cereus* *Brucella* spp. *Shigella* spp.	Sulphonamides Therapeutic drug residues Pesticides	
Canned fruits	*Salmonella* spp. *Staphylococcus aureus* *Clostridium perfringens* *Escherichia coli* (pathogenic type) *Bacillus cereus* Mycotoxins	Sulphonamides Pesticides	

Table 10.7 *(Continued)*

Ice cream	Mould spores *Escherichia coli* (pathogenic type) *Listeria monocytogenes* *Staphylococcus aureus*	Non-food chemical vapours (e.g. oil, ammonia) Allergens (e.g. peanuts, tree nuts)	
Mould ripened cheese	Mycotoxigenic fungi		
Raw milk	*Salmonella* spp. *Listeria monocytogenes* *Staphylococcus aureus* Staphylococcal enterotoxin *Clostridium perfringens* *Escherichia coli* (pathogenic type) *Yersinia* spp. *Campylobacter* spp. *Bacillus cereus* *Brucella* spp. *Shigella* spp.	Antibiotics Pesticides Sulphonamides	Insects Soil Glass fragments Wood slivers Metal fragments Foreign material
Nuts	Mycotoxigenic fungi *Salmonella* spp.	Fumigants Allergens	Insects Foreign material (e.g. shell fragments)
Packaging materials	Through damaged materials From airborne pathogens From condensate		Foreign material
Salt		Non-food grade impurities	Foreign material
Sweetening agents (dry)	*Listeria monocytogenes* Mould spores		
Sweetening agents (liquid)	*Listeria monocytogenes* *Clostridium botulinum* spores (e.g. honey, corn syrup) Mould spores		
Whey (dry, liquid and concentrate)	*Salmonella* spp. *Listeria monocytogenes* *Escherichia coli* (pathogenic type) Staphylococcal enterotoxin		Extraneous material

It is obvious that objects over 25 mm are not likely to be undetected and eaten or consumed, if present in a food. Therefore, physical objects that are larger than 25 mm (~1 in.) do not meet the above criteria and are classified as 'undesirable contaminants', but not 'physical hazards'.

10.6.4 *Potential hazards identified by source*

The data shown in Table 10.7 are not intended to be complete or comprehensive, but provide the reader with extensive examples of the hazards specific to certain dairy products or their ingredients.

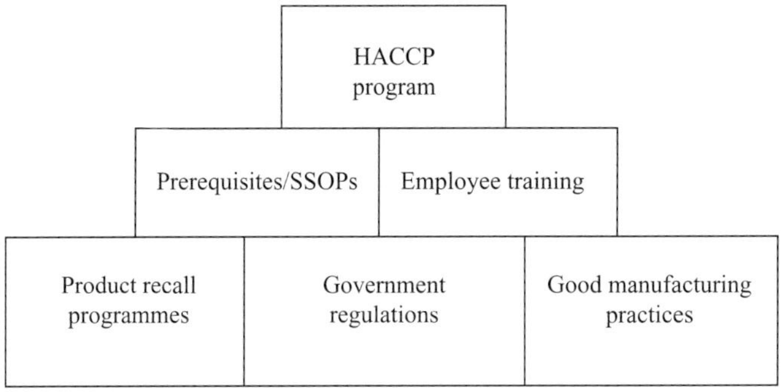

Fig. 10.5 An illustration of an HACCP pyramid.

10.7 Prerequisite programme

10.7.1 *Background*

Prior to the developing an HACCP plan, there is a requirement for dairy plants to have developed, documented and implemented programs to control factors that may not be directly related to product safety, but serve as a foundation of an HACCP system (see Figure 10.5). These programmes together are known as the 'prerequisite programme' (PP). A prerequisite programme should be written, effectively checked, documented and managed before attempting to develop the HACCP plan, and Figure 10.6 shows the pest management element of a prerequisite programme check.

Fig. 10.6 Prerequisite programme check of pest management.

Fig. 10.7 Prerequisite programme check of cross-contamination.

HACCP is not a stand-alone programme, but is part of a larger control system. The prerequisite programme is the universal procedure used to control the plant environment and operating conditions that contribute to the production of a safe, wholesome dairy product. They represent the sum of programmes, practices and procedures that must be applied to design, produce and distribute safe products in a clean, sanitary environment. Many of the conditions and practices are specified in federal, state and local regulations and guidelines. Plant SSOPs and GMPs programmes that serve as the basis for, and are necessary in order to build, a solid prerequisite programme (see Figure 10.7).

An outline for the development and documentation of a prerequisite programme include the following:

- Identify mandatory or key prerequisites
- Write brief description (see Tables 10.8, 10.9 and 10.10)
- Identify hazards reduced or eliminated
- Identify records maintained to verify reduced or eliminated hazard
- Identify staff responsible for maintaining records
- Written brief description of corrections

A prerequisite programme should address common public health concerns, which may be slightly different depending on the specific region of the world and specific dairy product. One example of government-based mandatory prerequisite or SSOP programmes is the USA

Table 10.8 Summary of the required monitoring documentation when preparing the prerequisite programme (PP#1).

PP#1
Safety of the water that comes into contact with food or food contact surfaces (including steam and ice)

Deep South Milk Company
Page 1 of 1, 15 February 2008
Approved by: ....

Goals:

- Water for milk plant purposes, including re-circulated cooling water, shall be from a system that is properly constructed, protected and operated, and shall be accessible, adequate and of a safe sanitary quality.
- Whenever steam is used in contact with milk or milk products, it shall be of culinary quality.
- Ice is produced from municipal potable water.

Procedures:

- Test well water and re-circulated cooling water for coliforms semi-annually, for example (a) monitoring – water sample log (laboratory) and (b) other documentation – water sampling reports.
- Use only approved boiler water and re-circulated cooling water additives, for example (a) monitoring – boiler chemical addition log (maintenance engineer) and (b) other documentation – additives with suppliers' certifications.
- Construction requirements for wells, the plant water distribution systems, the re-circulated cooling systems and the steam generation system will be evaluated upon installation and after major changes to the system.

Summary of required monitoring documentation:

Monitor	Frequency[a]	Document	Verification	Retention
Laboratory manager	Semi-annual	1.1.a. Water and re-circulated cooling water sample log	Quality manager	Laboratory – 3 years
Laboratory manager	Semi-annual	1.1. Water and re-circulated cooling water sampling reports file	File only	Laboratory – 3 years
Boiler maintenance engineer	Semi-annual	1.2.a. Boiler chemical addition log	Laboratory manager	Laboratory – 1 year
Laboratory manager	Semi-annual	1.1.a. Additives suppliers' certification file	File only	Laboratory – 3 years

[a] Sampling should be considered if the water system has been breached by construction.

Corrections:

- Corrections will be taken as needed at each step and noted on the monitoring form.
- Any correction that cannot be accomplished immediately will be reported to the appropriate supervisor to assess whether the non-conformity presents a significant weakness or has a potential impact on the ability to produce a safe product. In addition, the supervisor will assess whether immediate short-term measures are needed to minimise the effect of the problem on our product or operation.

Corrections that cannot be addressed immediately will be given a timeline for correction in the Scheduled Correction Log and will be reassessed on a weekly basis.

Table 10.9 Summary of the required monitoring documentation when preparing the prerequisite programme (PP#4).

PP#4 Maintenance of hand washing and sanitising, and toilet facilities	Old North State Milk Company Page 1 of 2, 15 February 2008 Approved by: ....

Goals:

- Toilet rooms shall not open directly into any room in which milk/milk products are processed. Toilet rooms shall be completely enclosed and shall have tight-fitting, self-closing doors. Dressing rooms, toilet rooms and fixtures shall be kept in a clean condition, in good repair and shall be well ventilated and well lighted. Sewage and other liquid wastes shall be disposed of in a sanitary manner.
- Convenient hand-washing facilities shall be provided, including hot and cold running water and or warm water, soap and individual sanitary towels. Hand-washing facilities shall be kept clean and in good repair.

Procedures:

- Restroom facilities and hand-washing areas will be inspected and maintained by the last powder packer of the shift, at the close of each shift.
- A contract cleaning service will clean and perform scheduled maintenance on restrooms weekly.
- Monthly plant audits will assess repair, ventilation and lighting.
- Annual verification will assess adequacy and convenience of hand washing facilities.

Note: Prerequisite programmes are broad areas of emphasis that need to have procedures established in order to identify specific actions that are to be taken. A programme may have multiple procedures that fix responsibilities and assign frequencies.

Summary of required monitoring documentation:

Monitor	Frequency	Document	Verification	Retention
2nd Shift sanitation employee	Daily	Sanitation checklist PS#11355	3rd Shift sanitation supervisor, daily	QC PP verification files
		Sanitation checklist PS#11355	QC supervisor, weekly	QC PP verification files
Supervisory team – different each month	Monthly	Monthly supervisory checklist – facilities inspection section: assess repair, general construction and maintenance of hand-washing and toilet facility.	Plant manager, monthly	QC PP verification files
HACCP team	Annually	Annual Verification Report	HACCP team leader	QC Plant verification files

Corrections:

- Corrections will be taken as needed at each monitoring step and noted on the monitoring form.
- Any correction that cannot be accomplished immediately will be reported to the appropriate supervisor to assess whether the non-conformity presents a hazard, a potential impact on our ability to produce a quality product or a legal violation. In addition, the supervisor will assess whether temporary measures are needed to minimise the effect of the non-conformity on our product or operation.
- Corrections that cannot be addressed immediately will be given a timeline for correction in the 'Scheduled Correction Log' and will be reassessed at weekly supervisor's meetings until corrected.

Note: All corrections cannot be accomplished immediately, and some corrections have a higher urgency than others. No identified non-conformance should be noted without a carefully considered correction plan and a timeline to get it accomplished. Have a procedure to reassess the timeline regularly and adjust as necessary to make sure the due date does not pass without some action, if only to extend the timeline as necessary.

Date: .... Signature: ....
Supersedes: 10 June 2007.

Table 10.10 Summary of the required monitoring documentation when preparing the prerequisite programme (PP#10).

PP#10 Temperature control programme			Page 1 of 1, 15 February 2008 Approved by: .... Signature: .... Date:	
Goals: To have adequate means of establishing, maintaining and monitoring temperatures • Raw materials/ingredients used in production. • Finished product to ensure safe storage of the food. • Distributed products. Procedures: • Raw materials/ingredients temperature control, for example (a) monitoring – silo storage temperature recording charts and (b) monitoring – incoming milk temperature log. • Finished product temperature control, for example (a) monitoring – finished product temperature recording charts, (b) monitoring – pasteurisation recording charts and (c) other documentation – daily temperature checks. • Pre-shipment transport requirements, e.g. monitoring – truck inspection/refrigeration checks.				
Monitor	Frequency	Document	Verification	Retention
Production supervisor	Daily	1.1.a Silo, temperature charts	Production manager – weekly	Production department – 1 year
Milk receiver	Daily	1.b Milk receiving temperature log	Production manager – weekly	Production department – 1 year
Warehouse supervisor	Weekly	2.a Finished product temperature charts	Warehouse manager – weekly	Warehouse department – 1 year
Production supervisor	Daily	2.b Pasteurisation temperature charts	Production manager – weekly	Production department – 1 year
Warehouse supervisor	Daily	2.c Daily temperature charts	Warehouse manager – weekly	Warehouse department – 1 year
Warehouse supervisor	Daily	3.a Transport inspection log	Warehouse manager – weekly	Warehouse department – 1 year
Corrections: • Corrections will be taken as needed at each step and noted on the monitoring form. • Any correction that cannot be accomplished immediately will be reported to the appropriate supervisor to assess whether the non-conformity presents a significant weakness or has a potential impact on the ability to produce a safe product. In addition, the supervisor will assess whether immediate short-term measures are needed to minimise the effect of the problem on our product or operation. • Corrections that cannot be addressed immediately will be given a timeline for correction in the 'Scheduled Correction Log' and will be reassessed on a weekly basis.				

National Conference on Milk Shipments (NCIMS) and the Food & Drug Administration's Juice HACCP regulation. Both programmes are based on a minimum of eight required prerequisites/SSOPs. There are eight *mandatory* prerequisites, which are addressed within the associated general prerequisites as follows (see Figure 10.8):

- Safety of water that comes into contact with food or food-contact surfaces, including ice (see Figure 10.9 and Table 10.8).

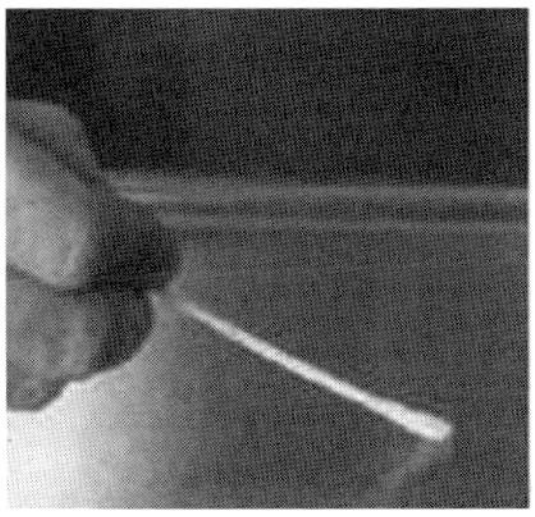

Fig. 10.8 Prerequisite check of condition and cleanliness of processing equipment.

Fig. 10.9 Water prerequisite – cooling water reservoir.

- Condition and cleanliness of the food contact surface.
- Prevention of cross-contamination between food, food-packaging material, food-contact surfaces and unsanitary objects; this includes contact between raw and processed product (includes allergen prevention programme).
- Maintenance of the hand washing and toilet facilities (see Table 10.9).
- Protection from adulteration with lubricants, fuel, pesticides, cleaning compounds, sanitising agents, condensate and other chemical, physical or biological compounds.
- Proper labelling, storage and use of toxic compounds.
- Control of employee health conditions.
- Exclusion of pests.

The intent of this section is to provide guidance to plant management and the HACCP team in evaluating the plant areas falling under the mandatory prerequisite programme (see above list) or identifying hazards that need to be addressed by new prerequisite programmes. The importance of the prerequisite programme cannot be overstated; it is the foundation of a comprehensive HACCP system and must be effective. If any hazard is not controlled, according to the hazard analysis, then the prerequisite programme may have to be adjusted or expanded to address such a hazard. A written and documented prerequisite programme that is a part of the complete dairy HACCP system will address many hazards identified in the hazard analysis. In summary, comprehensive, effective prerequisite programmes will simplify HACCP plans, and will ensure that the integrity and comprehensiveness of the HACCP programme is maintained and that the manufactured product is safe.

This section outlines examples of prerequisite areas including key environmental and operational areas. A dairy plant's prerequisite programme may include some or all of the areas listed below, depending on the plant, product and hazard analysis. A corresponding checklist useful in reviewing the areas listed below can be found in the section of the manual on 'evaluating and revising' the HACCP systems.

Premises – They include (a) outside property, (b) building and personnel traffic patterns and (c) sanitary facilities (mandatory PP#4 – see Table 10.9).

Water/steam/ice safety (mandatory PP#1 – see Table 10.8) include the following: (a) potable water supply, (b) steam supply, (c) ice supply, (d) cooling waters and (e) reclaimed water.

Receiving/storage/transportation consists of (a) supplier control, (b) receiving of raw materials, ingredients and packaging materials (i.e. specifications and storage conditions) (mandatory PP#6), (c) temperature control programme and (d) transportation programme.

Equipment performance and maintenance programme include (a) general equipment design, (b) equipment installation and (c) equipment maintenance and calibration.

Personnel training programme and employee hygiene involves (a) training, (b) hygienic practices (mandatory PP#7), (c) infectious disease policy (mandatory PP#7), (d) injury/open wound policy (mandatory PP#7), (e) controlled access policy (mandatory PP#7) and (f) personnel safety programme (mandatory PP#7).

Environmental and processing equipment hygiene programme involves manufacturing controls.

Cleaning and sanitation programme (mandatory PP#2) takes into account (a) interior facility cleaning, (b) processing equipment cleaning and sanitation and (c) pest control (mandatory PP#8 – see previous item 'G').

Recall programme involves (a) product traceability and (b) product recall system.

Cross-contamination prevention programme include the following: (a) allergens (mandatory PP#5), (b) general adulteration (mandatory PP#5) and (c) cross-contamination (mandatory PP#3).

10.7.2 *Prerequisite examples and forms*

Complete examples of three brief written prerequisite programmes, one for water safety (Table 10.8), one for maintenance of hand washing and toilet facilities (Table 10.9) and one for product temperature control (Table 10.10) should be used as a reference when developing written mandatory or voluntary prerequisite and/or SSOP programmes.

10.8 The principles of an HACCP plan

There are seven principles in establishing an effective HACCP plan, which is used in conjunction with the prerequisite programme and other associated parts of the entire HACCP programme; these principles are

- Conduct a hazard analysis
- Determine CCPs
- Establish critical limits
- Establish monitoring procedures
- Establish corrective actions
- Establish verification procedures
- Establish record-keeping and documentation procedures

During the evaluation of the seven principles of HACCP, the hazard analysis (Table 10.11) and the HACCP plan summary (Table 10.12) are working documents that can be utilised to capture decisions and information by the HACCP team for the HACCP plan. The information in these documents serves as the justification and basis for a dairy plant's HACCP plan.

10.8.1 *Principle 1 – conduct hazard analysis*

A thorough hazard analysis addressing food safety hazards is the key to preparing an effective HACCP plan. If the hazard analysis is not done correctly and the hazards warranting control within the HACCP system are not identified, the plan will not be effective regardless of how well it is implemented.

The HACCP team conducts a hazard analysis and identifies appropriate control measures using two separate but related steps called 'hazard identification' and 'hazard evaluation'. The purpose of hazard identification is to develop a list of potential hazards, based on historical operations, HACCP team experience, scientific literature and governmental requirements. This can be regarded as a brainstorming session. During this stage, the HACCP team reviews the ingredients used in the product, the packaging, the activities conducted and

Table 10.11 Summary of hazard analysis of a modified dairy HACCP tree[a].

Stage	Identify potential hazard	Q1 to Q6 of the dairy HACCP tree[a]		
Process step or ingredient input	Biological, physical or chemical	Q1 – Is the hazard identified at this step of sufficient likelihood of occurrence to warrant its control? If 'yes', then proceed to Q3. If 'no', stop and document at Q2.	Q2 – Identify the prerequisite programme or procedure which reduces the likelihood or severity of the hazard to ensure that control at this step is not necessary.	Q3–Q6. Does a control measure exist at this step to prevent, reduce or eliminate the likely occurrence of a hazard to an acceptable level? If 'yes', document as a CCP. If 'no', indicate where control exists.
Product description:				
Firm name: .... Firm address:				
Method of storage and distribution:				
Intended use and consumer				
Signature: .... Date:				

[a] Refer to Figure 10.10 for further information.

Table 10.12 HACCP plan summary – Model CCP[a] of continuous flow (HTST and HHST)[b] pasteurisation.

			Monitoring						
CCP	Hazard(s)	Critical limits	What	How	Frequency	Who	Corrective actions[a]	CCP verification[b,c]	Records
Milk and milk products pasteurisation	Biological (vegetative pathogens, i.e. non-sporeformers)	Time and temperature	Temperature at the exit of the holding tube	Continuous temperature recorder chart	At least once per shift by the operator	Pasteuriser operator	Manually divert flow of product	Record review	Pasteuriser charts
		Note: Assuring that the minimum holding times are met in systems which use a sealed timing pump would be CCP verification, i.e. equipment calibration	Residence time in the holding tube in continuous flow pasteurisers with magnetic flow meter based timing systems	Flow recorder chart	Continuous during operation	Pasteuriser Operator	Isolate the affected product Evaluate and determine disposition of the product (reprocess or disposal) Document actions	Pasteuriser charts verified Equipment function checks Operator performs required daily tests and record on the temperature charts Authorised plant person (supervised by regulatory when required) conducts checks listed in the *Milk Plant Equipment Test Report* (FDA Form 2359b) Seals: Verify required regulatory seals daily	Corrective action records CCP verification-records, including equipment-testing records

Product Description: .... Method of storage and distribution: ....

Intended use and consumer: ....

Signature: .... Date: ....

CCP, critical control point; HTST, high temperature and short time; HHST, high heat – short time.

[a] A properly operating HTST or HHST pasteurisation system will divert raw product to the balance tank when predetermined set points are not met.

[b] Every particle of milk or milk is heated in a properly designed, calibrated and operated pasteuriser, to one of the temperature and time combinations specified in the current Grade A Pasteurised Milk Ordinance (PMO).

[c] Pressure in the regenerator of continuous flow pasteurisers, and in the case of HHST pasteurisers as required in the holding tubes, across steam injectors and within infusion chambers shall be addressed in the HACCP plan and managed as CCP verification(s).

Table 10.13 An example of biological hazards and associated control measures that might be considered.

Hazard	Measures
Presence of pathogens	Eliminated or reduced to acceptable levels by pasteurisation
Post-pasteurisation contamination by pathogens	Managed by GMP and prerequisite on equipment construction, cleaning, and sanitising
Growth of pathogens and toxin production	Managed by prerequisite programme and cooling

the equipment used at each step in the process, product storage, distribution, the intended use and consumers of the product. Based on this review, the team develops a list of potential biological, chemical or physical hazards, which may be introduced, increased or controlled at each step in the production process and for each ingredient and packaging type. Common sense should be used to identify likely hazards.

The next step in the hazard analysis is 'hazard evaluation'. The HACCP team needs to decide which potential hazards must be addressed in the HACCP plan. During this stage, each potential hazard is evaluated on the basis of the severity of the potential hazard and its likelihood of occurrence. Severity is the seriousness of the consequences of exposure to the hazard. During the evaluation of each potential hazard, the food, its method of preparation, transportation, storage and persons likely to consume the product should be considered to determine how each of these factors may influence the likely occurrence and severity of the hazard being controlled. The team must consider the influence of likely procedures for manufacturing and storage and whether the intended consumers are susceptible to a potential hazard.

Hazard evaluation can be conducted using the decision tree found later in this section, and the results recorded on the hazard analysis (Table 10.13) also found later in this section. Hazards that are not severe or not reasonably likely to occur (see definition) would not require further consideration within an HACCP plan.

The hazard identification and hazard evaluation accomplish three objectives:

- The hazards are identified at each processing step and for each ingredient and material used.
- The hazards are evaluated to determine their severity and likelihood of occurrence.
- The hazard analysis provides a basis for determining management or control measures, such as prerequisite programmes/SSOPs or CCPs in Principle 2.

Upon completion of the hazard analysis, the significant hazards associated with each step in the flow diagram should be listed along with any control measures for each hazard. For example, if an HACCP team was to conduct a hazard analysis for the manufacture of yoghurt, possible pathogens in the raw milk would be identified as a potential hazard. Thus, heat treatment of the milk base would be the control measure (Table 10.14). Table 10.13 illustrates an example of biological hazards (in this case pathogenic bacteria) and the associated control measures that might be considered.

The following questions may be a useful tool in assessing hazards. Appendix C of the 1997 NACMCF guideline (www.fsis.usda.gov/ophs/hacmcf/past/jfp0998.pdf) on

Table 10.14 Examples of how the stages of hazard analysis are used to identify and evaluate hazards.

Stage	Hazard analysis	Incoming raw milk
Stage 1 – hazard identification	Determine potential hazards associated with product	Enteric pathogens (i.e. *Escherichia coli* O157:H7 and *Salmonella* spp.)
Stage 2 – hazard evaluation	Assess severity of health consequences if potential hazard is not properly controlled Determine likelihood of occurrence of potential hazard if not properly controlled Using information above, determine if this potential hazard is to be addressed in the HACCP plan	Epidemiological evidence indicates that these pathogens cause severe health effects including death among children and elderly. Un-pasteurised milk has been linked to disease from these pathogens *Escherichia coli* 0157:H7 is both likely to occur in raw milk as well as other enteric pathogens The HACCP team decides that enteric pathogens are hazards for this product. Sanitation and temperature control will not destroy these pathogens. The microbiological hazard must be controlled in the plan

Note: For illustrative purposes only. The potential hazards identified may not be the only hazards associated with the products listed; however, the responses may be different for different establishments.

HACCP provide useful questions that can be used by the plant's HACCP team to evaluate hazards:

- Does the dairy product contain any ingredients that may present microbiological hazards, chemical hazards or physical hazards?
- Does the dairy product permit survival or multiplication of pathogens and/or toxin formation in the dairy food during processing?
- Will the dairy product permit survival or multiplication of pathogens and/or toxin formation during subsequent steps in the dairy food chain?
- Are there other similar dairy products in the marketplace? What has been the safety record for these dairy products?
- Does the process include a controllable processing step that destroys pathogens?
- Is the dairy product commercially sterile?
- Does the microbial population change during the normal time the dairy food is stored prior to consumption?
- Will the processing equipment provide the time–temperature control that is necessary for safe dairy food?
- Can the processing equipment be sufficiently monitored and controlled so the process will be within the tolerances required to produce a safe dairy product?
- Is there a chance for dairy product contamination with hazardous substances?
- Are there metal detectors, thermometers, sifters, filters, screens, etc., used to enhance consumer safety?
- What is the likely use of the end product?

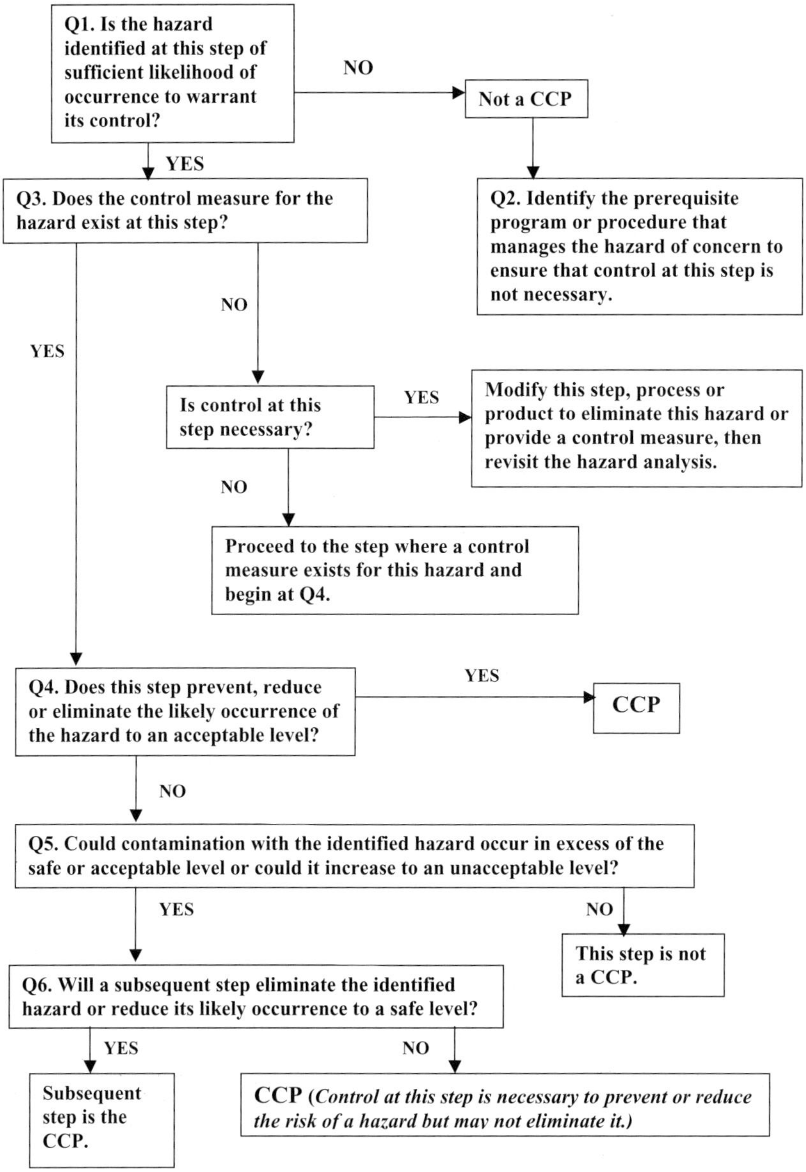

Fig. 10.10 Modified decision tree for HACCP.

A decision tree can be a helpful tool during the hazard evaluation process. There are many hazard analysis decision trees. One example is shown in Figure 10.10 where this decision tree has the advantage of being tied directly with the hazard analysis form, allowing the HACCP team the ability to answer questions raised in the decision tree about a specific hazard and recording the answers to the questions directly on the hazard analysis form. A summary of the hazard analysis form is shown in Table 10.11, and Figure 10.10 illustrate a modified decision tree for HACCP.

10.8.2 *Principle 2 – determine CCPs*

A CCP is defined as a step at which control can be applied, and is essential to prevent, eliminate or reduce a food safety hazard to an acceptable level. The hazard analysis conducted under Principle 1 (see Section 10.8.1) has identified areas where it is necessary to implement control measures. The prerequisite programme may be used to control many of the identified hazards. Any hazards not controlled through prerequisite programmes must be identified as CCPs. CCPs may vary depending on the hazard analysis, plant, product and production method.

Information developed during the hazard analysis should enable the HACCP team to identify which steps in the process are CCPs. Identification of each CCP can be facilitated by the use of the hazard analysis decision tree. Although application of the hazard analysis decision tree can be useful in determining if a particular step is a CCP for a previously identified hazard, it is merely a tool and not a mandatory element of HACCP. A hazard analysis decision tree is not a substitute for expert knowledge.

Different facilities preparing the same dairy product can differ in their hazards and the points, steps, or procedures, at which are CCPs. This can be due to differences in each facility layout, equipment, selection of ingredients (including raw versus pasteurised milk) or the process that is employed.

Once the CCP has been identified, it should be transferred to column 1 on the HACCP plan summary (see Table 10.12). The hazard should be transferred to column 2 on the same form.

10.8.3 *Principle 3 – establishments of critical limits*

A critical limit is a maximum and/or minimum (i.e. scientifically based) numerical value to which a biological, chemical or physical parameter must be controlled at a CCP to prevent, eliminate or reduce to an acceptable level the occurrence of a food safety hazard. A critical limit is used to distinguish between safe and unsafe operating conditions at a CCP. Critical limits should not be confused with operational limits, which are established for reasons other than food safety. Critical limits must be met to ensure the safety of the dairy product, and may be derived from different sources, such as regulatory standards and guidelines, scientific literature searches, experimental studies and experts. Critical limits are parameters, which may be established as control measures and include:

- Temperature
- Time
- Water activity (a_w)
- pH
- Titratable acidity
- Safe or tolerance levels of drug residues

Example of a very common CCP in the dairy industry is pasteurisation, and 'time' and 'temperature' are the critical limits in pasteurisation (see Table 10.12).

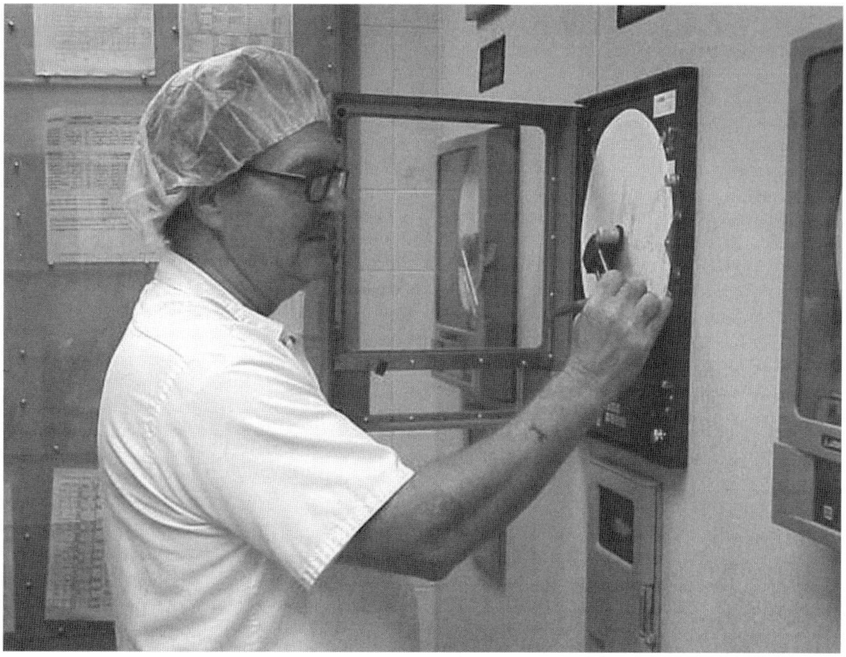

Fig. 10.11 Operator monitoring CCP of a pasteuriser.

10.8.4 *Principle 4 – establish monitoring procedures*

Monitoring is a planned sequence of observations or measurements used to assess whether a CCP is under control, and an accurate record exists for future use in verification. Figure 10.11 shows an operator monitoring CCP of a pasteuriser; however, monitoring serves three purposes:

- Monitoring is essential to dairy product food safety management in that it tracks the system's operation.
- Monitoring is used to determine when there is loss of control and a deviation occurs at a CCP (i.e. exceeding the critical limit); corrective action must be taken.
- Monitoring provides written documentation for use in verification of the HACCP plan.

Because of the potentially serious consequences of a deviation, monitoring procedures must be effective. Continuous monitoring is possible with many types of physical and chemical methods, for example, the time and temperature of pasteurisation. Monitoring equipment must be carefully calibrated for accuracy (verification activity). When it is not possible to monitor a critical limit on a continuous basis, it is necessary to establish the monitoring interval, which will be reliable enough to indicate the hazard is under control.

Assignment of the responsibility for monitoring is an important consideration for each CCP. Specific assignments will depend on the number of CCPs, their critical limits and the complexity of monitoring. Those individuals monitoring CCPs must

- Be trained in the technique used to monitor each critical limit
- Fully understand the purpose and importance of monitoring
- Have ready access to the monitoring activity
- Accurately report the monitoring activity

The person responsible for monitoring must also inform management when a process or product does not meet critical limits so that immediate corrective action can be taken.

Most monitoring procedures for CCPs will need to be done rapidly because they relate to online processes, and there will not be time for lengthy analytical testing. Microbiological testing is seldom, if ever, effective for monitoring CCPs due to the time required to conduct tests. Therefore, physical and chemical measurements are preferred because they may be conducted rapidly and can indicate the conditions of microbiological control in the process.

The following areas must be addressed when considering monitoring/inspection:

Monitoring controls – Need to know correct control points (critical) *before* establishing monitoring programmes; designed to measure if the CCP is *in* or *out* of control.

Frequency – Will vary depending on likelihood and severity of the identified hazard. Continuous monitoring and recording are preferred in most situations

Responsibility, i.e. position title assigned and work station.

Once established, the monitoring procedures should be transferred to columns 4, 5, 6 and 7 on the HACCP plan summary (see Table 10.12). The name of the records utilised to monitor the critical limit(s) should be placed in column 10 on the HACCP plan summary table (see Table 10.12).

10.8.5 *Principle 5 – establish corrective actions*

Corrective actions are procedures to be followed when a deviation occurs. Because of variations in CCPs for different dairy products and the diversity of possible deviations, specific corrective action plans must be developed for each CCP. At a minimum, the production personnel shall have the responsibility and the authority to take corrective actions such as notifying a supervisor or shutting down the production line and putting the product distribution on hold. Corrective actions must demonstrate the CCP has been brought under control, i.e. within critical limits. Individuals who have a thorough understanding of the dairy process, product and HACCP plan are to be assigned responsibility for taking corrective action. Corrective action procedures must be documented in the HACCP plan (see Table 10.15)

Written corrective action plans may include

- Elimination of the actual or potential hazards created by deviation.
- Specific corrective actions for each CCP, i.e. halt production of the product, isolate the affected product, return process to control and determine the disposition of the product.
- Disposition of the dairy product involved.
- Methods to demonstrate that the CCP is brought under control, i.e. determination of the cause of the deviation.

Table 10.15 A typical form for a corrective action procedure.

Subject: Corrective Action Log **Plant name** **Address**	
Deviation #1	
Today's date **Date reported:** **Explain CCP critical limit deviation**:	**Date of incident:** **Reported by:**
Product/process involved • Product name and description: • Code date(s): • Date(s) of manufacture: • Production line number:	
Corrective action:	**Action taken**
1. Segregate and hold the affected product until items 2 and 3 are completed;	☐ Yes ☐ No Date: Comments
2. Perform or obtain a review to determine the acceptability of the affected product for distribution. The review shall be performed by an individual or individuals qualified by training or experience to perform such a review;	☐ Yes ☐ No Date: Comments
3. Take corrective action, when necessary, with respect to the affected product to ensure that no product is allowed to enter commerce that is either injurious to health or is otherwise adulterated as a result of the deviation;	☐ Yes ☐ No Date: Comments
4. Take corrective action, when necessary, to correct the cause of the deviation; and	☐ Yes ☐ No Date: Comments
5. Perform or obtain timely validation by a qualified individual(s), as required in Appendix K, to determine whether modification of the HACCP plan is required to reduce the risk of recurrence of the deviation, and modify the HACCP plan as necessary.	☐ Yes ☐ No Date: Comments
Disposition of product	**Root cause of deviation**

Another approach to establishing corrective action plans in advance is to utilise the five steps when an HACCP deviation occurs. These steps are widely recognised as being adequate for the HACCP team to use in documenting a deviation, and they are:

- Segregate and hold the affected product.
- Perform or obtain a review to determine the acceptability of the affected product for distribution.
- Take corrective action, when necessary, with respect to the affected product to ensure that no injurious or adulterated product enters commerce.
- Correct the cause of the deviation.
- Perform or obtain timely validation as required by a qualified individual(s), to determine whether modification of the HACCP plan is required to reduce the risk of recurrence of the deviation, and modify the HACCP plan as necessary.

When the critical limit has been exceeded, decisions must be made on the basis of whether there are indications that

- Evidence or existence of a direct health hazard.
- Evidence that a direct health hazard could develop.
- Indications that a dairy product was not produced under conditions assuring safety.
- Evidence that a CCP is not under control.

Once the corrective actions have been established they should be transferred to column 8 on the HACCP plan summary (see Table 10.12).

10.8.6 *Principle 6 – establish verification procedure*

Verification is addressed in two ways in the HACCP plan. First, establishing verification on the HACCP plan summary (Table 10.12), and second, establishing internal verification monitoring. In both situations, documentation is critical.

Verification is the activity designed to insure that the monitoring programme is operating according to the requirements of the HACCP programme. The verification format at this point is determined by the HACCP team as to what actions will be taken in reference to the CCPs and is placed in column 9 on the HACCP plan summary (see Table 10.12). Once verification has been established, the internal verification monitoring programmes can be developed.

Verification reports should include (a) status of records associated with CCP monitoring, (b) direct monitoring data of the CCP while in operation, (c) calibration and testing of monitoring equipment, (d) deviations and corrective actions, (e) training and knowledge of individuals responsible for monitoring CCPs and (f) a check of chart to show the records have been verified.

Internal verification monitoring can be utilised to ensure that the HACCP plan is functioning effectively rather than relying on end product sampling. Firms must rely on frequent reviews of their HACCP plan for effectiveness, and examples of internal verification monitoring may include

- Establishment of appropriate verification monitoring schedules.
- Review of the CCP monitoring records at specified frequency.
- Visual verification of operations to observe if the CCPs are under control.
- Verify that changes have been implemented correctly after an HACCP plan has been modified.
- Review of consumer feedback records.

Verification of the HACCP programme should occur at least annually. At this point, if internal verification monitoring is not acceptable to resolve the issues, the HACCP team may have to validate the HACCP plan as described in the next section. One possible verification plan recording form is shown in Table 10.16.

Table 10.16 An example of verification plan recording form.

Specific activity	Classification of activity (i.e. CCP records, corrective actions, equipment calibration and/or other aspects)	Responsible verifier	Frequency	Outcome comments

Validating the HACCP plans

The HACCP team should conduct a validation of the HACCP plan annually or at any time when the process or product formulation is altered. The scientific or technical processes to validate the HACCP plan, CCPs and critical limits are satisfactory to guarantee that adequate controls are in place to reduce hazards. On a minimum basis, the HACCP plan must be validated by the HACCP team. In addition, the validation process may need to be conducted if the verification process does not correct and address the hazard. The HACCP team should validate the HACCP plan on an annual basis by reviewing the following items:

- The effectiveness of the process
- The accuracy of the flow diagram
- The completeness of the HACCP plan
- The soundness of the hazard analysis
- The appropriateness of the CCPs
- The scientific justification for the critical limits
- The comprehensiveness of the corrective actions
- The effectiveness of the monitoring programmes and the record keeping

In addition to the annual validation, some situations that require validations are

- New potential hazard for that dairy food, e.g. new pathogens and/or new CCPs.
- New scientific data available.
- Recall of dairy products.
- Response to new dairy product development, for example (a) raw materials change, (b) preparation and processing change, (c) formulation change, (d) packaging change and (e) new uses of dairy product by consumers.
- Response to manufacturing change, i.e. changes in dairy product flow in plant and/or equipment change.
- When dairy products have been implicated as a vehicle of foodborne disease, for example (a) based on consumer feedback, (b) based on product evaluations and (c) response to regulatory inspection/change.

An example of a form to record the validation information is shown in Table 10.17.

Table 10.17 An example of validation plan recording form.

Specific activity	Narrative of specific activity	Name of validator	Date of validation	Outcome and action items

10.8.7 *Principle 7 – records*

Records that are being used to monitor control points should be placed in column 10 on the HACCP plan summary (see Table 10.12). Records utilised in the total HACCP system may include the following, and they all must be documented. It is strongly recommended that an inventory (see example illustrated in subsequent section) of all HACCP records be made as soon at the HACCP written programme has been completed. This inventory summary is very useful in evaluating the HACCP programme to make sure all identified records have actually been created and utilised in the HACCP system.

The HACCP plan may include the following:

- Listing of the HACCP team and assigned responsibilities
- Statement of intent
- Description of the dairy product and its intended use
- Flow diagram for the entire dairy manufacturing process indicating CCPs
- Monitoring system
- Corrective action plans for deviations from critical limits
- Procedures for verification of HACCP system
- Records for all CCPs
- Hazard analysis
- Procedures for validation

The following records must be available for review:

- *Training* – documentation showing that all individuals have been properly trained in their role in monitoring CCP(s) identified in the HACCP plan summary (Table 10.14).
- *Processing* – records showing that the monitoring of the CCP is in place.
- *Deviation or corrective action log* – documentation to be completed in the event of a deviation from a critical limit for a CCP; a centralised deviation/corrective action log is strongly recommended.
- *Verification and validation records* – these include (a) records showing that validation has occurred a minimum of annually or as necessary and (b) records showing routine verification of control charts.
- *Records to show any major changes* to the HACCP plan.

In order to provide the HACCP team with an overall view of all records utilised and referenced to document the written HACCP programme, it is strongly recommended that each HACCP programme contain a centralised list of HACCP programme records (see Table 10.18). The purpose of Table 10.18 is to assist the plant HACCP team in demonstrating that

Table 10.18 An example of centralised HACCP record list.

Centralised list of HACCP programme records Plant name Address	Supersedes	Issue Date:	Page
Record	Available (√ = yes)	Most Current Version (√ = yes)	Comments
Required HACCP documents including forms are dated or identified with current version number. Each page is marked with a new date or version number whenever that page is updated. Most current versions used.			NA
Table of Contents			
Centralised list of HACCP programme records			
Document change log			
Process flow diagram(s)			
Product description(s)			
Written hazard analysis(s) for each product			
CCP, HACCP plan summary(s) for each product.			
CCP monitoring documents			
Centralised deviation log			
HACCP system verification documentation (including calibration of CCP monitoring equipment (i.e. pasteurisation equipment checks); review of CCP monitoring records, corrective action records and calibration records; and plant signatures and date on these records)			
HACCP system validation documentation (annually or when changes are made in raw materials or source of raw materials; product formulation; processing methods or systems, including computers and their software; packaging; finished product distribution systems; or the intended use or intended consumers of the finished product and consumer complaints)			
Prerequisite programme #1 – Safety of Water			
• Monitoring records related to this PP (list records by name)			
• Nonconformity correction records related to this PP			
Prerequisite programme #2 – condition and cleanliness of food contact surfaces			
• Monitoring records related to this PP			
• Nonconformity correction records related to this PP			
Prerequisite programme #3 – prevention of cross-contamination			
• Monitoring records related to this PP			
• Nonconformity correction records related to this PP			
Prerequisite programme #4 – maintenance of hand-washing and sanitising and toilet facilities			
• Monitoring records related to this PP			
• Nonconformity correction records related to this PP			
Prerequisite programme #5 – protection from adulteration			
• Monitoring records related to this PP			
• Nonconformity correction records related to this PP			
Prerequisite programme #6 – proper labelling, storage and use of toxic compounds			
• Monitoring records related to this PP			
• Nonconformity correction records related to this PP			
Prerequisite programme #7 – control of employee health condition			
• Monitoring records related to this PP			
• Nonconformity correction records related to this PP			
Prerequisite programme #8 – exclusion of pests			
• Monitoring records related to this PP			
• Nonconformity correction records related to this PP			
Other prerequisite programmes that are relied upon in the hazard analysis to reduce the likelihood of a potential hazard (list each separately, add rows as needed)			
• Monitoring records related to this PP			
• Nonconformity correction records related to this PP			
Other Applicable NCIMS requirements – Appendix K (list each separately, add rows as needed)			
• Monitoring records related to these requirements			
• Nonconformity correction records related to these requirements			

PP, prerequisite programme; NCIMS, US National Conference on Milk Shipments.

those records normally required. This checklist may also serve as a tool for internal and external HACCP auditors.

10.9 Internal and external auditing in the HACCP process

The audit process is a planned, independent, documented assessment that determines whether requirements for food safety are being achieved. An effective audit programme will assess compliance, system effectiveness and identify opportunities for continuous improvement. There are three types of audits necessary to support the HACCP system, and they are

- Partial or complete HACCP programme audits made by internal groups or the HACCP team.
- Partial or complete HACCP programme audits made by outside third parties or government employees.
- HACCP verification and/or validation audit performed by internal groups or the HACCP team.

Fundamental to the audit process is the belief that they benefit management, are performed by qualified individuals, are based on standards (dairy plant's written HACCP programme) and conclusions are only drawn from facts based on direct observations.

Floor supervisors, lead individuals or internal auditors can perform production floor audits. These individuals must be familiar with basic audit techniques and have process/system knowledge. Methods include record review, observation, sampling and use of short checklists. Results should be reported internally and used to develop corrective action plans if deficiencies are noted. The audit frequency will vary depending on results.

The plant HACCP team or internal auditing group can perform HACCP programme audits. Like the production floor audit, this is not an independent assessment of the HACCP system, but does provide a more in-depth verification of effectiveness. Training requirements for the HACCP team include intermediate auditing techniques and knowledge of the facility's internal management structure, HACCP programme and processing system. Methods include review of records, sampling and testing results, observation, interviews and checklists. These audits should be conducted at least quarterly with the results reported internally using formal documentation. This process can also be used to verify corrective actions have been successfully implemented.

A corporate team or external party performs HACCP verification and validation audits. Training for these individuals includes advanced audit techniques, process/system knowledge and food safety expertise. Methods include reviewing records, management controls, observation and interviews of operators, supervisors and management, sampling and testing, and comprehensive checklists. This independent audit is formally documented for distribution to plant management and the HACCP team and should be conducted annually.

Audits are used to assess HACCP system health and identify opportunities for improvement. If gaps are noted in a facility's HACCP system, the root cause of the deficiency must

Table 10.19 Table of contents for USA HACCP programme.

Section and subsections	Page number
1. Introduction	
a. Letter of management commitment	4
b. Principles of HACCP	5
c. Background documents including page on company history including current products covered under the HACCP programme – news stories	6
d. Staff organisational chart for the plant operation	8
e. Front plant view plus overheat view of the plant (floor layout)	9
f. Product flow diagram	10
g. Company policy on hold and release, and returned product disposition	11
h. Name of each HACCP team member with a brief bio for each. Include a log of team meetings, who attended, date and time and agenda of meeting	12
2. Plant GMPs and prerequisite programme (PP)	
a. Written summary of ice cream plant GMPs	14
b. Written description of ice cream plant prerequisites including a brief statement on document control (will provide some examples):	
i. Water safety	31
ii. Condition and cleanliness of processing equipment – address pre-operational start-up inspection, master sanitation schedule and limited inclusion of preventative maintenance to specific processing equipment preventative maintenance	32
iii. Cross-contamination – micro-based hazards, raw to pasteurised product	35
iv. Adulteration	
1. Chemical-based micro hazards, allergens	
2. Cleaning chemical and toxic compound labelling, use and storage	37
v. Temperature control – identify target temperatures for receipt of ingredients, storage of ingredients, storage of final product(s) and distribution of final product(s)	39
vi. Personnel training, employee, visitor and contractor hygiene – see PMO Section 13 and 14 for suggested items to be included in a company employee manual or policy on this subject. Identify training frequency as part of documentation supporting this prerequisite	40
vii. Receiving/storage/transport – address ingredient and package receiving, specifications, storage and distribution of final product(s)	42
viii. Product tracing and market withdrawals – include section on customer feedback summary information and mock recall procedures	44
ix. Pest control – include floor diagram and location of all pest stations, electrocutors, or bait stations. Address frequency of review for outside contractor's records	49
x. Maintenance of hand-washing, employee break-room and bathroom facilities	53
xi. Facility maintenance – includes grounds and outside maintenance (buildings, fence and everything between) and inside processing facility maintenance	54
xii. Air safety – focused primarily at air entrainment into the ice cream mix during freezing	55
c. Ingredient/packaging assessment	56

Table 10.19 *(Continued)*

3. Plant HACCP plan	
a. Product description & flow diagrams	57
i. Product A	59
ii. Product B	61
iii. Product C	63
iv. Product D	65
v. Product E	
b. Hazard analysis for each product based on those hazard advance from the hazard identification	67
c. HACCP plan summary table recording details of CCPs	72
d. Corrective action blank form with five standard steps – see IDFA recommended form	73
e. Verification and validation narrative and forms	74
4. Appendix	
a. Summary of all records identified in the written HACCP programme	79
b. Document change log that identifies date and items changed, with signature	86
c. Pre-operational start-up checklist	87
d. Procedure and record for the metal detector	88
e. Daily cleaning and sanitation log	89
f. Monthly internal inspection report form	90
g. Processing equipment maintenance programme and form	91

be determined and corrections actions identified. The three levels of auditing discussed can be used to enhance the effectiveness of the facility's corrective action programme.

10.10 Overview and summary

At this point, the dairy plant's HACCP team should have covered all important points in development of a written HACCP programme. In order to organise all the information and forms developed by the HACCP team, an example of a written HACCP programme 'table of contents' is shown in Table 10.19, which contains all the important information that should be included in any dairy plant's written HACCP programme.

11 Sensory Profiling of Market Milks

E. Molina, L. Amigo and A. Quirós

11.1 Introduction

Today's consumers are discerning, demanding and more knowledgeable about food and expect products which are safe, good value and of high sensory quality. Therefore, knowing consumers' preferences and perceptions of the sensory characteristics of food/drink products is very important for food manufacturers and retailers alike.

In the food industry, sensory analysis has proven valuable over years to test the characteristics of foods, and it is considered as a part of the total quality control of a product. Sensory evaluation is a science that studies a product from the point of view of how it is perceived by the consumer by their senses. The sensory attributes of taste, appearance, aroma, sound and texture are key factors in determining the acceptability of a food and its success in the highly competitive marketplace. The appeal of eating a food product plays a major role in the consumer's decision to make initial and repeat purchases.

Milk is a pleasant and satisfying food after correct production and processing. For this reason, the milk is analysed by sensory analysis in:

- *On-farm milk production* – control of the raw milk (effect of the forages used in cow's feed, health, environment of the herd).
- *Processing* – effects of processing methods on the finished product (temperature of heating and time), effect of packaging materials, inter-batch consistency.
- *Marketing* – assesses consumer preferences, understand how a product performs against competitors' products in relation to consumer perceptions and/or sensory characteristics, know consumer preferences and relate them to the sensory characteristics of the product, determine whether or not consumers can detect differences between products, develop innovative value-added food products.

As shown, judging of the milk starts on the farm so corrections can be made and finishes with retail quality control. In most cases, sensory analysis is done by specially trained and unbiased experts, called panellists, selected for their keen sense of taste, smell, sight, sound and touch to provide consistent and accurate results. Tests are conducted under controlled conditions using tests specifically designed to obtain the best answered, questions and the resulting information is analysed using the latest statistical techniques (Figure 11.1). Their conclusions are a valuable tool in tailoring a product for a specific market. In other cases, as in the consumers' preferences analysis, the panellists are inexperienced in sensory analysis, and preferably from the environment intended for the product (Carpenter *et al.*, 2000a).

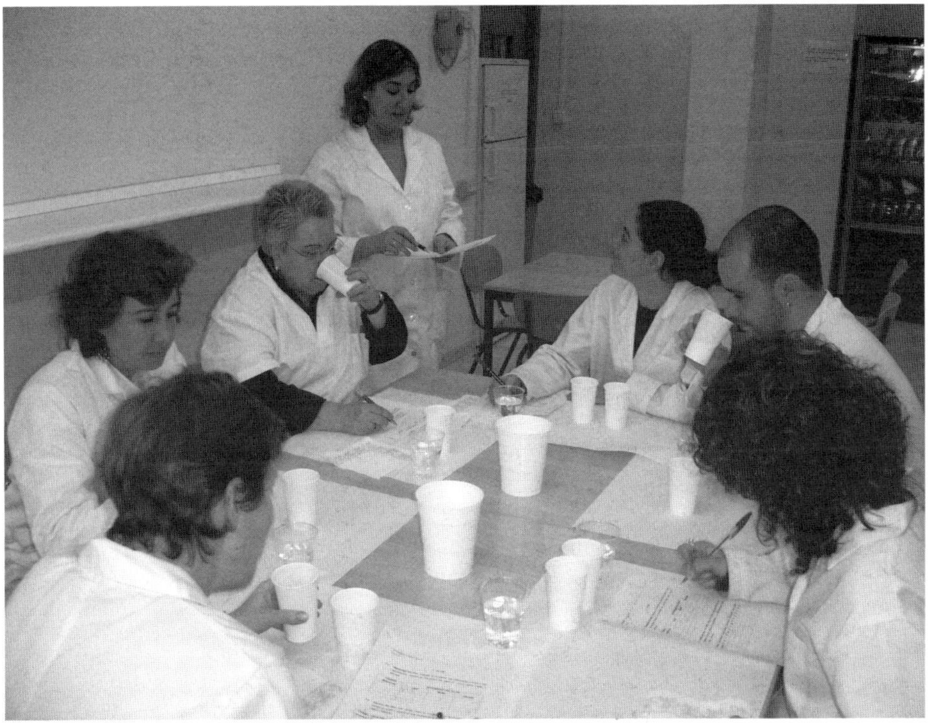

Fig. 11.1 Typical sensory analysis session by an experienced panel in a laboratory.

Different laboratory instruments can measure almost any attribute of food, but instruments cannot tell you if consumers will like your product. Using people, it is possible to measure the overall acceptability or quality of a product, or to know about factors like perceived sensory pleasantness and healthfulness that have been shown to be important in food choice.

11.2 Sensory properties of market milks

11.2.1 *Sensory evaluation of milk*

For dairy products, the most important senses are taste, smell and sight, as appearance, flavour (a combination of odour and taste) and texture are the three primary attributes that are considered to give an overall impression of the sensory quality of milk. The ability to critically evaluate dairy products can be learned, if close attention is directed to those senses with which practically everyone is endowed (Bodyfelt *et al.*, 1989).

The sense of sight is very important in the initial assessment of milk, allowing evaluation of colour and texture. It enables similarities and differences between products to be detected.

Smell is detected before and during drinking. Initially, by breathing or sniffing, volatiles come into contact with the uppermost regions of the nasal cavity and the odour is detected.

Then, during swallowing, a small vacuum is created in the cavity at the back of the nose, and part of the aroma is drawn in to it.

Taste is detected in the whole mouth, but most of the taste receptors are located in specific regions on the upper surface of the tongue. At the sides of the tongue, the sour taste is noted, saltiness along the side and tip, sweet at the tip and bitter at the base. Therefore, the sample must be moved all around the mouth.

In addition to the four basic tastes, the mouth can also receive other perceptions, such as astringency, pungency and warmth, called the trigeminal sensations. Moreover, the mouth also feels some tactile characteristics, such as smoothness and chalkiness.

Finally, the blend of sensations from the nose and mouth are taste, odour, aroma, trigeminal sensations and touch in the mouth is called flavour. It should be noted that the flavour of milk is a crucial sensory concept that determines acceptability of milk and it is the key to its popularity (Drake, 2004).

Condition of the evaluator

Evaluators should be in correct physical and mental condition and free from distractions (i.e. hunger). Health factors, such as colds or toothaches, should be noted on the score sheet. Evaluators should refrain from smoking, eating or drinking beverages other than water at least one and half hour before judging milks (Carpenter *et al.*, 2000b).

Before tasting the milk, it is necessary to condition the palate. Unsalted crackers may be used to cleanse the palate. A defect-free milk (i.e. control) is useful for conditioning the palate before and as a reference while evaluating test samples. The mouth should always be rinsed after a severe defect.

Condition of the sample

The correct sensory evaluation of a milk sample requires the sample to be presented in special conditions. Milks should be tempered to 15–21°C to allow flavours to become more volatile. A cold sample (i.e. below 7°C) chills the mouth and makes it difficult to distinguish certain flavours. If milk is tasted cold or too warm, a note of this should be made on the score sheet. Milks should be free of curd or films. If coagulated or otherwise visually unacceptable, milk should not be tasted. In addition, milk with objectionable odours should not be tasted.

Determining the flavour (odour and taste)

Odour and taste play a major role in quality determination at a factory-receiving platform. Certain flavours can be easily detected and nearly all can be prevented. For this reason, they should be identified and corrected as soon as possible by the producer. But consumer acceptance of milk also depends largely on a pleasing flavour.

Before tasting a milk sample, the odour should be noted. Smell the open container of milk, rather than the individual cup, since the greater volume in the bottle makes it easier to detect the odour. In contests, of course, it may not be possible for each contestant to follow this procedure. In this cases, swirl the milk, hold the cup up to the nose and remove the lid (if it has been fixed) to allow volatile compounds to evaporate.

Frequently, odour alone is sufficient to classify milk properly. Tasting permits further not volatile substantiates of the flavour to be determined. To taste a milk, take a generous sip of milk, roll in the mouth, note the flavour sensation and expectorate. Do not swallow the milk. Flavours can be enhanced by drawing in a breath of fresh air through the mouth followed by a slow exhale through the nose. Objectionable flavours, such as metallic, oxidised and rancid, can be readily recognised and are scored according to intensity. Other flavours, such as chemicals or fly sprays and sanitisers, are highly objectionable and should be quickly recognised. In learning to determine flavour, you should study the description of flavours and their causes outlined above.

Sensory judging of milk

Sensory analysis is a compilation of different measuring and evaluating tools of different food properties. Selection of the appropriate tool or test for a specific objective is required to obtain appropriate and optimal results. There are different approaches to the sensory judging of milk. Affective sensory tests will inform us about likes, dislikes and preferences of consumers, difference tests will determine whether two products are perceived as being similar or different and descriptive tests will characterise sensory profiles of milk (qualitative and/or quantitative). In any case, for the reliability of these analyses, it is necessary to be objective and normalise vocabulary and conditions of the tests in order to obtain quantitative and reproducible data. There are various standards or methods for measuring milk quality.

The sensory judging of milk for the general purpose of quality and shelf life evaluation is most often based on defining milk flavour, and it is performed under the established guidelines of the American Dairy Science Association (ADSA) (Shipe *et al.*, 1978). Milk flavour identification and evaluation are done on score cards prepared for this purpose. A scorecard was one of the earliest forms used for evaluating and recording quality. Milks are graded on a score of 1–10 on the basis of the lack or severity of perceived defects. No criticisms correspond to a score of 10, and unacceptable to a score less than 6. Criticisms are also classified with the intensity of the defects (slight, definite or pronounce) (see Table 11.1). Flavours usually considered in the score cards are acid, astringent, barny, bitter, burnt, caramelised, chemical, cooked, cowy, feed, flat-watery, foreign, fruity, garlic/onion, lack of freshness, malty, medicinal, metallic, papery, putrid, rancid, salty, unclean and weedy.

The 'ideal' natural flavour of milk should be kept in mind for comparison. The mental image will help in distinguishing between samples which show varying characteristics and types of flavour. Each sample should be scored on its own merits in comparison with the 'ideal'. However, the scores depend on an individual's own judgment, and do not let the comments or facial expressions of others influence you.

11.2.2 *Sensory attributes of flavour defects*

Milk of a good quality is a bland food with a slightly sweet taste, very little odour and a smooth, rich feel in the mouth that leaves only a clean, pleasing sensation and, if an odour or aftertaste is detected, the milk has a flavour defect. Appearance is characterised as whiteness, glossiness and transparency of the liquid milk (Phillips *et al.*, 1995). However

Table 11.1 Milk flavour criticisms and suggested flavour scores (out of 10) for milk with designated intensities of flavour defects.

	Flavour intensity of defect		
Attributes	Slight	Definite	Pronounced
Astringent	8	7	5
Barny	7	5	3
Bitter	7	5	3
Cooked	9	8	6
Cowy	6	4	1
Feed	9	7	5
Flat	9	8	7
Foreign	5	3	0
Garlic/onion	5	3	1
High acid	3	1	0
Bacterial	5	3	0
Lack of freshness	7	5	3
Malty	7	5	3
Oxidised	7	5	3
Rancid	7	5	3
Salty	8	6	4
Unclean	7	5	3

Goff (1995).

milk can often acquire or develop an undesirable colour (yellowness, brownish) and flavours (off-flavours).

Good quality raw milk is required to make good quality market milk. Processing and storage conditions have a great effect on the quality of milk, and because of the perishable characteristic of milk and the nature of milk production and handling procedures, the development of off-flavours/odours is not uncommon. Once raw milk is defective, it cannot be improved during processing, and defects often become more pronounced (Mounchili *et al.*, 2005). Sometimes the application of certain nutritional programmes or management practices on the farm can cause off-flavour problems in milk. These off-flavours are a major control problem for the industry, as consumer acceptance can be greatly affected. To prevent flavour/odour defects in milk, proper milk-handling procedures from the farm to the consumer are essential (Ishler & Roberts, 1991).

Major flavours defects are caused, in general, by five factors: health of the cow, feeds consumed by the cow, bacteriological action, chemical changes and absorption of foreign flavours after the milk is drawn (Ishler & Roberts, 1991). A number of off-flavours, together with their descriptors, are listed below, following the committee on off-flavour nomenclature and reference standards of the ASDA (Table 11.2). These off-flavours of milk can be divided into seven different categories according to the nature of the mechanism that produces them, and by understanding their origin, it is possible to have a better background to identify and recognise each of these particular flavours.

Transmitted flavours are caused by specific nutritional and management factors (Mounchili *et al.*, 2005). Cows are particularly bad for transmitting flavours through milk, and milk is equally susceptible to picking up off-flavours in storage (Goff, 1995). The feed a

Table 11.2 Off-flavours attributes that have characterised in milk according to the sensory scheme of the American Dairy Science Association (ADSA).

Off-flavour	Descriptors
Transmitted	Feed, weedy, garlic/onion, barny, cowy
Light-induced	Burnt sunlight, 'medicinal'-like flavour
Lipolysed	Rancid, butyric, bitter
Microbial	Acidic, bitter, fruity, malty, putrid unclean
Oxidised	Papery, cardboardy, metallic, oily, fish
Heated	Cooked, caramelised, scorched
Miscellaneous	Flat, chemical, foreign, lack of freshness, salty

Shipe *et al.* (1978).

cow eats may impart certain flavours to milk. Green grass, silage, turnips, alfalfa hay, silage or turnips can transfer flavours to the milk. The weedy flavour has a bitter characteristic varying with specific weeds of certain localities. It may include obnoxious flavours caused by plants such as ragweed, bitter weed, or peppergrass, and may become a very troublesome flavour defect. Garlic/onion imparted to milk when the cow eats garlic, onions or leeks is recognised by the distinctive taste and odour suggestive of its name. All these flavours are absorbed through the cow's system rather than directly into the milk (from the mouth or nose to lungs or digestive tract to blood to milk). Certain feeds can be detected in milk if fed to the cow even 15–30 min before milking, and some weed flavours persist for longer than 12 h after they are eaten. To avoid these off-flavours, it is necessary to keep cows away from feed or weed-infested pastures 2–4 h before milking (Bruhn, 1996).

Other transmitted flavours can be picked up in the milk if there is poor air quality, or the barn is not properly cleaned and cows breathe the air. These barny flavours are volatile so can be driven-off through vacuum de-aeration (Goff, 1995).

Transmitted off-flavours have major incidence in the milk industry. Mounchili *et al.* (2004), in a prospective case-control study with data on herd management, reported that the 69% of the cases milk off-flavours identified during the study period were due to 'transmitted' (feed) off-flavours, and 31% were due to other off-flavours.

In general, transmitted off-flavours can be prevented by taking cows off offending feeds or feeding cows after milking. Appropriate ventilation and cleaning in barns, adequate installations for avoiding exposure of milk to the farm environment, proper cleaning of equipment and good farm procedures (e.g. cooling practices) are also required. Milk from healthy cows should be used, and discard milk from old cows and late lactation period.

Light-induced defect can result when milk is exposed to ultraviolet rays from sunlight or fluorescent lighting catalysing oxidation in unprotected milk (Olsen & Ashoor, 1987). Photo-oxidation activates riboflavin which is responsible for catalysing the conversion of methionine to methanal (Bosset *et al.*, 1993). It is, therefore, a protein reaction rather than a lipid reaction. However, the end product flavour notes, such as burnt sunlight or 'medicinal'-like flavours, are similar but tend to diminish after several days storage.

Chapman *et al.* (2002) reported the results on sensory threshold of light-oxidised flavour defects in milk, and they suggest that the majority of milk on the market in light transmissible containers is vulnerable to development of detectable light-oxidised flavour defects as ~50%

of plastic containers remain in dairy cases for at least 8 h. To prevent light-induced off-flavours, it is necessary to avoid exposure of milk to light.

Lipolysed flavour is a defect due to the liberation short-chain fatty acids (particularly butyric acid) produced from the hydrolysis of milk fat as a consequence of a chemical development which continues until the milk is pasteurised (Santos *et al.*, 2003). It involves an enzyme called the lipoprotein lipase (LPL). LPL can be indigenous or bacterial. It is active at the fat/water interface but is ineffective unless the fat globule membrane is damaged or weakened (Goff, 1995).

The flavour defect is described as rancid (Shipe *et al.*, 1978). Lipolysed off-flavours are characterised by descriptors such as soapy, blue cheese-like aroma, slightly bitter, foul, pronounced aftertaste and do not clear up readily. Lipolysed off-flavour is closely associated with bitter flavour, but unlike the common bitter flavour, it has an odour resembling spoiled nut. A soapy-bitter taste is identifiable with rancidity.

The key to prevention is to have intact membranes around the milk fat globules as they protect fat from the action of the lipase. Causes of rancid flavour are any factors that result in weakened or broken milk fat globule membranes, such as freezing or over agitation in the bulk tank or homogenisation prior to pasteurisation (Cartier & Chilliard, 1990).

There are some other causes for rancid flavour related to air leaks in the pipeline of the milking installation, flooding of pipelines and receiver jars or pipes in the processing section of the factory or running pumps in a starved condition (Ishler & Roberts, 1991).

Microbial flavour defects of dairy products may be caused by bacteria, yeasts or moulds. Milk is an excellent growth medium for bacteria. It provides the nutrients and moisture and has a near neutral pH. Off-flavours are the result of bacterial growth (psychrotrophs) (Cousin, 1982). The type of bacteria is more important than the number. Psychrotrophs include many kinds of bacteria, all of which cause spoilage. Their ideal growth temperature is 18–21°C. Cold temperatures slow their growth, but do not kill them. At 7°C or above, bacterial growth is steady and off-flavours may be present in 2–3 days (Wilkes *et al.*, 2000).

In raw milk, the high-acid/sour flavour is caused by the growth of lactic acid bacteria (generally, *Lactococcus lactis* subsp. *lactis*), which ferment lactose (milk sugar). In both raw and processed milk, fruity flavours may arise due to psychrotrophs such as *Pseudomonas fragi*. Bitter or putrid flavours are caused by psychrotrophic bacteria which produce protease. It is the proteolytic action of protease that usually causes spoilage in milk.

Putrid flavours are the result of bacterial contamination and may originate in raw milk from contamination, and holding raw milk for 3 or 4 days after collection from farms, or from storage temperatures above 4–5°C. Occasionally, retail samples of milk are found with putrid flavours, but with excellent bacterial counts. Although more of a tactile defect, ropy milk is also caused by bacteria, specifically those which produce exopolysaccharides (Goff, 1995).

Milk from the bulk tank should be emptied and collected at least every other day. Longer time periods between pickup and partial emptying of a tank causes problems. Any milk left in a bulk tank for more than 3 days has the potential of high bacterial counts and off-flavours in fluid milk and manufactured products.

Sanitation is a key factor to preventing microbial flavours. All milk-handling equipment surfaces should be washed after each use, and sanitised just prior to reuse (Bramley & McKinnon, 1990). Certain bacteria from improperly cleaned equipment, especially milking machines, may contaminate the milk and cause the objectionable malty flavour. Malty flavours are caused by *L. lactis* subsp. *lactis* var. *maltigenes*, and are characterised by a corn flake-type flavour (Goff, 1995). It rarely develops in pasteurised milk.

The most common source of the problem is contamination in processing plants following pasteurisation. Psychrotrophs do not generally survive pasteurisation unless large numbers of the bacteria are present in the raw milk at the time of pasteurisation (Ishler & Roberts, 1991). Proper pasteurisation and packaging without contamination eliminates most of the psychrotrophs. Although all the bacteria are destroyed, the spoiled flavour remains. No heating, vacuum treatment or other processing procedure will lessen the flavour.

Oxidised flavour is the result of milk fat oxidation is catalysed by copper and certain other metals with oxygen and air. This leads to an auto-oxidation reaction. It is usually initiated in the phospholipids of the fat globule membrane. Propagation then occurs in triglycerides, primarily double bonds of unsaturated fatty acids. During propagation, peroxide derivatives of fatty acids accumulate. These undergo further reactions to form carbonyls, of which some, like aldehydes and ketones, have strong flavours. Dry feed, late lactation, added copper or other metals, lack of vitamin E (tocopherol) or selenium (natural antioxidants) in the diet all lead to spontaneous oxidation. Exposure to metals during processing can also contribute to this. It is characterised by descriptors, such as metallic, wet cardboard, oily, tallowy, chalky, and the mouth usually perceives a 'puckery' or astringent feel (Goff, 1995). The flavour can be detected in raw milk, but sometimes not until 2 days after collection. It can also be a problem in any pasteurised milk or dairy product that has not been flavoured. Causes are different than for the light-induced flavour of milk purchased at stores, although the taste is similar.

The oxidised flavour embraces many other flavours, which represent various stages of oxidation or partial changes in the fatty portion of milk. Papery or cardboard, sunlight and tallowy are forms of oxidised flavours with varying degrees of intensity. This is one of the most troublesome milk flavours and should be easily recognised (USDA, 2002).

Oxidised-metallic flavour is also caused when the milk comes into contact with corrodible metal, such as exposed copper on equipment or rusty milk cans or lids. It is very objectionable and may lead to further serious defects in certain dairy products, such as butter, when kept in storage. The intensity of the metallic flavour may increase in proportion to the extent of the milk's contact with metal. Metallic flavour is rough and puckery on the mouth and tongue (USDA, 2002).

To avoid these off-flavours, it is recommended to segregate susceptible milk, maintain equipment surfaces properly cleaned, eliminate copper and white metal and avoid exposure to light.

Heat-induced flavour as a result of the heat treatment may affect the sensory properties of the milk. Heat-induced flavours have been classified into the following types: (a) cooked or sulphurous, (b) heated, caramelised and (c) scorched (Jensen & Poulsen, 1992). Flavours caused by heat are a function of the time and the temperature of the heat treatment and the sensitivity of each milk component. Thus, pasteurised milk is less affected by heat than ultra-high-temperature (UHT) milk, and sterilised milk is the most affected. Many changes are related to any 'burn-on' action by heat of the whey proteins and subsequent interactions (Goff, 1995). Among them, cooked flavour is related to the release of sulfhydryl and hydrogen sulphide from β-lactoglobulin, and also the formation of small quantities of free sulphides and mercaptans giving the milk a burnt taste (Shipe *et al.*, 1978). The defect is most obvious immediately after heating, but dissipates and is not appreciated after 2 or 3 days of storage (Bodyfelt *et al.*, 1989).

Maillard reaction (i.e. a non-enzymatic browning reaction) is the other main change that occurs from heating milk. In its initial phase, condensation occurs between free amino groups of the casein, especially the lysine group, and the lactose, i.e. reducing aldehyde function(s). It leads to the formation of a Schiff base followed by the Amadori rearrangement and degradation into a brown pigment (melanoidins) (Walstra & Jennes, 1984). This reaction leads to a slight browning of the milk and the development of a caramel-like taste. The browning does not always appear during heat treatment since the Schiff bases so formed can become degraded slowly in dry, liquid or medium-moist milk products during storage.

Under the influence of intense heating, the fat produces lactone and methylketone compounds which, if highly concentrated, cause undesirable flavours as coconut flavour in concentrated and dried milk.

The more severe heat treatments are accompanied by changes in the tactile, or mouth-feel, characteristics of milk. Such flavours are typical of sterilised milks, and are objectionable to most consumers. Hence, extensive research has been aimed at sterilising milk without producing the objectionable flavour caused by conventional sterilising treatments (Bodyfelt *et al.*, 1989).

The use of certain additives like sodium bisulphite, sulphur dioxide and formaldehyde, or the presence of (–SH) groups in the environment can effectively help to prevent these defects. Recently, the ability of epicatechin to inhibit the thermal development of aroma compounds (i.e. Maillard reaction products) formed during UHT processing of cow's milk has been reported (Colahan-Sederstrom & Peterson, 2005; Schwambach & Peterson, 2006).

Although thermal treatment brings about changes in the flavour of milk, it has been effectively used for decades as a method to extend the shelf life of fluid milk and to eradicate pathogenic bacteria from it. Use of a non-thermal technology such as microwaves, high-pressure, pulsed electric fields (see Chapter 8) may represent a viable alternative to produce an extended shelf life (ESL) product, or to apply a second preservation treatment to thermally pasteurised milk without further altering its sensory and nutritional attributes (García-Risco *et al.*, 2000; Bendicho *et al.*, 2002; Clare *et al.*, 2005). To our knowledge, there is not enough information about the influence of these new technologies on the sensory properties of milk.

Miscellaneous defects

Other defects, which may be encountered in market milks, are briefly reviewed.

Flat

The source of this uncommon flavour is difficult to determine. The flavour may be described as tasteless. The characteristic flavour of normal milk is lacking, but the milk has no off-flavour. Flat-flavoured milk resembles normal milk that has been partially diluted with water, even though this may not have been done (USDA, 2002). Possible prevention is to dry out holding tanks completely.

Foreign
Any seriously objectionable flavour foreign to milk, such as fly spray, paint, oil, kerosene, creosote or a medicinal substance, will render the milk unpalatable or unfit for use. Such a flavour may either directly contaminate the milk or be absorbed. Sanitisers are included in this flavour category. The residue of sanitisers in milk, such as hypochlorite and iodophor, if left on dairy equipment, may be absorbed by the product and impart a foreign flavour. Phenolic compounds used in udder ointments may combine with iodophor or hypochlorite to form a highly objectionable foreign flavour, which is detectable at a very low concentration (USDA, 2002). These foreign flavours can be avoided by draining thoroughly the milking and bulk farm tanks, proper ventilation and cleaning the teats as well.

Salty
It is a taste, which may be present in milk from cows in the late stages of lactation, and is often characteristic of milk from cows infected with mastitis. It is not commonly found in herd milk or mixed milk received at a dairy plant. This defect cannot be detected by odour (USDA, 2002).

Unclean
Such as flavour is seldom found except in pasteurised milk that has been stored too long or at a slightly high refrigerator temperature. Unclean flavour often accompanies the bitter flavour. It may be caused by growth of bacteria in milk or from contact of milk with decomposed material on improperly washed or sanitised equipment or utensils (USDA, 2002).

11.2.3 *Other factors influencing the sensory properties*

Packaging materials

The sensory quality and the nutritional value of milk can be seriously affected by photo-oxidation which is caused by the combined effects of light and oxygen. The role of packaging is to provide a good protection that minimises photo-oxidative deterioration. Interactions between the packaging material and milk must be taken into account when choosing a packaging (Hotchkiss, 1997); permeability of gases, migration by polymers and absorption of flavour compounds are the main problems for dairy industries. A good barrier must retain the aroma and flavour of milk during storage (Gilbert *et al.*, 1983; Bosset *et al.*, 1993).

Dairy products are extremely sensitive to light due to the high amount of riboflavin (vitamin B_2). This strong photosensitiser can absorb visible and ultraviolet light and transfer this energy into highly reactive forms of oxygen like singlet oxygen and free radicals, such as superoxide O_{2-}. This may, in turn, induce a cascade of oxidation reactions leading finally to great losses of vitamins (A, B_2, C, D and E) and amino acids, lipid oxidation and discoloration; moreover, strong off-flavours are also produced (Borle *et al.*, 2001). Light oxidised flavour defects in milk are caused by the decomposition of lipid hydroperoxides and protein oxidation (Mortensen *et al.*, 2004). Volatile compounds like hexanal, pentanal, 1-octen-3-one, acetaldehyde, 1-hexen-3-one and dimethyl disulphide have been correlated with light-induced flavours in milk, some of them perceived as old vegetable oil, cardboard, goaty or metallic (Karatapanis *et al.*, 2006). Photodescomposition of methionine to methional and other sulphur compounds results in flavours characterised as burnt protein, burnt cabbage, cooked cabbage, mushroom, medicinal, or plastic-like (Chapman *et al.*, 2002).

Besides packaging, other factors that influence photo-oxidation are storage and milk-processing conditions. For example, whole milk is less light sensitive than low-fat and skimmed milks, due to its greater light-scattering properties; sterilised or UHT milks are less photosensitive than raw or low pasteurised milks due to their higher content of monodisulphide groups, which act as reducing agents (Borle *et al.*, 2001).

Chocolate milk has been reported to be less sensitive to light-induced off-flavours, probably due to a masking effect of the chocolate flavour itself (Chapman *et al.*, 1998). Also, extra protection is expected in chocolate milk against light by filtration. Moreover, cocoa and vanillin contain additional antioxidants, such as α-tocopherol, which may contribute to minimising light effects (Borle *et al.*, 2001).

Traditionally, milk has been packaged in glass bottles and laminated paperboard cartons. Over the past few years, new polymer packaging materials like polyethylene terephthalate (PET) and coextruded high-density polyethylene (HDPE), clear and pigmented, have been introduced into the milk market. In general, oxidation off-flavours increased more rapidly in light-exposed packages than in light protected packages (van Aardt *et al.*, 2001; Moyssiadi *et al.*, 2004; Mestdagh *et al.*, 2005). Pigmented PET and HDPE bottles reduce the side effects of light on milk and have considered as effective materials for protecting milk in terms of flavour and to extend its shelf life (Solano-López *et al.*, 2005; Karatapanis *et al.*, 2006). Several studies based on the sensory evaluation of milk in different types of packaging material (PET, HPDE and laminated paperboard carton) have been carried out. These studies show that multi-layer pigmented HPDE bottles followed by the paperboard carton and pigmented HDPE provides the best protection in terms of flavour (Moyssiadi *et al.*, 2004; Karatapanis *et al.*, 2006).

Storage

As mentioned elsewhere, storage conditions during the manufacture and distribution of milk can have a significant impact on its sensory attributes. Raw milk must have a high quality. Moreover, it is necessary to maintain strict control of sanitation and storage conditions, and temperature is an important limiting factor. Flavour defects developing in milk during storage can be related to lipolysed, oxidised and light-induced off-flavours (see Section 11.2.3, 'Packaging materials'). Lipolysed off-flavours like rancidity and bitterness occur when raw milk has a high bacterial count due to it being held for more than 3 or 4 days after collection from farms, after which the growth of psychrotrophic bacteria takes place. Although these bacteria are destroyed during heat treatment, thermo-resistant enzymes, proteases and lipases have already been formed, and can decrease the quality of milk during storage. This is an important problem in UHT milk because its storage is longer and, apart from the development of flavour defects, gelation of the product can occur (Kelly & Foley, 1997). Metallic off-flavours are due to the milk coming into contact with metal during processing and storage.

Light-induced off-flavours can be prevented by protecting the product with adequate packaging. But also, some storage characteristics like the type of light used in storage rooms as well as the exposure geometry and duration, play key roles in protecting the food against photo-oxidation. Mild, soft, light sources like 'warm white' fluorescent lamps, the shortest possible duration as well as the lowest possible intensity of light exposure and storage temperature are the main factors to reduce the effects of light on milk.

11.3 Different types of market milks

There is a wide range of liquid milk products and milk beverages available on the market worldwide, which are either raw or processed milks (Figure 11.2). The type of heat treatment is of major importance in the sensory properties of the milk, including some UHT processing combined with a specific packaging that increases the refrigerated shelf life of milk to 6 or more weeks, i.e. ESL. This product is more convenient for marketing and has longer distribution times (van Aardt *et al.*, 2005). It is also of interest to note that the level of fat in milk greatly affects the sensory properties of the product (Phillips *et al.*, 1995). On average, the fat content of whole milk is ~3.5 g 100 g^{-1} while in products with reduced fat its content ranges from <0.1, 1 to 2 g 100 g^{-1}. Sensory differences between non-fat and other types of milk are mainly found in the attributes of appearance, texture and mouth-feel (Tuorila, 1986; Phillips *et al.*, 1995).

Moreover, there is also an increased interest in enriched milks and very often functional ingredients are added and constitute part of the composition of a milk beverage (Hernández-Ledesma *et al.*, 2002). Other modifications, which can be applied to milk, are lactose-reduced milk or the addition of cocoa powder for the production chocolate-flavoured milks; these processes increase the number of possibilities for increasing the consumption of milk as beverage products. On the other hand, soymilk, which is not a dairy beverage, is briefly reviewed in this chapter, as normally it is used by consumers as an alternative option to cow's milk (or milk from other species of mammals) in cases of allergy or lactose intolerance. Soymilk consumption is quickly increasing in western countries, and it has a particular and much defined flavour.

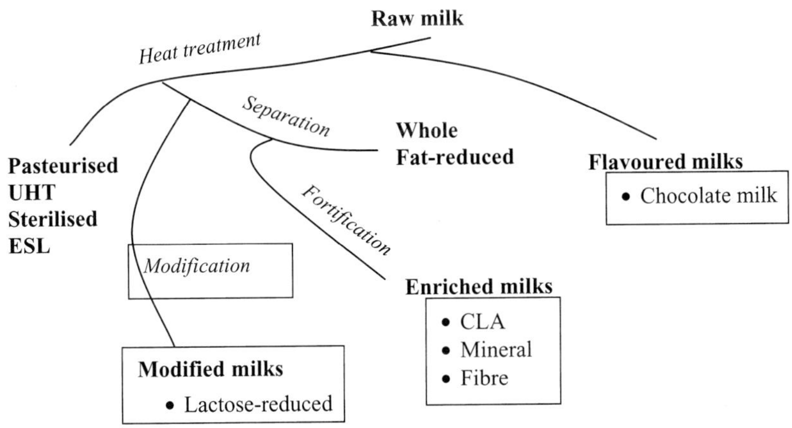

Fig. 11.2 Schematic diagramme of the some of the types of milk available at the market.

11.3.1 *Raw and heat-processed milk*

Raw cow's drinking milk represents only a small fraction of total milk consumption. The governments of the different countries, e.g. European Union, have implemented strict regulation on diary hygiene with rules prohibiting or restricting the placing on the market, within its territory, of raw milk for direct human consumption (FSA, 2005). Raw milk has been shown to have more 'taste and quality' than heated milk, maintaining especially the freshness. It is natural, creamy and has a high odour intensity. However, spoilage bacteria in milk produce foul smells and off-flavours.

Different heat treatments can be applied in the dairy industry (pasteurisation, uperisation, sterilisation, UHT), and these treatments provoke the development of heated off-flavours and a slight brown colour depending on the intensity of the heat and the duration of its application. Moreover, some other off-flavours related to the packaging and storage can be present in heated milk. Dairy industries have been developing different strategies to produce safe milk with a similar taste to fresh milk, and extension of the shelf life has been a prime objective (Goff & Griffiths, 2006). Ultrapasteurisation (UP) enables dairy processors to produce milk with an ESL similar to the UHT process, but with fewer flavour defects and resembling the fresh taste of pasteurised milk.

In addition, the higher the storage time, the higher the possibility for developing organoleptic defects. The flavour of UP milk has not been studied as extensively as UHT milks, but Mauri *et al.* (1995) reported that UP milk developed stronger painty and rancid flavours during storage than HTST milk. The extended storage time of ESL milk increases the opportunity for light-induced oxidation, for Maillard reaction or product/package interactions.

Table 11.3 summarise the typical sensorial profiles of raw and heated milk, and possible off-flavours that can be developed. Raw milk is principally fresh with no aftertaste. Although there has been considerable improvement with respect to the care and cleaning of the dairy cows, farms and milk collection and delivering systems, numerous off-flavours are usually developed at the farm. Moreover, microbial growth and metabolism shorten the shelf life of raw milk and, as a consequence, produce undesirable changes in aroma and taste attributes of milk; hence, milk is normally subjected to a heat treatment for hygienic reasons. As shown in the Table 11.3, during the heat treatment of the milk, other off-flavours may appear in milk. Their appearance and intensity are usually related to the time and temperature of the heat. Packages are designed to preserve the freshness and sensorial properties of the milk during storage, but sometimes can actually be directly responsible for causing flavour defects.

11.3.2 *Whole and fat-reduced milk*

Milk sensory properties of appearance, texture and flavour are largely determined by the fat concentration in the product. Fat molecules are too large to stimulate taste or smell, but do produce tactile sensations (i.e. oily, creamy). This is because the tactile sensations are mediated by the trigeminal nerve. Previous studies have also shown that taste perceptions of milk are influenced by the fat content (Richardson-Harman *et al.*, 2000).

Table 11.3 Principal defects of raw and heated milk – origin and prevention.

Type of milk	Typical sensory profile	Off-flavours/off-tastes/other defects	Possible origin of the defect	Prevention
Raw	Fresh milk (no aftertaste)	Salty, cowy, rancid, lack of freshness, medicinal-like, bitter, flat, oxidised Barny, unclean, foreign, feed Feed, cowy, weedy, garlic/onion, bitter High acid/sour, malty, fruity/fermented, bitter, putrid, ropy Medicinal, unclean plastic/rubber, rancid, metallic, high acid, foreign	Health/status of the cow Poor conditions of the barns Offending pastures short time before milk Bacterial growth Milking equipment	Healthy and young cows, avoid advanced lactation milk Ventilate and clean barns Control of the feed and the milking time Check bacterial counts and types, use clean and sanitise the equipment, cooling practices Proper use of the equipment, avoid copper and white metal
Pasteurised	Cooked flavour, but still fresh	Oxidised, metallic, cardboardy Rancid Unclean, putrid, bitter, fruity/fermented High acid Burnt, medicinal Cooked, sulphurous	Un-homogenised milk, milk fat interaction(s) Improper pasteurisation process Post-pasteurisation contamination Poor refrigeration Exposure of milk to light High heat treatment	Homogenise the milk, avoid copper and white metal and exposure to light Proper pasteurisation, and avoid using re-processed milk Check the cleaning and sanitising procedures Cool the milk properly Use proper packaging material and storage conditions Control of the pasteurisation process/temperature
UHT	Heated milk	Cooked, sulphurous, caramelised, bitter scorched, rancid, burnt Putrid Caramel-like taste and non-enzymatic browning Beany odour Bitter, rancid, gelation	Time and intensity of heat treatment, great storage time Organic sulphur compounds Maillard reaction Oxidation effects Enzymatic breakdown	Use enough heat to kill microorganisms, but not so high to induce chemical reactions Avoid excessive storage time Use of non-thermal technologies
ESL	Cooked flavour, but still fresh	Chemical off-flavours and non-enzymatic browning Rancid, bitter, cooked Papery, cardboardy, metallic Burnt, medicinal, plastic-like Foreign/painty	Maillard reaction Intense heat treatment Light-induced oxidation Exposure of milk to light Product/package interactions	Avoid excessive storage time Control temperature/time Proper packaging Proper packaging and storage Use proper package, and avoid excessive storage time

Whole milks are whiter and more viscous than skimmed milks. Also, the flavour of whole milk is more intense due to the naturally occurring fat flavour compounds. The fat content in the product determines the colour of milk, which is related to the perception of other sensory attributes like aroma and thickness (Phillips *et al.*, 1995). Low-fat milks are usually described with terms like boiled milk smell and flavour, transparency and blueness. The simple addition of a whitener to low-fat milk produces an increase in the sensory perception of whiteness, glass coating, thickness (visual and oral), creaminess, residual mouth-feel and total fattiness (Frøst *et al.*, 2001). However, the different levels of fat (<0.1–3.5 g 100 g^{-1}) content do not affect the sensory properties of milk in a linear manner (Figure 11.3). The sensory evaluation of milks with different fat contents have shown that whole milk has high score due to its taste, perceived as creamiest and mouth-feel perceptions, and there are relatively few sensory differences between whole and reduced-fat milk. However, the sensory differences between skimmed milk ($\sim$0.1 g 100 g^{-1}) and reduced-fat milk (1.3 g 100 g^{-1}) are greater (Frøst *et al.*, 2001). The consumption of a certain type of milk is not only influenced by taste, but other factors, such as health, cost, convenience, habits and nutritional knowledge, are also implicated. Health awareness among consumers has negative perceptions for whole milk with a high amount of fat because this is related to cholesterol and high-energy contents (Bus & Worsley, 2003). However, from a taste perspective, whole milk is more valuable than a low-fat product. A positive relationship between creaminess and preference has been found with higher fat products being perceived as the creamiest.

Some studies have been carried out to develop fat substitutes that enhance the sensory characterisation of low-fat milk. As the colour is a key attribute in the sensory perception of whole milk, it is important that an effective fat substitute makes the colour of milk whiter, less green and less blue; non-fat dry milk is sometimes added to skimmed and low-fat milks as a fat substitute. Their use increases the relative viscosity of skimmed milk, although it does not provide the product with the desirable sensory attributes of flavour, appearance, colour and thickness (Phillips *et al.*, 1995). Commercially available protein-based fat substitutes have also been unsuccessfully evaluated as an alternative to milk fat (Phillips & Barbano, 1997).

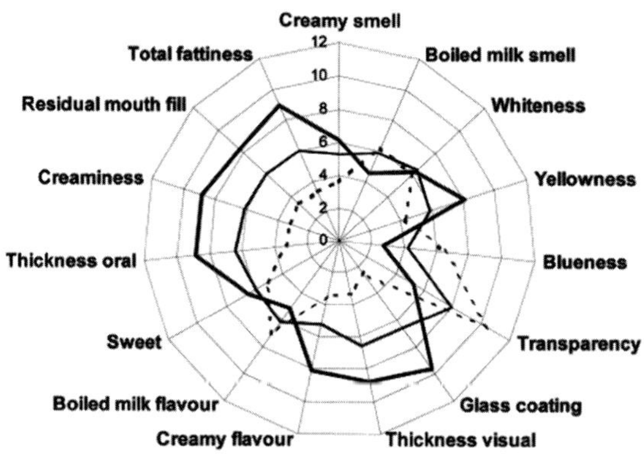

Fig. 11.3 Cobweb plot of sensory profiles of milks with three different fat levels. Note: –0.1 g 100 g^{-1}, –1.3 g 100 g^{-1} and –3.5 g 100 g^{-1}. Reproduced with permission from Frøst *et al.* (2001), and *Journal of Food Quality and Preference*.

11.3.3 *Enriched and modified milks*

Over the past few years, consumers' growing interest in healthy eating has led to the promotion and appearance on the market of many different types of modified milks. Apart from milks with different levels of fat, which have been on the market for a long time, there are other milks enriched with specific components (or modified), which are considered to have some health benefits. Some examples are milks fortified with vitamins, fibre, calcium, conjugated linoleic acids (CLA) or health-promoting microorganisms (Tamime *et al.*, 2005). There are also other milk products, such as cholesterol-reducing and lactose-free milk, and soymilk is an alternative to dairy fluid milk. Sometimes the sensory profile of these developed milks is very different to that of traditional fluid milk. Attempts are being made to describe the sensory characteristics of these enriched milks in order to better satisfy consumer demands.

CLA-enriched milks

Milk and dairy products are major sources of CLA in human diet. Growing medical evidence suggests that CLA helps to prevent cancer, and coincides with the recent advances in enhancing CLA levels in milk. The CLA content of milk can be increased in two ways: first, with direct fortification of milk with CLA oil (i.e. known as synthetic enrichment of CLA), and second, the content of CLA in milk can be enhanced naturally by modifying the cow's diet. The use of very high levels of synthetic CLA-enriched milks seems to reduce the overall acceptability, flavour and freshness perception of the product. Fortification of the milk with CLA produces a grassy/vegetable oil flavour, and decreases milk fat flavour. However, although the addition of high amounts of CLA to milk would help to obtain maximum health benefits, this is impractical because of cost constraints (Campbell *et al.*, 2003). In milk with naturally increased levels of CLA, the amount of these fatty acids are lower, and the results of several researchers suggest that consumer acceptability attributes of CLA-enriched milk are similar to products with low levels of CLA (Ramaswamy *et al.*, 2001; Khanal *et al.*, 2005).

Mineral-enriched milk

There are calcium-, magnesium- and iron-enriched milks available on the market. These types of milks are designed to supply the mineral necessities of mothers during breastfeeding and pregnancy or infants. Calcium and magnesium are involved in bone development and maintenance; its consumption decreases the risk of several diseases, such as osteoporosis, hypertension and colon cancer. The fortification of milk with iron can help to prevent and eradicate iron deficiency anaemia, which is one of the most widespread nutritional deficiencies in humans.

The sensory quality of mineral-enriched milks depends on the mineral source used for fortification. It is necessary to reach the optimal level of 'bio'-available mineral without affecting the taste and appearance of milk. In general, calcium is added to milk with other minerals (e.g. phosphorus, magnesium), vitamins (D, K), and more recently, phosphopeptides and probiotic microorganisms have been added in order to improve the absorption of minerals (Hilliam, 2003). There are several commercially available calcium salts used for enrichment of milk. These include inorganic salts (chlorides, carbonates, phosphates), milk minerals (calcium phosphate) and organic salts like tricalcium citrate, calcium lactate, calcium lactate gluconate and calcium gluconate. Some of them have a good solubility in milk but contain only low amounts of calcium, for example calcium gluconate or calcium lactate, and others like inorganic salts have a high quantity of calcium, but are poorly soluble. Regarding their influence on sensory properties, generally carbonates and phosphates impart astringent, bitter and chalky tastes to milk. Calcium lactate and calcium chloride promote bitter notes at high concentrations, whilst calcium carbonate may come across as soapy or lemony and calcium phosphate has a bland flavour, but imparts a gritty mouth-feel. Tricalcium citrate is a neutral tasting salt with a high calcium level and medium solubility; moreover, it is the most economic salt for calcium fortification, and these advantages make it now the first choice for enrichment of milk with calcium.

Iron is the most difficult mineral to add to foods and ensure adequate absorption, and because soluble iron has the highest 'bio'-availability, it often causes changes in flavour and appearance. On the other hand, insoluble forms, such as elemental iron powders, do not produce changes in the sensory characteristics of products, but may be very poorly absorbed. A metallic aftertaste and flavour defects caused by the oxidation-mediated rancidity of fats and colour changes resulting from interactions with anthocyanins, flavonoids and tannins have been described as negative effects produced in food fortified with iron (Mehansho, 2006). An example of an iron source is ferrous sulphate (i.e. a water-soluble iron compound), which has been successfully used to fortify infant formulae, but can produce a metallic aftertaste in liquid products (Hurrell, 2002).

Milk enriched with fibre

Fibre intake in the diet is continuously increasing due to its beneficial effects on health (risk reduction of coronary heart disease, high blood cholesterol and cancer, attenuation of levels and fluctuations of blood glucose and insulin, and control of weight due to its satiety effect). When fibre is added to a food, the taste and textural properties can be modified.

A small quantity of soluble fibre has been successfully added to skimmed milk. It produces an improvement in the texture of the product, which becomes similar in taste to reduced-fat or whole milks. Fibre-enriched milk improves in palatability and tolerance compared to other skimmed milk products, making it recommendable for people on a diet or who need a low-fat diet, but do not like the taste of skimmed milk.

Lactose-reduced milk

Consumers, who are lactose intolerant, do not produce enough of the enzyme lactase, which breaks down lactose to glucose and galactose. This lead to poor digestion of the carbohydrate in the intestine and, as a consequence, bacterial activity in the colon produces gas, bloating, pain and sometimes diarrhoea. Most technological approaches to obtain lactose-free milk products consist of adding lactase to milk to hydrolyse the lactose to glucose and galactose. Since the main carbohydrate is glucose instead of lactose, the taste of this milk is sweeter than normal milk, and many consumers do not like it. To resolve this handicap, a technological alternative based on a chromatographic process to separate lactose from skimmed milk has been developed (Jelen & Tossavainen, 2003). This process permits lactose-free milk to be produced, which does not taste any different to ordinary milk so consumers with lactose intolerance can enjoy the good taste of milk.

11.3.4 *Other types of milks*

Chocolate-flavoured milk

Chocolate-flavoured milk is a popular and widely consumed dairy product in many countries. There are two common modes of consumption: (a) adding a cocoa powder formulation to milk, either hot or cold and (b) chocolate milk ready to drink. Basically, these flavoured milks are made from milk, sucrose and cocoa powder, and some hydrocolloids are added to improve the consistency and prevent sedimentation of cocoa particles in the product (Yanes *et al.*, 2002). In some instances, dairy solids are added to the base milk, for example roller-dried and spray-dried whole milk powders, high-fat powders, butter-milk powders, whey powders and skimmed milk powder sprayed with anhydrous milk fat or cream. Alvarez *et al.* (2003) have demonstrated that acid whey could be used successfully to partially replace skimmed milk (i.e. 10–20%) in chocolate-flavoured beverages.

According the European Union Directive 2000/36/EC (EU, 2000), milk chocolate has to contain (g 100 g^{-1}) at least 25 dry cocoa matter, 2.5 defatted cocoa, 14 total dry milk solids, 3.5 milk fat and 25 fat from cocoa or milk (Bolenz *et al.*, 2006). The specific characteristics and different levels of these ingredients used during the manufacture of chocolate milks lead to a wide variability in the sensory character of the product. Different levels of sugar and fat produce very significant differences in the sensory properties of chocolate milk. Low-sugar products are more bitter, gritty and have roasted flavour notes, while chocolate milks with high sugar contents are more milky/dairy, vanilla/caramel and sweeter. Samples with high-fat levels are faster melting in the mouth. Milk chocolate drinks with low levels of sugar and fat are usually associated with viscous, mouth coating, fatty/oily, cocoa and darker notes (Guinard & Mazzucchelli, 1999).

Milk powders can also have a significant impact on the physical and organoleptic properties of these flavoured milks. Of these properties, the fat-free milk powder available to mix with cocoa butter in the chocolate and the particle characteristics are arguably the most important. The former component has a significant influence on the rheological properties of the manufactured product, whilst the particle size of the powder is correlated with the sensory property of smoothness (Liang & Hartel, 2004). The amount of added cocoa is correlated with the perceived thickness of the product in the mouth, and is also responsible for the typical bitter and cocoa flavour of these milks. The amount of stabiliser(s) used (i.e. hydrocolloids) is also responsible for the thickness attribute in chocolate-flavoured milk.

In the evaluation of product quality, the lack of defects is very important. Current defects in the appearance of chocolate milks are undissolved chocolate on surface, chocolate particles in suspension, chocolate precipitate, white spots on the surface and mottled surface.

Instrumental measurements of colour, viscosity, particle size, snap force, melting temperature and hardness are usually carried out by most chocolate manufactures to monitor the quality of these ingredients. The evaluation of chocolate milk sensory quality is normally done on the appearance, texture, aroma, taste and flavour. Sensory languages to document these properties have been developed and include using terms such as milky, vanilla, carob and roasted (Guinard & Mazzucchelli, 1999).

Consumer preferences of commercial chocolate-flavoured milks have been determined by preference mapping tools. The procedure requires an objective characterisation of product sensory attributes achieved by descriptive analysis, which is then related to preference ratings for the product obtained from a representative sample of consumers (Murray *et al.*, 2001). These studies show that cocoa aroma is a major driver influencing the acceptability of chocolate-flavoured milks. Other flavour attributes included are cooked, eggy and malty, and they positively influence acceptability within certain market segments.

Soymilk

Soymilk consumption is growing in popularity. It is being marketed as an alternative to dairy milk, and as a good way to consume soybean, which has been attributed with several health benefits including the prevention of cancer, osteoporosis and coronary heart disease. Although soymilk has different nutritional properties, it is marketed as though it is a milk product with similar nutritional characteristics. Despite this marketing concept, the consumption of soymilk is limited in western countries due to soy aroma, astringency and bitterness. The astringent compounds are thought to include polyphenols and saponins. Hexanal and other aldehydes, ketones and alcohols produced by three lipoxygenase isozymes contained in soybean also contribute to off-flavour(s) in the milk. The necessity of enhancing the acceptance of soymilk has led to the development of descriptive vocabularies or lexicons to describe its characteristics of flavour, aroma and taste (Torres-Penaranda & Reitmeier, 2001; N'Kouka *et al.*, 2004). The term 'beany' has often been chosen by many researchers to describe the flavour properties of soybean products. More specific terms such as green, raw soy, grassy, painty, rancid, astringent, and bitter have also been used to describe soymilk flavour. More recently, there have been some technological and formulation approaches to reduce the objectionable flavours of these products. For example, the perception of astringency decreases with the addition of sucrose or by increasing soymilk viscosity (Courregelongue *et al.*, 1999). Moreover, sugar addition produces a decrease in bitterness and reduces the attributes related to 'beaniness' (Torres-Penaranda & Reitmeier, 2001). The genetic removal of the lipoxygenase isozymes from soybeans led to the production of soymilk, which had less cooked beany aroma, cooked beany flavour and astringency than soymilk made with normal soybeans (Torres-Penaranda *et al.*, 1998).

11.4 Conclusion

Milk is a highly nutritious beverage, and has a pleasant flavour and taste, i.e. it constitutes a perfect drink. In order to preserve the satisfying experience of drinking milk and maintaining the freshness of milk, it is of primary importance that flavour defects should be avoided. Off-flavours in milk are caused by different mechanisms, and at different moments in the marketing chain, i.e. from farm to consumer. New processing techniques and packaging materials have been developed to minimise these undesirable off-flavours. Moreover, a wide range of milk products with different flavours or with different added compounds is available in the market to diversify the tastes and flavours in order to meet the demand of consumers.

References

Alvarez, V.B., Ji, T. & Balakrishnan, S. (2003) Acid whey as an ingredient in a chocolate-flavored beverage. *Milchwissenschaft*, **59**, 524–527.

Bendicho, S., Barbosa-Cánovas, G.V. & Martín, O. (2002) Milk processing by high intensity pulsed electric fields. *Trends in Food Science & Technology*, **13**, 195–204.

Bodyfelt, F.W., Tobias, J. & Trout, G.M. (1989) Principles of sensory perception: an overview. *The Sensory Evaluation of Dairy Products*, pp. 8–35, Van Nostrand Reinhold, New York.

Bolenz, S., Amtsberg, K. & Schäpe, R. (2006) The broader usage of sugars and fillers in milk chocolate made possible by the new EC cocoa directive. *International Journal of Food Science and Technology*, **41**, 45–55.

Borle, F., Sieber, R. & Bosset, J.O. (2001) Photo-oxidation and photoprotection of foods, with particular reference to dairy products – an update of a review article. *Sciences des Aliments*, **21**, 571–590.

Bosset, J.O., Gallman, P.U. & Sieber, R. (1993) Influence of light transmittance of packing materials on the shelf-life of milk and dairy products – a review. *Lait*, **73**, 3–49.

Bramley, A.J. & McKinnon, C.H. (1990) The microbiology of raw milk. In: *Dairy Microbiology*, Vol. 1, 2nd edn (ed. R.K. Robinson), pp. 163–208, Elsevier Applied Science, London.

Bruhn, J.C. (1996) *The Dairy Research & Information Centre*, The University of California, CA. Available at http://drinc.ucdavis.edu/about.htm

Bus, A.E.M. & Worsley, A. (2003) Consumers' sensory and nutritional perceptions of three types of milk. *Public Health Nutrition*, **6**, 201–208.

Campbell, W., Drake, M.A. & Larick, D.K. (2003) The impact of fortification with conjugated linoleic acid (CLA) on the quality of fluid milk. *Journal of Dairy Science*, **86**, 43–51.

Carpenter, R.P., Lyon, D.H. & Hasdell, T.A. (2000a) Para qué se utiliza el análisis sensorial? *Análisis Sensorial en el Desarrollo y Control de la Calidad de Alimentos*, pp. 1–10, Editorial Acribia, Madrid.

Carpenter, R.P., Lyon, D.H. & Hasdell, T.A. (2000b) Quiénes son las personas adecuadas para el análisis sensorial? *Análisis Sensorial en el Desarrollo y Control de la Calidad de Alimentos*, pp. 67–86, Editorial Acribia, Madrid.

Cartier, P. & Chilliard, Y. (1990) Spontaneous lipolysis in bovine milk: combined effects of nine characteristics in native milk. *Journal of Dairy Science*, **73**, 1178–1186.

Chapman, K.W., Rosenberry, L.C., Bandler, D.K. & Boor, K.J. (1998) Light-oxidized flavor development and vitamin A degradation in chocolate milk. *Journal of Food Science*, **63**, 930–933.

Chapman, K.W., Whited, L.J. & Boor, K.J. (2002) Sensory threshold of light-oxidized flavor defects in milk. *Journal of Food Science*, **67**, 2770–2773.

Clare, D.A., Bang, W.S., Cartwright, G., Drake, M.A., Coronel, P. & Simunovic, J. (2005) Comparison of sensory, microbiological, and biochemical parameters of microwave versus indirect UHT fluid skim milk during storage. *Journal of Dairy Science*, **88**, 4172–4182.

Colahan-Sederstrom, P.M. & Peterson, D.G. (2005) Inhibition of key aroma compound generated during ultrahigh-temperature processing of bovine milk via epicatechin addition. *Journal of Agricultural and Food Chemistry*, **53**, 398–402.

Courregelongue, S., Schlich, P. & Noble, A.C. (1999) Using repeated ingestion to determine the effect of sweetness, viscosity and oiliness on temporal perception of soymilk astringency. *Food Quality and Preference*, **10**, 273–279.

Cousin, M.A. (1982) Presence and activity of psychrotrophic microorganisms in milk and dairy products: a review. *Journal of Food Protection*, **45**, 172–207.

Drake, M.A. (2004) ADSA foundation scholar award: defining dairy flavors. *Journal of Dairy Science*, **87**, 777–784.

EU (2000) Directive of 23 June 2000/36/EC relating to cocoa and chocolate products intended for human consumption. *Official Journal of the European Communities*, **19**, L197.

Frøst, M., Dijksterhuis, G. & Martens, M. (2001) Sensory perception of fat in milk. *Food Quality and Preference*, **12**, 327–336.

FSA (2005) Historical background to the different control requirements for raw drinking milk and raw cream in the different countries of UK, 24–6-2005.

García-Risco, M.R., Olano, A., Ramos, M. & López-Fandiño, R. (2000) Micellar changes induced by high pressure. Influence in the proteolytic activity and organoleptic properties of milk. *Journal of Dairy Science*, **83**, 2184–2189.

Gilbert, S.G., Hatzidimitriu, E., Lai, C. & Passy, N. (1983) Study of barrier properties of polymeric films to various organic aromatic vapors. In: *Instrumental Analysis of Foods*, Vol. 1 (eds J. Crank & G.S. Parker), p. 405, Academic Press, New York.

Goff, H.D. (1995) *Dairy Science and Technology Education*, University of Guelph, Canada. Available at www.foodsci.uoguelph.ca/dairyedu/home.html.

Goff, H.D. & Griffiths, M.W. (2006) Major advances in fresh milk and milk products: fluid milk products and frozen desserts. *Journal of Dairy Science*, **89**, 1163–1173.

Guinard, J.X. & Mazzucchelli, R. (1999) Effects of sugar and fat on the sensory properties of milk chocolate: descriptive analysis and instrumental measurements. *Journal of the Science of Food and Agriculture*, **79**, 1331–1339.

Hernández-Ledesma, B., Recio, I., Ramos, M. & Amigo, L. (2002) Preparation of ovine and caprine β-Lg hydrolysates with ACE-inhibitory activity – identification of active peptides from β-Lg hydrolysed with thermolysin. *International Dairy Journal*, **12**, 805–812.

Hilliam, M. (2003) Future for dairy products and ingredients in the functional foods market. *Australian Journal of Dairy Technology*, **58**, 98–103.

Hotchkiss, J.H. (1997) Food-packaging interactions influencing quality and safety. *Food Additives and Contaminants*, **14**, 601–607.

Hurrell, R.F. (2002) Fortification: overcoming technical and practical barriers. *Journal of Nutrition*, **132**, 806S–812S.

Ishler, V. & Roberts, B. (1991) Guidelines for preventing off-flavors in milk. The Dairy Practices Council, Pennsylvania, PA.

Jelen, P. & Tossavainen, O. (2003) Low lactose and lactose-free milk and dairy products – prospects, technologies and applications. *Australian Journal of Dairy Technology*, **58**, 161–165.

Jensen, G.K. & Poulsen, H.H. (1992) Sensory aspects. *Bulletin of the International Dairy Federation*, **271**, 26–31.

Karatapanis, A.E., Badeka, A.V., Riganakos, K.A., Savaaidis, I.N. & Kontominas, M.G. (2006) Changes in flavour volatiles of whole pasteurized milk as affected by packaging material and storage time. *International Dairy Journal*, **16**, 750–761.

Kelly, A.L. & Foley, J. (1997) Proteolysis and storage stability of UHT milk as influenced by milk plasmin activity, plasmin/β-lactoglobulin complexation, plasminogen activation and somatic cell count. *International Dairy Journal*, **7**, 411–420.

Khanal, R.C., Dhiman, T.R., Ure, A.L., Brennand, C.P., Boman, R.L. & McMahon, D.J. (2005) Consumer acceptability of conjugated linoleic acid-enriched milk and cheddar cheese from cows grazing on pasture. *Journal of Dairy Science*, **88**, 1837–1847.

Liang, B. & Hartel, R.W. (2004) Effects of milk powders in milk chocolate. *Journal of Dairy Science*, **87**, 20–31.

Mauri, J.L., Christen, G.L., Melton, S.L., Penfield, M.P. & Baird, C.S. (1995) Effect of processing method and storage temperature and time on chemical and sensory attributes of low-fat milk. *Journal of Dairy Science*, **78**, 146.

Mehansho, H. (2006) Iron fortification technology development: new approaches. *Journal of Nutrition*, **136**, 1059–1063.

Mestdagh, F., Meulenaer, B.D., Clippeleer, J.D., Devlieghere, F. & Huyghebaert, A. (2005) Protective influence of several packaging materials on light oxidation of milk. *Journal of Dairy Science*, **88**, 499–510.

Mortensen, G., Bertelsen, G., Mortensen, B.K. & Stapelfeldt, H. (2004) Light-induced changes in packaged cheeses-a review. *International Dairy Journal*, **14**, 85–102.

Mounchili, A., Wichtel, J.J., Bosset, J.O., Dohoo, I.R., Imhof, M., Altien, D., Mallia, S. & Stryhn, H. (2005) HS-SPME gas chromatographic characterization of volatile compounds in milk tainted with off-flavour. *International Dairy Journal*, **15**, 1203–1215.

Mounchili, A., Wichtel, J.J., Dohoo, I.R., Keefe, G.P. & Halliday, L.J. (2004) Risk factors for milk off-flavours in dairy herds from Prince Edward Island, Canada. *Preventive Veterinary Medicine*, **64**, 133–145.

Moyssiadi, T., Badeka, A., Kondyli, E., Vakirtzi, T., Savvaidis, I. & Kontominas, M.G. (2004) Effect of light transmittance and oxygen permeability of various packaging materials on keeping quality of low fat pasteurized milk: chemical and sensorial aspects. *International Dairy Journal*, **14**, 429–436.

Murray, J.M., Delahunty, C.M. & Baxter, I.A (2001) Descriptive sensory analysis: present and future. *Food Research International*, **34**, 461–471.

N'Kouka, K.D., Klein, B.P. & Lee, S.-Y. (2004) Developing a lexicon for descriptive analysis of soymilks. *Journal of Food Science*, **69**, S259–S263.

Olsen, J.R. & Ashoor, S.H. (1987) An assessment of light-induced off-flavors in retail milk. *Journal of Dairy Science*, **10**, 1362–1370.

Phillips, L. & Barbano, D. (1997) The influence of fat substitutes based on protein and titanium dioxide on the sensory properties of low fat milks. *Journal of Dairy Science*, **80**, 2726–2731.

Phillips, L.G., McGiff, M.L., Barbano, D.M. & Lawless, H.T. (1995) The influence of fat on the sensory properties, viscosity, and color of lowfat milk. *Journal of Dairy Science*, **78**, 1258–1266.

Ramaswamy, N., Baer, R.J., Schingoethe, D.J., Hippen, A.R., Kasperson, K.M. & Whitlock, L.A. (2001) Short communication: consumer evaluation of milk high in conjugated linoleic acid. *Journal of Dairy Science*, **84**, 1607–1609.

Richardson-Harman, N.J., Stevens, R., Walker, S., Gamble, J., Miller, M., Wong, M. & McPherson, A. (2000) Mapping consumer perception of creaminess and liking for liquid dairy products. *Food Quality and Preference*, **11**, 239–246.

Santos, M.V., Ma, Y., Caplan, Z. & Barbano, D.M. (2003) Sensory threshold of off flavors caused by proteolysis and lipolysis in milk. *Journal of Dairy Science*, **86**, 1601–1607.

Schwambach, S.L. & Peterson, D.G. (2006) Reduction of stale flavour development in low-heat skim milk powder via epicatechin addition. *Journal of Agricultural and Food Chemistry*, **54**, 502–508.

Shipe, W.F., Bassette, R., Deane, D.D., Dunkley, W.L., Hammond, E.G., Harper, W.J., Kleyn, D.H., Morgan, M.E., Nelson, J.H. & Scanlan, R.A. (1978) Off flavours of milk: nomenclature, standards and bibliography. *Journal of Dairy Science*, **61**, 855–869.

Solano-López, C.E., Taehyun, J. & Alvarez, V.B. (2005) Volatile compounds and chemical changes in ultrapasteurized milk packaged in polyethylene terephthalate containers. *Journal of Food Science*, **70**, C407–C412.

Tamime, A.Y., Saarela, M., Korslund Søndergaard, A., Mistry, V.V. & Shah, N.P. (2005) Production and maintenance of viability of probiotic micro-organisms in dairy products. In: *Probiotic Dairy Products* (ed. A.Y. Tamime), pp. 39–72, Blackwell Publishing, Oxford.

Torres-Penaranda, A.V. & Reitmeier, C.A. (2001) Sensory descriptive analysis of soymilk. *Journal of Food Science*, **66**, 352–356.

Torres-Penaranda, A.V., Reitmeier, C.A., Wilson, L.A., Fehr, W.R. & Narvel, J.M. (1998) Sensory characteristics of soymilk and tofu made from lipoxygenase-free and normal soybeans. *Journal of Food Science*, **63**, 1084–1087.

Tuorila, H. (1986) Sensory profiles of milks with varying fat contents. *Lebensmittel Wissenschaft und Technologie*, **19**, 344–345.

USDA (2002) *Judging and Scoring Milk and Cheese*, Farmer's Bulletin No. 2259, United States Department of Agriculture, Washington DC.

van Aardt, M., Duncan, S.E., Marcy, J.E., Long, T.E. & Hackney, C.R. (2001) Effectiveness of polyethylene terephthalate and high-density polyethylene in protection of milk flavor. *Journal of Dairy Science*, **84**, 1341–1347.

van Aardt, M., Duncan, S.E., Marcy, J.E., Long, T.E., O'Keefe, S.F. & Nielsen-Sims, S.R. (2005) Aroma analysis of light-exposed milk stored with and without natural and synthetic antioxidants. *Journal of Dairy Science*, **88**, 881–890.

Walstra, P. & Jennes, R. (1984) *Dairy Chemistry and Physics*, Wiley and Sons, New York.

Wilkes, J.G., Conte, E.D., Kim, Y., Holcomb, M., Sutherland, J.B. & Miller, D.W. (2000) Sample preparation for the analysis of flavors and off-flavors in foods. *Journal of Chromatography A*, **880**, 3–33.

Yanes, M., Durán, L. & Costell, E. (2002) Rheological and optical properties of commercial chocolate milk beverages. *Journal of Food Engineering*, **51**, 229–234.

Index

acetylcholinesterase (AchE) inhibition test, 78
Achromobacter spp., 49, 59
Acinetobacter spp., 57
additive framework Directive 89/107/EEC, 111
aerobic plate count (APC), 157
aerobic spore-formers, 58
Aeromonas, 49
agricultural chemicals, 147–148
Alcaligenes spp., 49, 52, 59
  A. tolerans, 54
alkaline phosphatase (ALP), 37
allergen, 255
ampicillin, 153
animal feed, contamination of, 146, 148–149
animal health management, 9
APV Pure-Lac™ system, 197
aseptic packaging, 188–190
Association of Official Analytical Chemists (AOAC), 74
'athermal' effects of microwaves, 207

Bacillus spp., 42, 50, 53, 58, 60, 73, 177
  B. abortus, 145
  B. badius, 56
  B. cereus, 1, 3, 11–12, 42, 48, 50, 53, 56, 58, 177, 181, 221
  B. circulans, 53, 181
  B. coagulans, 53, 56, 58
  B. firmus, 181
  B. licheniformis, 50, 53, 56, 58, 178, 181
  B. megaterium, 50, 56
  B. melitensis, 145
  B. mycoides, 53
  B. polymyxa, 56
  B. pumilis, 58, 181
  B. sphericus, 56
  B. sporothermodurans, 3, 56, 181, 240
  B. subtilis, 50, 181, 215, 224, 240
  B. subtilus, 56
Bacillus stearothermophillus var. *calidolactis* spores, 78
bacterial contamination, prevention strategies for
  addition of carbon dioxide, 62
  application of bio-preservatives, 62–64
  cleaning and sanitation, 61–62
  cooling during storage, 62
  multitarget attack/integrative approaches, 65
  non-thermal treatments, 65
  thermal treatments, 64
Bactofuge™, 181
bactoscan, 245
batch heating, 174
batch pasteurisation time, 200
benzalkonium chloride, 52
bifidobacteria, 26
Bifidobacteria spp., 11
biologically derived contaminants, 147
bio-preservatives, 65
blood serum albumin, in cow's milk, 32
booster pump, 176
bovine serum albumin (SA), 32
bovine spongiform encephalopathy (BSE), 149–150
Brevibacillus brevis, 58
Brucella spp., 54
  contamination, of milk, 145
buffering capacity, of milk, 38–39
building construction, and hygiene practice, 241, 258
Buna rubber gaskets, 51
butyric acid bacteria, 3, 15
California mastitis test (CMT), 77
Campylobacter spp., 54, 153, 237
  C. coli, 141
  C. fetus, 141
  C. jejuni, 1, 3, 50, 77, 177, 207
  contamination, of milk, 141
capillary electrophoresis (CE), 86, 88, 90
caprylic acid, 152
caramelisation, 27
carbon dioxide (CO_2), in milk preservation, 62
Carnobacterium piscicola CS526, 144
β-casein, 34, 89
γ-casein, 34
κ-casein, 34–35, 39, 86
casein content analysis, 82
casein micelles, 34–36, 169
caseinomacropeptide (CMP), 86
caseins, in cow's milk, 33–36
CCPs decision tree, 255
chemical contaminants and residues
  biologically derived contaminants, 147
  industrially derived contaminants, 146
  pesticides and residues of plant health agrichemicals, 147–148
  residues of animal remedies, 148–149
chloramphenicol, 154
chromatographic techniques, 83
chromatography, 82
Chromobacterium, 49
CIP systems, 243
Cl. perfringens, 55

Clostridium spp., 50
C. beijerinckii, 7
C. botulinum, 180–181
C. butyricum, 7
C. perfringens, 55
C. sporogenes, 7
C. tyrobutyricum, 7, 10
Cloxiella burnetii, 131
Codex Alimentarius. *See* legislation, for market milks
coliforms, 49
colloidal calcium phosphate (CCP), in milk, 29
colloidal salts, in milk, 29
colostrum, 25
computer models, of processing technologies milk, 226–228
concentration/enrichment, of microorganisms, 76
contagious mastitis organisms, 4
contamination, sources of
pasteurised milk
air, 61
packaging materials, 61
processing equipment, 60–61
raw milk
exterior of the udder, 59
interior of the udder, 58–59
milking equipment, 60
water, 59–60
UHT milk, 61
control, defined, 255
control point (CP), 255
cooling of milk, for preventing contamination, 8
corrective action, 255
Corynebacterium-like gram-positive non-spore-forming rods, 57
Corynebacterium spp., 54
C. bovis, 4
cows. *See also* cow's milk
health status of, 25
infection of, 1
transmission of contagious pathogens, 4
cow's milk
composition and properties of, 24–25
various breeds, 25
constituents
indigenous milk enzymes, 36–37
lactose, 25–28
lipids, 29–21
milk salts, 28–29
proteins, 31–36
heat-induced changes
acid-induced coagulation properties, 40
buffering capacity, 38–39
coagulation properties, 40–41
creaming or gravity-driven separation of, 39
pH, 37–38
rennet coagulation properties, 39–40
mineral content, 28
proteins, 31–36
triacylglycerols in, 30
Coxiella burnetii, 54, 121
Creameries Acts 1928–1934, 119
creaming, of raw milk, 39
Creutzfeldt-Jakob disease, 149
critical limit, 255
Crohn's disease, 150
Cryptosporidium spp., 54
culling, of chronically infected cows, 9

Dairies (General) Regulations 1959, 115
Dairy Produce Acts 1924–1984, 119
deficiency, HACCP definition, 255
denaturation, of milk proteins, 85, 209
denaturation, of whey proteins, 32–33
density, 199
differential scanning calorimetry (DSC), 33
dioxins, 146
direct and indirect UHT heating systems, 189–190
direct microscopy (DMSCC), 78
dirt, transmission of, 5
disodium phenyl phosphate, 173
DNA-based methods, 77
DNA-based subtyping methods, 77
Drinking Milk Regulations 1998, 116
dry cow antibiotic therapy, 9

electrophoresis, 82
emission monitoring, 146
1,2-enediol, 26
energy units, 199
enolisation reaction, 26
ω-3-enriched milks, 91
Ent. faecalis, 60
Enterobacteriaceae, 49, 52
Enterobacter spp., 49
E. sakazakii, 151–152
Enterococcus spp., 54
Enterococcus faecium F58, 144
environmental mastitis organisms, 4
enzyme assays, 84
enzyme-linked immunosorbent assays (ELISA), 145
enzymes, in raw milk, 174
ermC gene, 155
ermX gene, 155
Escherichia coli (VTEC) contamination, of milk, 141–142
Escherichia spp., 49
E. coli, 3–4, 6, 50–51, 177, 207, 214–215, 238
E. coli O26, 54, 154
E. coli O157, 54, 154
E. coli O157:H7, 6, 76–77
E. faecium strains, 155
E. sakazakii, in PIF, 152
ethanol stability test, 194–195
EUR-LEX website, 102
European Communities (Standards for Heat-Treated Milk in Intra-Community Trade) Regulations 1990, 119
European Hygienic Engineering Design Group (EHEDG), 241
European Union Statutory Instruments (SI), related to milk composition. *See* legislation, for market milks
extended shelf life (ESL) milk, 102, 195–197
microflora of, 57–58

facility hygiene, 11–12
fat globules, of milk, 31
fatty acids, in cow's milk, 30
Feed Hygiene Regulation, 11
'flat sour' defect, 55–56
Flavobacterium spp., 49, 52, 59, 243
fluorimetric detection, 78
fluorimetry, 84
fluoro-optical sensors, 78
Food and Drugs Act 1955, 115
Food Hygiene (England) Regulations 2006, 116

Food Hygiene (Scotland) Regulations 2006, 116
Food Standard Agency (FSA) website, 117
Fossomatic method (FSCC), 78
Fourier transform technique (FT-MIR), 80
fresh bedding, microorganisms in, 5
furosine, 85, 89

galactosamine, 34
gamma irradiation, 152
General Agreement on Tariffs and Trade (GATT), 126
Geobacillus stearothermophilus, 56, 181, 192, 240
γ-glutamyltransferase, 212
glycosylation, 83
good dairy farming practices, 8
Grade A (Certified) Milk, 115
Grade 'A' Pasteurised Milk Ordinance, 122
gram-negative psychrotrophic microorganisms, 238
GSUDT. *See* legislation, for market milks

hazard analysis of critical control points (HACCP) systems, 1, 8, 11, 143, 244
- audit process, 291–293
- benefits of, 254–255
- commitment by management, 259
- concepts of, 253
- definitions, 255–257
- flow diagram for process, 263
- hazard components, 263–269
- implementation and maintenance of, 257–260
- letter of endorsement of, 259–260
- prerequisites, 269–277
- principle 1, 277–283
- principle 2, 283
- principle 3, 283–284
- principle 4, 284–285
- principle 5, 285–287
- principle 6, 287–288
- principle 7, 289–291
- product description form, 262
- in quality control, 73
- steps to implementation, 260–263
- team, 256, 260–262
- validation of plans, 288

hazards, in milk
- antimicrobial resistance, 152–156
- bovine spongiform encephalopathy (BSE), 149–150
- chemical contaminants and residues, 147–149
- *Enterobacter sakazakii*, 151–152
- *M. avium* subsp. *paratuberculosis* (MAP), 150–151
- microbial hazards, 140–145

headspace volume, 193
heat exchanger design, 200
heat film coefficient, 199
heating effects, on milk
- buffering capacity, 38–39
- in caseins and casein micelles, 25–26
- coagulation, 40–41
- creaming, 39
- in lactose, 26–28
- in milk fat and fat globules, 31
- pH, 37–38
- rennet coagulation properties, 39–40
- in the salt balance, 29
- in whey proteins, 32–33

heating processes, 170–171
heat-resistant bacteria, of pasteurised milk, 53
heat-resistant enzymes, 57
heat-resistant spore-forming bacteria, 191–193
heat transfer, principles of, 170–171
high-performance liquid chromatography (HPLC), 78
high pressure (HP) processing technologies, milk
- effects on properties, 210–212
- operating principle, 210
- state of the art and opportunities, 212–213

high-temperature, short-time (HTST) process, 39, 172
holder process, 174
homogenisation, before pasteurisation, 176
Hull method, 57
human microbial pathogens, in milk, 1
hurdle technology, 214
hygiene practices, in liquid milk dairies
- building construction, 241
- cleaning, 243–244
- monitoring and control
  - in-line detection of microorganisms, 245
  - optimisation of CIP procedures, 245–246
- packaging materials, 244
- pasteurisation of milk, 238–240
- processing aids, 244
- processing equipment, 242–243
- quality of the raw milk, 237–238
- sterilised milk, 240
- zoning, 241–242

hypothiocyanite, 62
hypothiocyanite and its conjugate acid (HOSCN), 37

immnoglobulins (Ig), in cow's milk, 32
immunoassay-based biosensor systems, 76
immunodetection methods, 76
in-bottle sterilisation, 101
in-container sterilisation, 181–183
innovative steam injection processing technology, milk
- effect on product properties, 221–222
- operating principle, 221
- state of the art and opportunities, 222

International Dairy Federation (IDF), 74
International Standards Organisation (ISO), 74
ionic calcium, role in sedimentation, 194
Ireland Statutory Instruments (SI), related to milk composition. *See* legislation, for market milks
isoelectrofocusing, 86
isomerisation, 26

Jersey cows, 29
Johne's disease, 15, 54
Johne's disease, in cows, 150

Kjeldahl method, 79, 84
Klebsiella spp., 4, 49

Lac. lactis subsp. *lactis*, 64
α-lactalbumin, 26, 89
α-lactalbumin (α-la), in cow's milk, 32
β-lactams, 78
lactate-oxidising yeasts, 10
lactation stage and fatty acid composition, 30
lactic acid bacteria, 50, 59, 64
Lactobacillus spp., 54, 197
Lactococcus lactis subsp. *lactis*, 155, 225
Lactococcus spp., 2
lactoferrin, 63
β-lactoglobulin (β-lg), in cow's milk, 32, 185, 195, 212
lactoperoxidase inactivation, 178
lactoperoxidase (LPO), 37, 62–63
lactoperoxidase system (LPS), 177
lactose
  concentration, 25
  primary structure of, 26
  synthesis of, 26
β-lactose, 26
lactose-derived compounds, 84
lactulose, 84
lactulose concentrations, 26
late-blowing defect, 7
latent heat, 199
Le Chatelier–Braun principle, 210
legislation, for market milks. *See also* hazard analysis of critical control points (HACCP) systems
  Codex Alimentarius, 173, 179
    Commission, 127
    concept, 125
    content, 127
    food additives and, 132
    hygiene codes of practice, 129–131
    membership and structures, 125–126
    standards, 126–128
    standards for direct human consumption, 128–129
  European Union
    amendment to compositional requirements, 105
    Article 12 of Regulation 178/2002, 117
    CN Code 04.01, 103
    compositional requirements, 102–105
    criteria for raw milk supply, 112
    dairy hygiene Directive 92/46, 117
    Directive 45/2007, 120
    Directive 85/397, 105–106, 119
    Directive 88/388, 111
    Directive 90/496, 105
    Directive 92/46, 106–107
    Directive 94/36, 111
    Directive 89/107/EEC, 111
    hygiene and food safety requirements, 105–115
    Hygiene Directive 92/46, 119
    hygiene requirements of milk and milk product production, 108
    labelling requirements, 112–113
    packaging requirements, 113
    Regulation 539/94, 103
    Regulation 852/2004, 107, 110, 117, 119
    Regulation 853/2004, 107–109, 120
    Regulation 854/2004, 108
    Regulation 1153/2007, 116
    Regulation 1411/71, 105, 119
    Regulation 2074/2005, 108–109
    Regulation 2597/97, 103, 105, 120
    shell life requirements, 114–115
    size requirements, 113–114
    website, 102
  Ireland
    background, 119
    compositional requirements, 120
    on hygiene, 119–120
    range of packaging sizes for fresh drinking milk, 120
  United Kingdom
    background, 115
    code of practice for HTST pasteurisation, 118
    compositional requirements, 115–116
    on hygiene, 116–117
    permitted range of packaging sizes for milk, 118
  United States of America
    background, 120–121
    Code of Federal Regulations, 124
    definitions and standards, 122–124
    historical details, 121
    Pasteurised Milk Ordinance (PMO) and the Code of Federal Regulations, 122
    sale of raw milk, 124–125
lipids, of milk, 29–30
lipolysis, 36, 84, 175, 180
lipopolysaccharides (LPS), 63
lipoprotein lipase (LPL), 36
Listeria spp., 51, 54, 155, 214
  L. innocua, 143, 214
  L. monocytogenes, 1, 3, 6, 8, 50, 75, 77, 143–144, 156, 207, 224, 237–238
listeriosis, in cattle herds, 6

M. paratuberculosis, 77
Maillard reaction, 26–28, 73, 89, 170
  effects of microwaves, 208
Maillard reaction products, 85
market milks, types of
  chocolate-flavoured milk, 311–312
  CLA-enriched milks, 309
  enriched and modified milks, 309
  lactose-reduced milk, 310–311
  milk enriched with fibre, 310
  mineral-enriched milk, 309–310
  raw and heat-processed milk, 306
  soy milk, 312
  whole and fat-reduced milk, 306–308
mastitis organisms, 2–5, 12, 25, 51
  contamination, 148
medium infrared spectroscopy (MIR), 80
melanoidins, 28
membrane filtration, 15
metabolite testing, 76
methicillin-resistant *Staphylococcus aureus* (MRSA), 155
methylene blue test, 172
micellar calcium phosphate (MCP), in milk, 29
Microbacterium spp., 54
microbial contamination, of milk
  control, 8–15
    future strategies for mass production, 15–16
  hazards, in milk, 264–266

Brucella spp., 145
Campylobacter spp., 141
E. coli (VTEC), 141–142
L. monocytogenes, 143–144
Mycobacterium bovis, 144–145
route of entering by microorganisms, 140
S. aureus, 142–143
Salmonella spp., 140–141
significance for public health, 1
significance to animal welfare, 1
sources
aerial contamination, 2
exterior of the teats, 2
farm environment, 5–7
growth during milk storage, 8
mastitis bacteria, 2–5
microorganisms, 3
psychrotrophic microorganisms, 2
via the milking equipment, 7–8
Micrococcus-like gram-positive non-spore-forming rods, 57
Micrococcus spp., 54
microfiltration method processing technology, milk
effect on product properties, 219–220
operating principle, 219
state of the art and opportunities, 220
microwave method processing technology, milk
drawback, 207
effects on properties, 207–209
operating principle, 207
Milk and Dairies Acts 1935–1956, 119
Milk and Dairies (Skimmed and Semi-skimmed Milk) (Heat Treatment and Labelling) Regulations 1973, 115
milkborne antimicrobial residues, 149
milk composition, of dairying species, 24, 169–170
milk concentration, at the dairy farm, 15–16
milk fat globule membrane (MFGM), 31
Milk Hygiene Directive 92/46/EEC, 105, 238
milking equipment, contamination of, 7–8
milking operations, hygienic, 12–13
milk products, reconstituted, 197–198
milk sickness, hepatotoxicity syndrome of, 147
Milk (Special Designations) Regulations 1977, 115
milk yield, relationship with composition, 25
Minispec (Bruker), 80
Ministry Food (Continuance) Act of 1920, 115
mislabelling, of functional processed milks, 90–91
multi-residue analysis, 79
mutarotation process, 26
Mycobacterium africanum, 144
Mycobacterium avium subsp. *paratuberculosis* (MAP), 54, 118, 150–151, 173, 178
Mycobacterium bovis, 144–145
Mycobacterium paratuberculosis, 3, 75
Mycobacterium tuberculosis, 172
complex (MTC), 144
mycotoxins, 78, 147

N-acetylneuraminic acid residues, 34
NACMCF guidelines, 280
National Antimicrobial Resistance Monitoring System (NARMS), 153
nisin, 64
NIZO Premia software, 227, 246
non-conformity, HACCP definition, 256
non-spore-forming bacteria, 56–57
non-thermal treatments, of milk, 65
nuclear magnetic resonance (NMR) spectrometry, 80

4-*O*-β-D-galactopyranosyl-D-glucopyranose. *See* lactose
Office International des Epizooties (OIE), 150
Office of Public Sector Information (OPSI) website, 115
Official Journal of the European Union, 102
online monitoring, of tank milk, 16
online monitoring systems, 14
OPTI-CIP, 246
orthophosphate, 83
overall heat transfer coefficient (OHTC), 174, 199

Paenibacillus lactis, 56, 191
Paenibacillus spores, 191
Panton-Valentine leukocidin (PVL)-negative staphylococcal cassette chromosome (SCC)*mec*-type IVa strain, 155
paratuberculosis, 150
Pasteur, Louis, 101
pasteurisation, 16, 101, 139, 150, 238–240
Codex Alimentarius, 173
factors affecting the quality of, 177–180
high-temperature, 195–197
methods, 174–177
principles, 171–174
pasteurised milk, microflora of
endospore-forming genera, 53
gram-negative psychrotrophic bacteria, 51–52
heat-resistant bacteria, 52–53
Lactobacillus spp., 52
Lactococcus spp., 52
pathogenic organisms, 54–55
vegetative bacteria, 53–54
pasture preservation, 156
pathogenic microorganisms
in pasteurised milk, 54–55
in raw milk, 50–51, 54
in thermised milk, 54
penicillin G, 154
performance standards, HACCP definition, 256
pesticide contamination, 147–148
pH, of milk, 37–38, 169–170
phosphoglycerides, 83
phosphorylated carbohydrates, 83
phosphorylation, 83
phosphoserine, 83
phytotoxins, 147
plasmin activity, impact on milk, 36, 90
plate heat exchanger (PHE), 174
polychlorinated biphenols (PCBs), 146
polymerase chain reaction (PCR), 143
post-milking teat disinfection, 9, 12

post-pasteurisation contamination (PPC), 179
potentially hazardous food, HACCP definition, 256
powdered infant formula (PIF), contaminated, 151–152
pre-incubation (PIC) test, 110
prerequisite programme (PP), of HACCP, 269–277
processing technologies, milk
  combined
    effect on product properties, 224
    preservation in the presence of antimicrobials, 224–225
    state of the art and opportunities, 225–226
  computer models, 226–228
  high pressure (HP)
    effects on properties, 210–212
    operating principle, 210
    state of the art and opportunities, 212–213
  innovative steam injection
    effect on product properties, 221–222
    operating principle, 221
    state of the art and opportunities, 222
  microfiltration method
    effect on product properties, 219–220
    operating principle, 219
    state of the art and opportunities, 220
  microwave method
    drawback, 207
    effects on properties, 207–209
    operating principle, 207
  pulsed electric fields (PEF)
    effect on properties, 214–215
    operating principle, 213–214
    state of the art and opportunities, 215
  ultrasound method
    effects on product properties, 216–217
    operating principle, 216
    state of the art and opportunities, 217–218
protease-producing bacteria, 57
protein stabilisation step, 185
proteolysis, 84
proteose peptones, in cow's milk, 32
pseudomonads, 49, 60
Pseudomonas spp., 3, 7, 11, 42, 49, 52, 59, 77, 243
  P. aeruginosa, 52, 178
  P. fluorescens, 49, 52, 207, 225
  P. fluorescens NCDO 2085, 64
  P. fragi, 49, 52
  P. putida, 52
Psychrobacter, 49
'psychro fast' test, 52
psychrotrophic bacteria, 59–61
  growth of, 190–191
psychrotrophic microorganisms, 16
pulsed electric fields (PEF) processing technology, milk
  effect on properties, 214–215
  operating principle, 213–214
  state of the art and opportunities, 215
pulse-field gel electrophoresis (PFGE), 77
  subtyping, 143
pyrolysis-mass spectrometry, 89
pyrrolizidine, 147

Q fever, 121
quality control. *See also* hazard analysis of critical control points (HACCP) systems
  analysis of fraudulent addition of ingredients and authentication
    mislabelling of fortified and functional milks, 90–91
    non-dairy proteins, 89
    reconstituted milk, 89–90
    substitution of milk from a different species of mammals, 90
    substitution of milk with different feeding or geographical origin, 90
    water, 86
    whey protein, 86–89
  components analysis
    degradation products formed during milk storage, 84
    fatty acid composition, 83
    heat load, 84–86
    lactose, 79
    milk components, 83
    MRI standardised methods, 80
    nuclear magnetic resonance (NMR) spectrometry, 80–81
    protein concentration, 79, 82–83
    Raman spectroscopy, 80
    reference methods, 79–80
    spectroscopic methods, 80
  methods of analysis
    abnormal milk, 75
    mastic milk, 77–78
    microbial analysis, 75–77
    for residue compounds, 78–79
  of processed milk, 73
  of raw milk, 72
quality of raw milk and dairy products, 41–42
quaternary ammonium compounds (QAC), 154

raw milk
  hygiene practices for, 237–238
  microflora of
    gram-negative psychrotrophic bacteria, 48–49
    gram-positive psychrotrophic bacteria, 49–50
    pathogenic microorganisms, 50–51, 54
reaction kinetic parameters, 170
reasonably likely to occur, HACCP definition, 256
regeneration efficiency, 200
rehydration during storage, impacts, 10
rennet coagulation properties, of milk, 39–40
residue risks, in milk, 148–149
risk management strategies
  harvest phase management of dairying, 156–157
  post-phase management of dairying, 157–158
  pre-harvest management of dairying, 156
roughages, microbiological quality of, 10
rybotyping, 77

S. thermophilus, 178
safety, of raw milk. *See also* hazard analysis of critical control points (HACCP) systems

chemical contaminants and residues
biologically derived contaminants, 147
industrially derived contaminants, 146
pesticides and residues of plant health agrichemicals, 147–148
residues of animal remedies, 148–149
emerging hazards
antimicrobial resistance, 152–156
bovine spongiform encephalopathy (BSE), 149–150
Enterobacter sakazakii, 151–152
M. avium subsp. *paratuberculosis* (MAP), 150–151
implication of milk in human disease, 139–140
microbial hazards in milk
Brucella spp., 145
Campylobacter spp., 141
E. coli (VTEC), 141–142
L. monocytogenes, 143–144
Mycobacterium bovis, 144–145
S. aureus, 142–143
Salmonella spp., 140–141
risk management strategies
harvest phase management of dairying, 156–157
post-phase management of dairying, 157–158
pre-harvest management of dairying, 156
Salmonella spp., 1, 3, 50, 54, 72, 75–77, 177, 237–238
contamination, of milk, 140–141
S. Dublin, 140
S. enterica, 6
S. typhimurium, 140, 207
Salmonella ser. typhimurium DT104, 51
salt balance, in milk, 29
salt composition, of milk, 28–29
sanitation standard operating procedure (SSOP), 256–257
sensitive ingredient, HACCP definition, 257
sensory properties, of market milks
attributes of flavour defects, 297–303
evaluation, 295–267
factors influencing, 303–305
serovars, 140
severity, HACCP definition, 257
silage contamination, 7
simulated milk ultrafiltrated (SMUF), 214
single-strand conformational polymorphism (SSCP), 144
somatic cell count (SCC), 42
specific enthalpy, 199
specific gravity, 199
specific heat, 199
spectinomycin, 154
spoilage microorganisms, of milk and dairy products, 237
of pasteurised milk, 51–53
spore concentrations
in faeces, 5
in silage, 10
spore-forming bacteria, 7, 49–50
standard plate count (SPC), 76
staphylococcal food poisoning, 142–143
Staphylococcus aureus, 2–4, 50–51, 72, 75, 177–178, 207, 224, 238
steam infusion process, 186
steam injection (ISI) process, 184, 186
sterilisation, of milk, 180–181, 240
in-container, 181–183
Str. dysgalactiae, 51
Str. uberis, 51
streptococci species, 4
Streptococcus spp., 54
S. agalactiae, 4, 51, 155
S. dysgalactiae, 4
S. thermophilus, 3, 53, 178, 242
S. uberis, 4–5
streptomycin, 154
sub-clinical mastitis, 4
sulphonamide, 154
sulphydryl–disulphide interchange reactions, 30
surface plasmon resonance biosensor (Biacore), 85

tandem mass spectrometry, 79
target levels, HACCP definition, 257
teat cleaning methods, 12
tetracycline (ACSSuT), 154
thermal conductivity, 199
thermal diffusivity, 199
thermal treatments, of milk, 64
thermisation, 16
of raw milk, 171
thermistor cryoscope method, 86
thermoduric bacteria, effects of, 172
thermophilic organisms, 52
thermoresistent streptococci (TRS), 245
thin-layer chromatography (TLC), 78
toxic compounds, HACCP definition, 257
transmissible spongiform encephalopathies (TSEs), 149
trichloroacetic acid (TCA), 79
Trp fluorescence, 85
tuberculosis bacillus, 172
tubular heat exchanger (THE), 174
turbidity test, 182
tyndallisation, of raw milk, 171

udder contamination, 58–59
udder health programme, 9
UK Dairy Industry Federation Code of Practice, 118
UK Statutory Instruments (SI), related to milk composition. *See* legislation, for market milks
ultra-high temperature (UHT) milk, 101, 168
factors affecting, 190–193
methods, 185–190
microflora of
heat-resistant enzymes, 57
heat-resistant spore-formers, 55–56
post-sterilisation contaminants, 56–57
principles, 183–184
quality control of, 73
sediment formation and fouling, 193–195

ultrasound method processing technology, milk
effects on product properties, 216–217
operating principle, 216
state of the art and opportunities, 217–218
USA Statutory Instruments (SI), related to milk composition. *See* legislation, for market milks

validation, HACCP definition, 257
vegetative bacteria, 53–54
verification, HACCP definition, 257, 263
viruses, in milk, 140
viscosity, 175, 199

Wales Food Advisory Committee (WFAC), 117
water contamination, 59–60
Weights and Measures (Miscellaneous Foods) Order 1988, 118

whey protein denaturation, 85
whey proteins, in milk, 31–33
effects of microwaves, 208
world milk production, 23–24

Yersinia spp., 54
Y. enterocolitica, 50, 177, 207, 238

zoning, and hygiene practice, 241–242